Quantum Dots in Bioanalytical Chemistry and Medicine

Detection Science Series

Editor-in-chief:
Mike Thompson, *University of Toronto, Canada*

Series editors:
Subrayal M. Reddy, *University of Surrey, UK*
Damien Arrigan, *Curtin University, Perth, Australia*
Mengsu (Michael) Yang, *City University of Hong Kong, Hong Kong*

Editorial advisor:
Electra Gizeli, *IMBB-FORTH, Greece*
Thiago Paixão, *University of São Paulo, Brazil*

Titles in the Series:
1: Sensor Technology in Neuroscience
2: Detection Challenges in Clinical Diagnostics
3: Advanced Synthetic Materials in Detection Science
4: Principles and Practice of Analytical Techniques in Geosciences
5: Microfluidics in Detection Science: Lab-on-a-chip Technologies
6: Electrochemical Strategies in Detection Science
7: Peroxynitrite Detection in Biological Media: Challenges and Advances
8: Biological Fluid-Surface Interactions in Detection and Medical Devices
9: Advanced Environmental Analysis: Applications of Nanomaterials, Volume 1
10: Advanced Environmental Analysis: Applications of Nanomaterials, Volume 2
11: Quenched-phosphorescence Detection of Molecular Oxygen: Applications in Life Sciences
12: Carbon-based Nanomaterials in Analytical Chemistry
13: Forensic Analytical Methods
14: Immunosensors
15: Analytical Electrogenerated Chemiluminescence: From Fundamentals to Bioassays
16: In Situ Analysis of Cellular Functional Molecules
17: Confining Electrochemistry to Nanopores
18: Detection Methods in Precision Medicine
19: Analytical Strategies for Cultural Heritage Materials and their Degradation
20: Challenges in Detection Approaches for Forensic Science
21: Disposable Electrochemical Sensors for Healthcare Monitoring
22: Quantum Dots in Bioanalytical Chemistry and Medicine

How to obtain future titles on publication:
A standing order plan is available for this series. A standing order will bring delivery of each new volume immediately on publication.

For further information please contact:
Book Sales Department, Royal Society of Chemistry, Thomas Graham House, Science Park, Milton Road, Cambridge, CB4 0WF, UK
Telephone: +44 (0)1223 420066, Fax: +44 (0)1223 420247
Email: booksales@rsc.org

Visit our website at www.rsc.org/books

Quantum Dots in Bioanalytical Chemistry and Medicine

Edited by

Michael Thompson
University of Toronto, Canada
Email: m.thompson@utoronto.ca

and

Zahra Ramezani
Ahvaz Jundishapur University of Medical Sciences, Iran
Email: zahramezani@gmail.com

ROYAL SOCIETY OF CHEMISTRY

Detection Science Series No. 22

Print ISBN: 978-1-83916-714-0
PDF ISBN: 978-1-83916-956-4
EPUB ISBN: 978-1-83916-957-1
Print ISSN: 2052-3068
Electronic ISSN: 2052-3076

A catalogue record for this book is available from the British Library

The Royal Society of Chemistry is a charity, registered in England and Wales, Number 207890, and a company incorporated in England by Royal Charter (Registered No. RC000524), registered office: Burlington House, Piccadilly, London W1J 0BA, UK, Telephone: +44 (0) 20 7437 8656.

For further information see our website at www.rsc.org

Printed in the United Kingdom by CPI Group (UK) Ltd, Croydon, CR0 4YY, UK

Preface

Semiconductor nanocrystals, often referred to as quantum dots (QDs), are particles with dimensions of a few nanometers that possess properties associated with quantum mechanical behavior. The physical nature of such nanomaterials confers electronic and optical features that differ from those of macroscopic particles. The absorption and emission of radiation characteristics for QDs reflect transitions between discrete quantum mechanically allowed electronic energy levels. In this respect, they are considered to behave in an intermediate fashion between individual atoms and bulk materials. Aside from fundamental studies of the physics of QDs, the particles have received intense investigation in terms of practical applications. Among these possibilities for quantum computing, the design of solar cells (photovoltaics), photodetectors and semiconductors, and LED and laser device fabrications have figured prominently. With regard to more chemical aspects, employment as components of thin films in Langmuir-Blodgett layers, coating technology, catalysis, and ink-jet printing have been introduced. In contrast, the present text considers applications of QDs in the world of bioanalytical science, including medically oriented aspects. Overall, the text proceeds logically from describing the production of QDs and their surface modification to several aspects of their application in medical science, such as detection in diagnosis. This is followed by a discussion of their implementation in more general situations, such as environmental chemistry and food safety. The text includes a level of deliberate overlap in the material in the chapters, with a view to each being self-contained with respect to the readership.

Chapter 1, following an introduction to the general characteristics of the QDs, opens with a detailed comparison of the luminescence properties of QDs compared to more conventional organic fluorophores. The particles display a number of strong advantages over the latter, which include superior

Detection Science Series No. 22
Quantum Dots in Bioanalytical Chemistry and Medicine
Edited by Michael Thompson and Zahra Ramezani

Published by the Royal Society of Chemistry, www.rsc.org

chemical-, potential thermal- and photostability; high fluorescence lifetime; high extinction coefficient and brightness; and *in vivo/vitro* optical trackability. The discussion then looks at the various types of QDs; these include semiconductor materials composed of Groups I–VI and II–VI elements, followed by Groups III–V, IV–VI, and IV QDs, and finally other materials such as carbon, graphene, and perovskite. After a comparison of semiconductor versus carbon-based QDs (CQDs), the section proceeds to an examination of detection approaches based on resonance energy transfer (RET, fluorescence, bioluminescence, and chemiluminescence). Chapter 2 deals with the issue of the surface attachment of probes to the particle so that the system can be used to detect the presence of analytes via interaction with the selective binding probe. Much of this bioconjugation chemistry reflects that developed with respect to conventional biosensor technology. It is generally the case that modification of CQDs is more facile than for semiconductor materials due to the ease of generating surface-linking moieties. As for biosensors, the probe–linker interaction must be highly selective in nature and on the QD surface. Hopefully, this will be subject to minimal non-specific adsorption in the staple under study. The methods for probe surface attachment include carbodiimide, avidin-biotin, His-tag, and molecular self-assembly chemistry. The section also looks at the potential employment of QDs in the world of theranostics in addition to the opposing problematic area of particle toxicity.

Chapters 3 to 6 discuss the implementation of QDs in various aspects relevant to medical science. Chapter 3 concentrates on the potential use of the particles in clinical diagnosis. As expected, the role played by the surface chemistry involved is critical in this area. Following a discussion on the surface binding of probes and the physical properties of dots, Chapter 3 evaluates the use of QDs in point-of-care, bio-imaging, biosensors, and theranostics detection approaches. This includes a description of efforts to detect biomarkers for genetic and infectious diseases, imaging of tumor cells together with therapy, and optical and electrochemical biosensor technology. This discussion concludes with a critical examination of a number of negative issues with QD-based technology, such as potential cytotoxicity and the all-pervasive non-specific surface adsorption or fouling problem. Chapter 4, in addition to a further detailed look at imaging, investigates the potential for QDs in targeted drug delivery. This involves chemotherapy, phototherapy, and gene therapy, among a number of other applications. The concept is the attachment of a probe for tissue-specific recognition combined with a drug for therapy, with the QD acting as a nano-carrier. This area includes QDs employed in cell apoptosis, autophagy, and ferrotosis, although, as mentioned above, the ubiquitous cytotoxicity issue exists. The detection of biological particles such as bacteria and viruses using dots follows in Chapter 5. QD labeling strategies for virus tracking are often categorized into major groups depending on the virus's labeling site, such as internal or external components. For this type of study, the hope is that virus activity will be maintained regardless of the conjugation strategy employed. Using fluorescence-based detection, a number of viruses, such as influenza

H5N1 and SARS-COV2, have been tracked via attachment to particle surface proteins. Analogous strategies have been used for the assay of bacteria such as *S. aureus, E. coli*, and *P. aeruginosa.* This section is completed with a further discussion of the introduction of biosensor technology for the specific detection of particles outlined above. Chapter 6 deals with biological and medical aspects and describes intracellular and extracellular pH measurements using the QD–fluorescence approach. This parameter is known to be critical with regard to cell apoptosis, drug resistance, ion transport, cell growth, and endocytosis. CQDs, graphene-based QDs (GQDs), and silicon nanocrystals have been surface modified with pH-sensitive moieties where the fluorescence characteristics are altered by changing the hydronium ion concentration. In terms of the future, research on pH-sensitive fluorescent QDs is being focused on the development of robust nanocomposites with higher pH resolution, stability, and sensitivity.

Chapters 7 to 9 discuss the application of QDs in general environmental science and food safety, with an emphasis on the detection of contaminants such as toxins. With respect to the former area, QDs have been used to detect species such as heavy metal ions, organic pollutants, polycyclic aromatic hydrocarbons (PAHs), pesticides, and drug residues. All of these constitute serious health hazards. An assay of mercuric, ferric, cupric, and lead ions has been achieved, for the most part, via quenching of the fluorescence of CQDs produced from ‘green’ natural materials. PAHs have been detected through surface-modified dots by probes that are selective binding agents for these pollutants. Various measurement strategies have been utilized for this purpose, such as square wave electrochemistry and “on–off” fluorescence spectroscopy. Similar methods are described for detecting pesticides such as triticonazole and nicosuluron; interestingly, the latter is via ZnS QDs modified by molecularly imprinted polymeric (MIPs) material. With regard to drug residues, an analogous MIPs approach with QDs was used for the assay of ceftazidime (a broad-spectrum antibiotic) in milk samples.

Chapter 8 discusses the QD-based assay of toxins associated with pathogens in food products and beverages. Screening for foodborne pathogens is clearly critically important from a public health perspective, and there is also the specter of bioterrorism in terms of national security. Toxins of several types originate from bacteria such as *S. botulinum* and *S. aureus,* fungi, and algae. A particularly interesting example of pathogen detection involving QDs was research on using GQDs in conjunction with micromotors as mobile sensors for toxins produced by enterobacteria, enabling the detection of contaminants in food. As in many previous studies, “on-off” fluorescence was the mechanism employed for the assay. GQDs have also been applied to the electrochemical and fluorescence detection of mycotoxin, ochratoxin A, via an aptamer probe attached to the QDs. The section includes a comprehensive list of QDs utilized for the detection of both bacterial and marine toxins, which includes QD and toxin types, technique, analytical criteria, and the nature of the examined sample. Chapter 9 describes how CQDs produced from waste materials using green synthesis procedures can be used in various food safety applications, including detection and wastewater disinfection. Various synthetic protocols to generate CQDs from beetroot, lemon peel waste, perennial grass, seeds, waste from coffee grounds, and fruit shells are evaluated. The discussion incorporates a detailed look at the dispersion techniques used, such as microwave radiation, ultrasound, extraction, laser ablation, arc discharge, and calcination. In a similar vein to that outlined above, CQDs synthesized by these procedures have seen use in detecting metal ions, pesticides, food colorants, and antibiotics, usually by luminescence-based methodology.

Finally, we would like to express our deep appreciation to all the book's authors for contributing their knowledge and sharing their experience with regard to QDs and their applications, especially as they pertain to detection science in bioanalytical chemistry. Zahra Ramezani would like to thank the past and present members of her research group at Ahvaz Jundishapur University of Medical Sciences, especially Dr. Iman Saeedi, her first student working on QD synthesis and applications. Mike Thompson would like to express his sincere appreciation to the members of the Thompson Biosensors Group at the University of Toronto for their many valued research contributions and unfailing support. We wish to thank Dr. Brian De La Franier and Janine Harper of the University of Toronto for their efforts with regard to the preparation of the manuscript, and the Royal Society of Chemistry Publishing, especially Matt Caffrey, for the dedicated and expert work involved in the production of the text.

Zahra Ramezani and Mike Thompson

Contents

Detection Science Series No. 22
Quantum Dots in Bioanalytical Chemistry and Medicine
Edited by Michael Thompson and Zahra Ramezani

Published by the Royal Society of Chemistry, www.rsc.org

CHAPTER 1

Classifications of Quantum Dots and Their Detection Principles in Sensing

ZAHRA RAMEZANI*[a,b], FATEMEH KIANI GHALEHSARDI[b] AND SIAMAK NOORIZADEH[c]

[a]Toxicology Research Center, Medical Basic Sciences Research Institute, Ahvaz Jundishapur University of Medical Sciences, Ahvaz, Iran; [b]Nanotechnology Research Center, Faculty of Pharmacy, Ahvaz Jundishapur University of Medical Sciences, Ahvaz, Iran; [c]Chemistry Department, College of Sciences, Shahid Chamran University of Ahvaz, Ahvaz, Iran
*E-mail: zramezani@ajums.ac.ir; Zahra.ramezani@utoronto.ca

1.1 Introduction

Our universe is made up of molecules in different physical states. These molecules range from single atoms (like noble gases) that are far too small to be seen with the naked eye to huge macromolecules (such as polymers and proteins) with macroscopic sizes. Therefore, molecules' dimensions range from a few angstroms to around one billionth of a meter. These molecules can aggregate or bind together to form different particles and materials. In industry, the materials are mostly used in the form of solid species, mainly conductors, semiconductors, and insulators. Conductors allow energy (electricity, heat, and sound) to easily flow through them; semiconductors allow a moderate flow of energy; insulators exhibit low conductivity.

Detection Science Series No. 22
Quantum Dots in Bioanalytical Chemistry and Medicine
Edited by Michael Thompson and Zahra Ramezani

Published by the Royal Society of Chemistry, www.rsc.org

Key characteristics of individual particles include material properties (elastic deformation, physical adsorption, heat conduction, magnetization, thermal radiation, *etc.*) and geometric properties (shape, size, size distribution, *etc.*). In particulate samples, particle size is the most important physical property, and many properties of bulk solids (such as electronic, optical, magnetic, mechanical, and catalytic) are significantly altered with a reduction in particle size. Nanoscale materials are substances that are smaller than 100 nanometers in at least one dimension. These materials can be categorized based on their sizes into zero, one, two, and three dimensions. In 0-D nanomaterials (such as nanoparticles), all dimensions are measured at the nanoscale (<100 nm): 1-D (nanotubes, nanorods, nanowires, *etc.*), 2-D (graphene, nanofilms, nanocoatings, *etc.*), and 3-D (fullerenes, bulk powders, *etc.*). Nanomaterials are substances that are not confined to the nanoscale in one, two, or three dimensions, respectively. Nanomaterials can also be classified into inorganic-based (nanosilver, nanogold, nanocopper, *etc.*), carbon-based (fullerenes, graphene, nanodiamonds, *etc.*), organic-based (cyclodextrin, dendrimers, micelles, liposomes, *etc.*), and composite (metal-organic frameworks) types.

QDs are quasi-zero-dimensional nanosystems consisting of light-emitting spherical semiconductor nanocrystals with radii of 1–10 nm. Although QDs contain between 100 and 100,000 atoms, they exhibit properties as though they were composed of a single atom. However, an atom-like spectrum is rarely seen in the case of these nanostructures, as reported QDs are typically 2–20 nm in size. These nanostructures are formed in various shapes, such as spherical (CdSe/ZnS), cylindrical (GaAs/AlAs), pyramidal (core–shell InGaAs/GaAs), conical (Au cones), and lens shapes (InAs).

Due to quantum effects, QDs display properties that differ significantly from bulk materials. In fact, the properties of QDs can vary depending on their shapes and sizes. For example, samples of a particular QD with different particle sizes have different colors and possess unique magnetic and electrical properties.[1,2]

Quantum dots were first detected in a glass matrix by Alexey I. Ekimov in 1981, who was the first to work on them.[3,4] Prior to the 1990s, traditional chemical methods (such as micro-emulsions and coprecipitation) were used for the synthesis of colloidal nanoparticles. Although these methods could control the size of nanocrystals, the prepared nanocrystals did not emit light, and their optical performance was very poor. In the early 1990s, CdSe nanocrystals were synthesized using a metal–organic-coordinating solvent and high temperature method from $CdMe_2$, a method that is expensive, explosive, and very toxic. At the beginning of the 21st century, $CdMe_2$ was replaced by non-toxic CdO for QD synthesis. In 2002, QDs became commercially available in life science. In 2014, high-performance and low-cost quantum dot light-emitting diodes (QLEDs) (electroluminescent QD-based devices) were synthesized using a solution-processed method. QLEDs possess the advantages of both GaN

LEDs and organic LEDs.[5] Researchers are still working to improve synthesis protocols in order to take advantage of QDs' small size in different fields, especially in biomedicine. This is done by decorating QDs' surfaces with different ligands and molecules or biomolecules through homogeneous and heterogeneous synthetic strategies.

QDs have a wide range of applications.[6] They can absorb high-energy light on one wavelength and efficiently convert it to low-energy light on other specific wavelengths to create more distinct red, green, and blue colors that are extremely pure with excellent color contrast. QDs' peak emissions can be tuned to within 1 nm, giving excellent control over the color spectrum output of LCD displays. This unique property makes them desirable for use in high-quality television screens; this technology has been very popular since 2014.[2,7]

QDs are used as photocatalysts, amplifiers, biological sensors, *etc.*[2,7–12] Additionally, they increase the efficiency of silicon photovoltaic cells and can be used in solar cells, fluorescent biomedical imaging, and spectroscopy. Many molecules and antibodies can be attached to a single dot due to the very high surface area of QDs, which may amplify the fluorescent signal used for targeted sensing. Accordingly, some applications related to photoluminescence are also proposed for QDs. For example, QDs containing cadmium sulfide and selenide, gallium arsenide, germanium, silicon, and zinc selenide in their structures show the ability to efficiently emit light in the visible or near-infrared (NIR) ranges at room temperature, which is a unique photoluminescent property.

QDs have many advantages over traditional organic dyes (see Table 1.1). They can act as multimodal/multifunctional probes because of their high specific surface area, superior chemical and photo-stability, high fluorescence lifetime, high extinction coefficient and brightness, large stock shifts, and *in vivo*/*in vitro* optical trackability.[14–16] The thermal stability of QDs is high and strongly depends on their type and surface chemistry.

1.2 Size-dependent Band Gap of SQDs

Based on quantum mechanical calculations, discrete energy levels are not observed in bulk materials; the quantized energy effect is unique to small nanoparticles such as QDs. In fact, nanoparticle sizes and particle wavelengths are similar in magnitude, and, as a result, they have discrete energy levels. QDs have both valence (occupied orbitals) and conduction (unoccupied orbitals) bands, and the gap between these bands can be tailored by simply altering size, shape, and charge potential. When an electron is excited from the valence band (*i.e.*, ejected from the ground state), a corresponding positively charged hole is left in its wake. When this occurs, electron flow begins in the conductive band of the QD. To re-establish the ground state, electrons and holes recombine and emit photons. This emission is effectively influenced by the excited state of the QD, and the

Table 1.1 Comparison of traditional organic dye fluorophores with QDs. Reproduced from ref. 13, https://doi.org/10.1080/14686996.2019.1590731, under the terms of the CC BY 4.0 license, http://creativecommons.org/licenses/by/4.0/.

Property	Traditional organic fluorophores	Quantum dots
Chemical properties	Chemical resistance is often poor	Resistant to chemical degradation; sensitivity to pH determined by coatings
Size scale	Molecular, <0.5 nm	Colloidal, 1.5 to 10 nm diameter
Hydrodynamic radius	Small, <0.6 nm	Variable, 1.4 to 40 nm
Absorption spectra	Discrete bands, FWHM, 35 to 80 nm to 100 nm	Strong and broad
Emission spectra	Broad, red-tailed, and asymmetric, FWHM, 35 to 70 to 100 nm	Narrow, symmetric, FWHM, 30 to 90 nm
Two-photon cross-section	10 to 500 GM	(2000 to 47 700) GM
Molar absorption coefficient	(10^3 to 10^5) cm^{-1} mol^{-1} L	(10^5 to 10^6) cm^{-1} mol^{-1} L
Quantum yield	Variable, 0.05 to 1.0	High, >20%
Fluorescence lifetime	Short, <5 ns, mono-exponential decay	Long, >10 ns, typically multi-exponential decay
Solubility or dispersibility	Control by substitution pattern	Control *via* surface chemistry (ligands)
Thermal stability	Dependent on dye class; can be critical for NIR-wavelength dyes	High; depends on shell or ligands
Photostability	Usually, poor	Excellent resistance to photobleaching; observation time of minutes to hours
Bioconjugation labels	Monovalent to multivalent labeling possible	Scaffolds: Monovalent conjugation can be challenging; distribution of multivalences often encountered
Applicability to single molecule analysis	Moderate; limited by photobleaching	Good; limited by blinking
Spectral multiplexing	Possible	Ideal for multi-color experiments; up to five colors demonstrated
Multifunctionality	Difficult and few	Great potential
Toxicity	Variable, based on dye	Related to heavy metal

wavelength of the emitted light is dependent on the particle size. Larger-sized QDs emit large wavelength photons (red emission), whereas small-sized QDs emit short wavelength light (blue emission).

From the optical point of view, removing an electron from a valence band and placing it into the conduction band creates an electron–hole pair (referred to as an exciton). However, the finite size of a dot means that talking about a free electron and a free hole is meaningless. The distance between the electron and the hole (known as the Bohr radius) is controlled by the size of the dot. By shrinking a QD, the size of the particle approaches the size of the Bohr radius; in this case, the exciton in the QD behaves nearly like a "particle in a box." The Brus equation can be used to describe the band gap energy of 3-D spherical QDs. Based on this model, the band gap of a QD is (ΔE) evaluated according to eqn (1.1).

$$\Delta E = E_g(bulk) + \frac{h^2}{8R^2}\left(\frac{1}{m_e^*} + \frac{1}{m_h^*}\right) - \frac{1.786\, e^2}{4\pi\varepsilon_0\varepsilon_R R^2} \tag{1.1}$$

where E_g (*bulk*) is the band gap energy of the bulk semiconductor; R is radius of the QD; ε_0 is the permittivity of vacuum; ε_r is the relative permittivity; and m_e^* and m_h^* are the effective masses of the excited electron and excited hole, respectively. According to this equation, the band gap is inversely proportional to the square of the QD's size. Thus, with a decrease in the size of a QD, the energy gap will increase (see Figure 1.1); consequently, the emission wavelength changes with the QD's size. The changes in the band gap that occur with changes in size are responsible for QDs' multi-color nature, as various methods of synthesis can produce QDs with a Gaussian-size distribution in the range of 2–10 nm. The size-dependent band gap of semiconductor QDs (SQDs) is a well-known quantum confinement effect, which is the special confinement of electron–hole pairs in one or more dimensions within a material. Such behavior is not observed in bulk materials. In fact, as particle sizes increase beyond the magnitude of the characteristic wave function, particles tend to behave freely and form a continuous state. Hence, bulk materials do not have a specific fluorescence size effect.

Despite the many applications that exist for QDs, some disadvantages limit their applications. For example, these materials pose potential risks to human health due to the leaching of toxic heavy metals contained in their core; QDs' size and surface chemistry also make them intrinsically toxic.[1,2] Research on QDs must, therefore, balance attempts to take advantage of their small size and other desirable features with efforts to resolve their safety concerns. An increase in the rate of publications with "Quantum Dots" in their titles in the period from January 2020 to September 2022 (see Figure 1.2) suggests a rising interest in these materials. Nanodots have, thus, attracted and continue to attract, considerable R&D interest from researchers all over the world; many

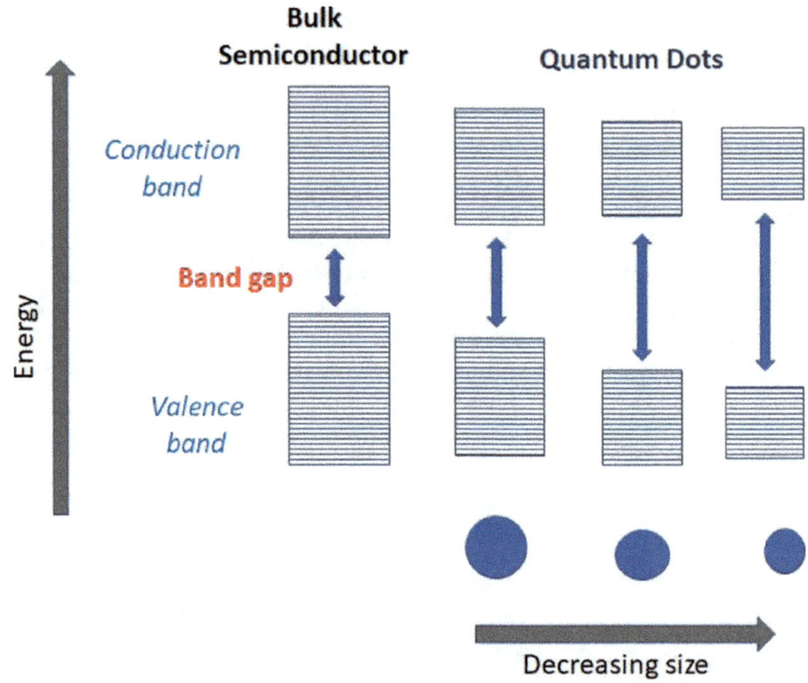

Figure 1.1 Variation of band gap with semiconductor sizes.

start-ups, laboratories, and companies have been founded to commercialize QD technology from the lab to the market.

This chapter provides a historical overview of the development of QDs, from their initial detection to present-day research. Their advantages and limitations in biomedicine and detection sciences are also discussed. Finally, we discuss different detection mechanisms in biomedical and chemical sensing using QDs as fluorescence probe.

1.3 Classification of Quantum Dots

The original QDs consisted of group IV and III–V compounds, with significant progress made thereafter on the incorporation of compounds from other groups.[17,18] QDs are categorized according to their compositions. A QD may be composed of a single element (*e.g.*, Si and Ge QDs), or it can be a compound semiconductor (*e.g.*, CdSe, CdTe, PbS, and PbSe QDs). Types of QD include core–shell, metal dichalcogenide, and metal oxides. Recently, carbon-based QDs (CDs) have been a focus of research due to the many advantages they have over the other QDs; CQDs are a shining star in the QD sky. The preparation of CQDs is simple, their precursors are inexpensive, and they are biocompatible.

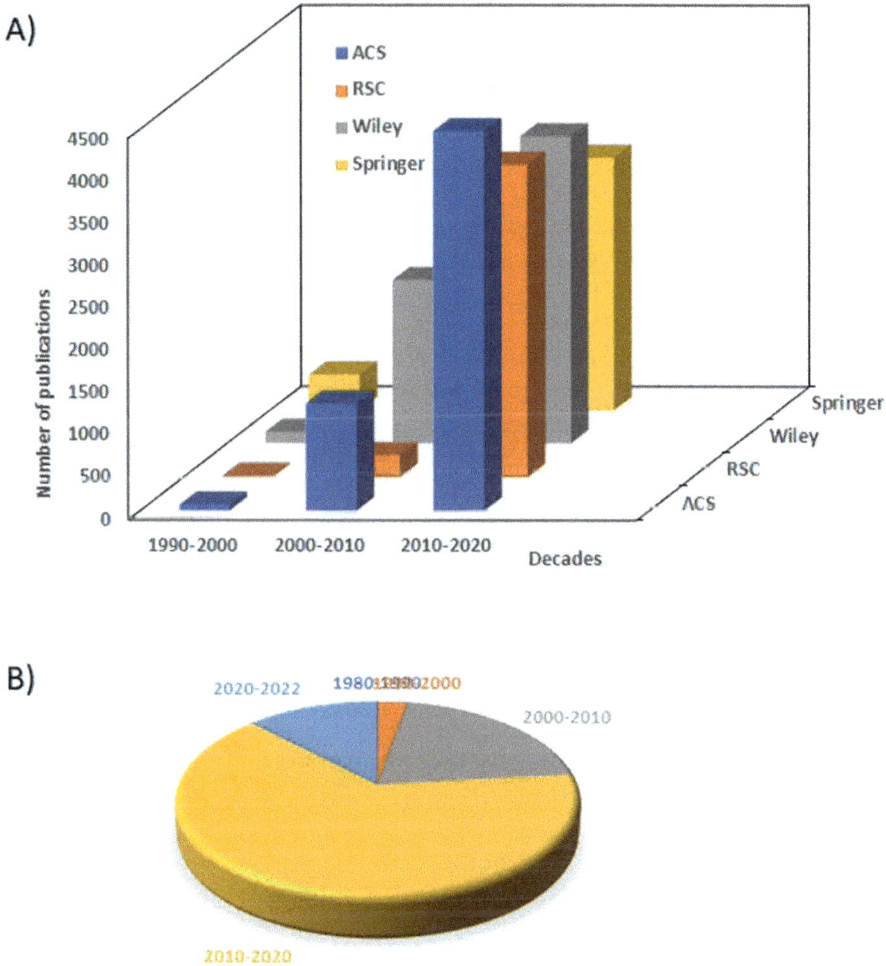

Figure 1.2 The number of published articles on QDs by four different publishers: (A) for three different decades starting from the first time they were invented, and (B) total number of published articles from 1980 to September 2022.

1.3.1 Semiconductor QDs

QDs prepared from semiconductor materials have achieved great interest because they emit light in a broad range of electromagnetic radiations (from UV to NIR[19–21]) with high quantum yield and excellent photostability. SQDs are categorized according to the names of the periodic table groups to which their elements belong, using Mendeleev's group numbering system (I–VIII). For example, Cu from group I (transition metal) and S from group VI (nonmetal) form Cu_2S, which is categorized as a group I–VI semiconductor. SQDs can be roughly classified as group I–VI (Cu_2S), group

I-VII (AgBr), group II–VI (ZnO, ZnS, ZnSe, ZnTe, CdS, CdSe, and CdTe,), group III–V (InN, InP, InAs, InSb, GaN, GaP, GaAs, and GaSb), group IV–VI (PbS, PbSe, and PbTe), and group IV (Si) according to this system. These semiconductor groups are briefly described below.

1.3.1.1 Group I–VI SQDs

Group I–VI elements are combined to prepare this kind of SQD. The most widely used SQDs of this group are Ag_2X (where X = S, Se, Te)[20] because they have a small band gap and are categorized as NIR-emitting SQDs. Compared with Pb, Cd, and Hg, Ag is less harmful to bioenvironments in SQDs, as only an ultra-trace amount of Ag is released to the surrounding media due to its very low water solubility ($k_{sp} = 6.3 \times 10^{-50}$). Surface passivation of Ag_2X is easily performed with plenty of functional groups, which makes Ag-based SQDs reliable for NIR applications in biomedical sciences through easy conjugation with biomolecules. Ag_2S, Ag_2Se, and Ag_2Te QDs have been prepared with emission wavelengths in the range of 700 to 1700 nm for targeted drug delivery and for the imaging of different cell lines and tissues. Ag_2Te has been introduced as a fluorophore for ultrafast brain imaging. The only point is that there are challenges in Ag_2Se synthesis as it is the most difficult to prepare.

1.3.1.2 Group II–VI SQDs

Group II–VI materials are used to prepare SQDs for photochemical studies. CdS was the first synthetic QD in this group; metal sulfides are easy to prepare, and the starting materials are readily accessible.[22–24] In 1983, researchers realized that the photoactivity of CdS increased enormously with a decrease in its particle size, and accordingly, they suggested the concept of colloidal QDs[25] as well as relationship between size and emission of the SQDs. This work opened a new window into future QD applications. In 1993, CdS, CdSe, and CdTe were prepared in an organic solvent and then different preparation methods were developed. Thereafter, ZnO QDs (Zn is in the same group as Cd) were prepared, and their photoluminescent properties were evaluated. From that time on, group II–VI QDs become one of the most widely used and well-characterized types of QDs. However, group II–VI QDs contain heavy metal elements, such as cadmium, which limit their application in nanomedicine due to toxicity issues. Therefore, scientists have tried to introduce reliable and bio-friendly QDs as alternatives *via* various methods of surface passivation.

1.3.1.3 Group III–V Quantum Dots

GaN, InSb, GaSb, InP, GaP, InAs, AlSb, InN, and GaAs QDs are examples from this group. Some members of this group are highly luminescent and have low toxicity.[20–22,26] These properties make them suitable for biomedical applications. The synthesis methods for this group of QDs are well documented with stable QDs being prepared compared to group II–VI. Therefore, significant research has been conducted on group II–V QDs. Spectroscopic evaluations of prepared QDs show that their emissions are size-dependent. Among these QDs, InP is used extensively in biomedical systems because it is biocompatible and lacks the biological toxicity associated with Cd ions. Additionally, some ternary SQDs are prepared from group III–V QDs, such as InGaAs and InGaP. The introduction of a third element to binary QDs plays a crucial role in their potential applications, as this addition alters their photochemical and chemical properties extensively. Ternary QDs are less toxic and emerging as a safer alternative to binary QDs while exhibiting similarly excellent luminescence. Ternary SQDs are also prepared with elements from other periodic table groups. For example, $CuInS_2$, $AgInSe_2$, and $AgInSe_2$ are group I–III–VI ternary QDs. They have the advantage of a tunable band gap, which makes it easy to adjust their fluorescence from the red to the NIR region, and they have a long fluorescence lifetime. They will be discussed in more detail in later sections.

1.3.1.4 Group IV–VI Quantum Dots

Group IV–VI salt nanostructures such as PbS, PbSe, and PbTe are more appropriate for research of confinement limit and accordingly provide a chance to study the effect of strong confinement due to approaching the limit of quantum confinement compared with the groups II–VI and III–VI. They are also reliable for studies of dimensional quantization systems. It is possible to prepare high-quality PbS QDs, and they can easily be grown on glass substrates with a tunable quantum binding energy gap from 1.1 to 2 μm. However, the toxicity of lead limits their applications in nanomedicine.

1.3.1.5 Group IV

The most important semiconductor from group IV of the periodic table is silicon (Si). It is abundant and widely used in microelectronics. Si bulk material has an indirect band gap and, hence, it is not suitable for light emitting devices, but it has efficient fluorescence when converted to QDs; it has a quantum yield of up to 60–75%.[22,27] In fact, SiQDs are used abundantly in various *in vivo* applications, owing to their low toxicity as well as advancements in preparation and surface functionalization There are two main issues related to SiQD preparation. First, it is of the utmost importance that colloidal SiQDs be both high quality and efficiently prepared. At least, the preparation method should produce SiQDs particles with a size range of 1–5 nm, a quantum yield above 10–15%, and an emission wavelength from blue to NIR. The second concern is the need to modify SiQDs' surfaces, which are normally prone to oxidation, which in turn decreases their fluorescence quantum yield. However, the synthesis of SiQDs *via* non-thermal plasma in the absence of oxygen and the passivation of their surface with organic ligands produce SiQDs with a high yield and a high quantum yield of 62% as well as a narrow size distribution. However, experimental conditions must be adjusted, and this process is time-consuming and costly.

1.3.1.6 Core–Shell QDs

The toxic effects of CdS or CdTe encouraged scientists to introduce ternary compounds known as core–shell QDs (CSQDs). In an ordinary QD, non-radiative recombination occurs because a large number of atoms are on the surface due to the compound's high surface area-to-volume ratio.[2] In a CSQD, the core is covered with semiconductor materials with different band gaps and band alignments. The coating stabilizes the core and reduces non-radiative recombination, resulting in luminescence with a high quantum yield. It also protects the core from oxidation and inhibits the production of free heavy metals, and it enables the easy localization of dopants such as Mn, Ag, Cu, and Ni. Metal dopants significantly improve the physical properties of QDs. The optical properties of CSQDs can be adjusted by changing the shell thickness and the coating element. They have more photostability, high extinction coefficient, lower width at emission half height, broad absorption spectra, and size-dependent emission. These improvements make it possible to design more sensitive, stable, and reliable biosensors with diverse applications.

CSQDs of various shapes (spherical, disk, wire, rod, hexagonal, *etc.*) have been prepared through different synthetic procedures. A spherical core covered with spherical semiconductor materials is the most common form. The optoelectronic properties of these CSQDs are both shape and size dependent; sensitivity and melting point are completely shape dependent, as are the optical and electronic properties. Accordingly, CSQDs are classified into two main sub-groups, as described below.

1.3.1.6.1 Classification of CSQDs Based on Band Alignment. CSQDs are classified based on the band gap and alignment of valence and conduction band edges for the shell and core materials (see Figure 1.3).[2] In Figure 1.3, E_g refers to the band gap and V_{VB} and V_{CB} represent the valence and conduction bond edges of the materials, respectively. Band alignments are classified as type I, inverse type I, type II, inverse type II, and quasi-type II. In type I (Figure 1.3A(a)), E_g is higher, but V_{VB} and V_{CB} are smaller for the shell compared with the core materials. These features provide comprehensive confinement to the carrier in the core region of these materials. In inverse type I, in contrast, everything is reversed (Figure 1.3A(b)): the band gap of the shell is smaller than the core materials and, therefore, the carriers are partially or completely confined within the shell materials depending on the structural dimensions. In type II (Figure 1.3A(c)), E_g, V_{VB}, and V_{CB} are smaller in the shell region compared to the core region; consequently, electrons and holes are mostly confined to the shell and

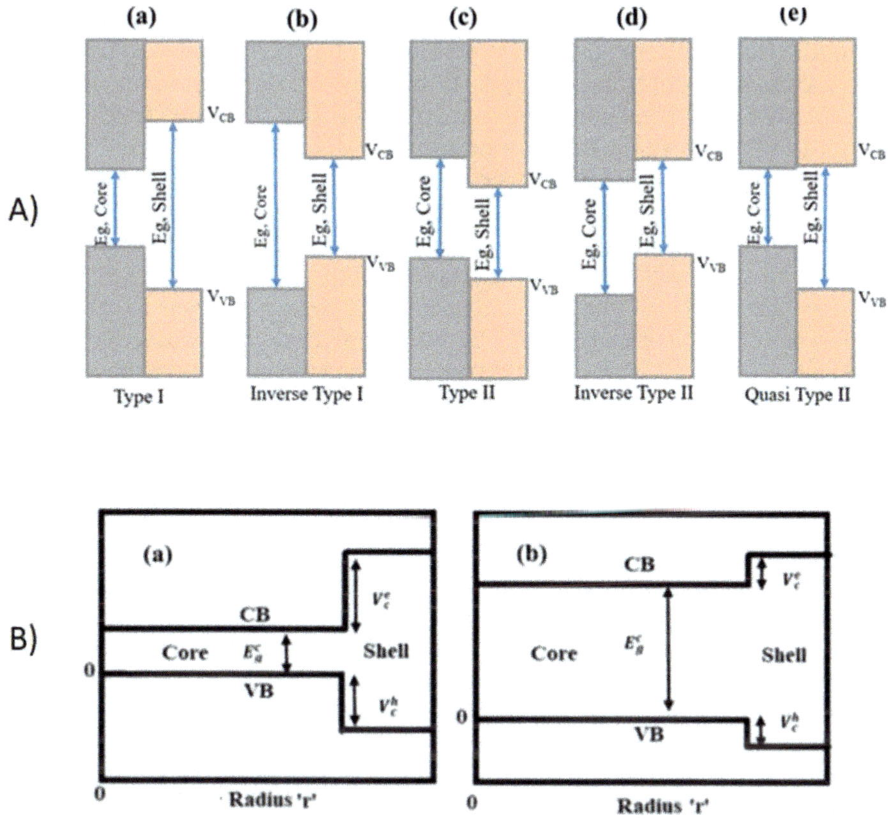

Figure 1.3 The classification of CSQDs based on (A) band gap band alignment and (B) confinement strength. Reproduced from ref. 2 with permission from Elsevier, Copyright 1991.

core regions. In inverse type II (Figure 1.3A(d)), the reverse is true: while the electrons are mostly confined to the core region, the holes are in the shell region. In this configuration, the leakage of the carrier in the shell region diminishes the chemical stability of CSQDs with increased Stoke's shift, which can be tuned by structural dimensions. Quasi-type II (Figure 1.3A(e)) is an intermediate type of CSQD in which V_{VB} offset of the core and shell are as in type I while V_{CB} of core and shell are quasi-aligned to each other. Therefore, the holes are confined to the core, while the electrons are delocalized over the whole CSQD.

It should be mentioned that different alignments have different applications. For example, quasi-type II and type II CSQDs are extensively used in solar cells. When determining which CSQD to use, properties such as the carrier's lifetime, stability, and leakage should be considered.

1.3.1.6.2 Classification of CSQDs Based on Confinement Strength. Confinement strength is defined as the strength of step potential at the core–shell interface.[2] Figure 1.3B shows the arrangement of band gaps of core and shell materials with respect to each other in this subgroup. In Figure 1.3B(a), the core has a narrow band gap and is encapsulated by the shell of a higher band gap material, as in PbS/CdS. As a result, a larger step potential is observed at the core–shell interface, which causes the carrier to be strongly confined in the core region. In these CSQDs, referred to as strong confined CSQDs, the step potential is much higher than the band gap of the core materials. In contrast, when a core with a wide band gap is encapsulated by a shell with slightly wider band gap materials (Figure 1.3B(b)) (as in ZnTe/ZnSe), the step potential is lower than the band gap of the core materials. This results in the weak confinement of the carrier, which affects the optoelectronic properties of the CSQD.

1.3.2 Carbon-based QDs

Carbon-based nanodots (CDs), also referred to as group VI QDs, are luminescent nanomaterials. The term is also a general one for all carbon-based nanomaterials that have at least one dimension less than 10 nm in size and contain sp^3/sp^2 carbon. These CDs may have nitrogen and oxygen atoms in their structures.[10,28] CDs are synthesized through the polymerization of small molecules or the pyrolysis of graphite-based materials. Graphite-based materials may exfoliate to CDs preparation *via* polymerization by small molecules producing cross-linked polymer clusters (CPDs) by dehydration or a well-defined π conjugation system graphitic structure that is CQDs. These CDs are bio-friendly and eco-friendly because they are mainly composed of carbon. In addition to their biocompatibility and environmental friendliness, CDs have the advantage of adjustability: the functional groups on their surface can be adjusted by choosing appropriate carbon sources and synthesis procedures. For example, nitrogen-doped CQDs are prepared *via* hydrothermal heating of a mixture of citric acid and ethylene diamine in a stainless-steel autoclave. Therefore, CQDs with different functional groups and doping elements can be prepared for a vast variety of applications. In addition, they have very low to no toxicity and are a suitable candidate for the production of fluorescence probes for biomedical applications. They have discrete energy levels, and their fluorescence originates from the π domain, surface defects, doping elements, quasi-molecules, and sub-fluorophores (see Figure 1.4).

1.3.3 Two-dimensional QDs (2-D QDs)

Research has shown that graphene illuminates when its size falls below 20 nm. Besides the nanoscale, graphene shows new characteristics such

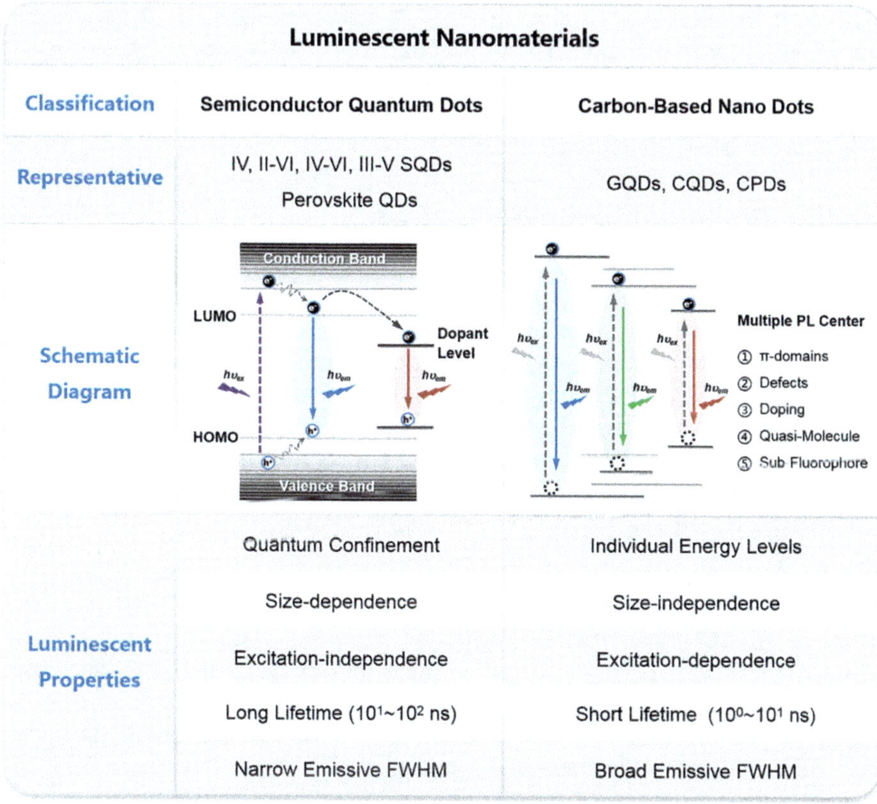

Figure 1.4 Photoluminescence behavior of SQDs *vs.* carbon-based QDs. Reproduced from ref. 28 with permission from American Chemical Society, Copyright 2019.

as a large surface area, higher solubility, and easier surface modification. Graphene-based QDs (GQDs) represent a new direction in graphene research as 2 D nanomaterials; they have attracted scientists' attention to other 2-D QDs such as graphitic carbon nitrite, hexagonal boron nitride (H-BN), and transition metal dichalcogenic.

1.3.3.1 *Graphene QDs*

GQDs are single-layer or multilayer graphene with dimensions of less than 20 nm. Compared with SQDs, GQDs are more chemically inert, have better solubility in water and polar organic solvents, and have a wider emission wavelength. They are also highly biocompatible due to their low toxicity.[29] Unlike their CD cousins, GQDs clearly possess a graphene structure inside the dots, regardless of the dot size, which endows them with some of the unusual properties of graphene.[30,31] The photoluminescence of GQDs is affected by their size, doped atoms, chemical moieties, and geometric shapes. It is, therefore, possible to prepare GQDs with appropriate properties for various applications by adjusting synthetic procedures. For these reasons, GQDs have attracted significant scientific attention; GQD chemistry is still a subject of active research.

1.3.3.2 Graphitic Carbon Nitride QDs

Graphitic carbon nitride (g-C_3N_4) QDs (CNQDs) are composed of covalently connected N and C atoms. This is considered as N-substituted graphene, but their physio-chemical properties are far different from those of N-doped GQDs. CNQDs can be prepared with simple procedures and have strong blue fluorescence and up-conversion characteristics. The quantum yields of CNQDs are usually high and can also be adjusted by changing the ratio of the precursors. CNQDs have attracted much attention for biomedical applications by virtue of their good biocompatibility, stable fluorescence, high quantum yield, and nontoxicity.[14]

1.3.3.3 H-BN QDs

In these materials, the C–C bonds of graphene QDs are completely replaced by B–N bonds. The B–N covalent bonds in h-BN QDS have the same electronic properties and structures as the C–C covalent bonds in graphene. These QDs exhibit excellent chemical and thermal stability, and they are easily synthesized[32]; for example, they can be produced through the exfoliation of h-BN flakes or from small starting materials through various methods. h-BN QDs also exhibit higher photophysical, chemical, thermal, and photochemical stability and are non-toxic and more biocompatible compared to conventional semiconductor materials. They are even more biocompatible than CQDs. Additionally, h-BN QDs are more pH tolerant in the range of 2–12 with no significant changes in their fluorescence, which makes them suitable fluorophores for use in biological environments.

1.3.3.4 Transition Metal Dichalcogenide QDs

The term transition metal dichalcogenide (TMD) refers to a class of materials with multifunctional properties. TMDs are composed of transition metal sandwiched between two hexagonal planes of chalcogen atoms. Some common examples of TMDs are WS_2, WSe_2, ReS_2, MoS_2, and $MoSe_2$. Like GQDs, they have a layered structure; weak van der Waals forces hold TMDs' layers together. TMD QDs can be prepared *via* ultrasonic agitation or solvent heat treatment. TMD QDs can be prepared from bulk materials *via* ultrasonic agitation at room temperature or *via* solvent heat treatment. The process of exfoliating bulk TMDs takes time; material scientists are therefore still conducting research to modify preparation methods. The intercalation of alkaline metals accelerates TMD QDs' preparation.

Recent microwave-assisted exfoliation of MoS_2 bulk powder in ammonia has produced highly luminescent MoS_2 QDs in about 30 min,[33] which is the lowest reported exfoliation time.[33] MoS_2 QDs are a type of transition metal disulfide nanomaterial with a unique 2-D structure similar to graphene.[7,34] These QDs have an adjustable emission spectrum and high QY, and they are easily functionalized. TMD QDs' good biocompatibility makes them suitable for tumor treatment platforms,[35] disease wound therapy,[36] and biosensors and fluorescence probes.[33] They are also suitable for use as catalysts in chemical reactions.

1.3.3.5 Atomic Black Phosphorus (Phosphorene) Quantum Dots

2-D black phosphorus (BP; phosphorene) is prepared from phosphorus atoms. Each phosphorus atom covalently binds to the other three phosphorus atoms and forms a folded honeycomb structure. The inherent direct band gap of the monolayer phosphene is 0.8–1.5 eV (theoretically calculated). In 2015, a simple solution method[37] was introduced for the preparation of BP QDs. As-prepared BP QDs had a size of 4.9 ± 1.6 nm and a thickness of 1.9 ± 0.9 nm (about 4.2 layers). Probe sonication and bath sonication methods[38] produce BP QDs with a transverse dimension of about 2.6 nm and a thickness of about 1.5 nm, with good light stability. The stability of BP QDs was enhanced in the medium after PEG modification, and no obvious toxicity to different types of cells was observed. Therefore, BP QDs are good candidates for biomedical and detection applications.

1.3.4 Perovskite Quantum Dots

Perovskite is the name for $CaTiO_3$ that a mineralogist discovered in the Ural Mountains.[22] Perovskite QDs are a rising star in scientific research, owing to their outstanding photoelectronic properties. Perovskite is a kind of ceramic oxide with the general formula ABX_3, in which X is an oxygen atom and the A ion is usually an alkaline or rare earth metal element that is capable of coordinating with 12 oxygen atoms to form a dense cubic structure that stabilizes the perovskite structure. The B ion is a transition metal element (such as Co, Fe, or Mn) that has a coordination number of six and can form an octahedral structure. The variable valence state of the transition metals determines many properties of the Perovskite structure. Perovskite's properties remain constant with changes in the A site and B site ions, and it has the same radius elements; together, these factors make it a reliable catalyst. Two general kinds of perovskite QDs are available: hybrid organic–inorganic (HOIPs) and all inorganic (AIPs) perovskites. Various AIPs and HOIPs have been synthesized recently, but their biomedical applications are limited due to the toxicity of some of their structural elements (*e.g.*, Pb) as well as their low stability. Material scientists are, therefore, working to fabricate more biofriendly and stable perovskite QDs for nanomedical applications.

1.3.5 Semiconductor QDs *Versus* Carbon QDs

CQDs and SQDs are small nanomaterials that fall within the field of quantum mechanics. The key difference between these QDs is that CDs are small carbon nanomaterials and SQDs are small semiconductor materials. CDs (especially CQDs) have recently attracted more scientific attention in biomedical research compared with SQDs for multiple reasons: they are more highly soluble in aqueous media, they are more eco- and bio-friendly, their starting materials are inexpensive, their synthesis is economic, and their physio-chemical properties are highly controllable.[11,39] Figure 1.4 compares the photoluminescent behavior of carbon-based dots and SQDs.[28] Although the photoluminescence mechanisms and luminescence performance of these QDs are different (Figure 1.4), both show a size-dependent emission. SQDs have a very high quantum yield (QY%): usually more than 90% with high chemical stability and photostability as well as a long lifetime. The QY% of carbon-based QDs is in the range of 3% to 90%, with relative stability depending on their carbon sources and methods of synthesis. The advantages of CQDs are low preparation cost and biocompatibility due to low or no toxicity while SQDs are very toxic to humans and the environment. In SQDs, photoluminescence (PL) arises from electron–hole recombination; PL is therefore mainly due to the quantum confinement effect, especially when the QDs' sizes are smaller than their exciton Bohr radius. Moreover, SQDs with uniform size distribution exhibit excitation-independent PL and have a narrow full width at half emission maximum (FWHM). Conversely, CQDs have discrete energy levels and multiple PL centers; π domains, defects, different surface states, doping, quasi-molecules, and subfluorophors provide various individual energy levels rather than a continuous band gap. CQDs therefore mostly have excitation-dependent emission, a wider FWHM, and a shorter fluorescence lifetime compared to SQDs.

1.4 Detection Mechanism

Due to their improved fluorescence properties and photostability, QDs offer tremendous opportunities for analyte detection. Most QDs-based sensors' signals are produced by either quenching the QDs' fluorescence emissions in the presence of the analyte or increasing fluorescence by suppressing the quenching effect.[40] Multiple quenching processes support detection mechanisms in QD-based systems, including static quenching, dynamic quenching, resonance energy transfer (RET), photo-induced electron transfer (PET), inner filter effect (IFE), and aggregation-caused signal changes (ACSC). To a lesser extent, enhanced QDs' emission is the signal in the response to an analyte. Recently, QDs have also acted as quenchers in biosensing devices to improve the sensitivity of the measure-

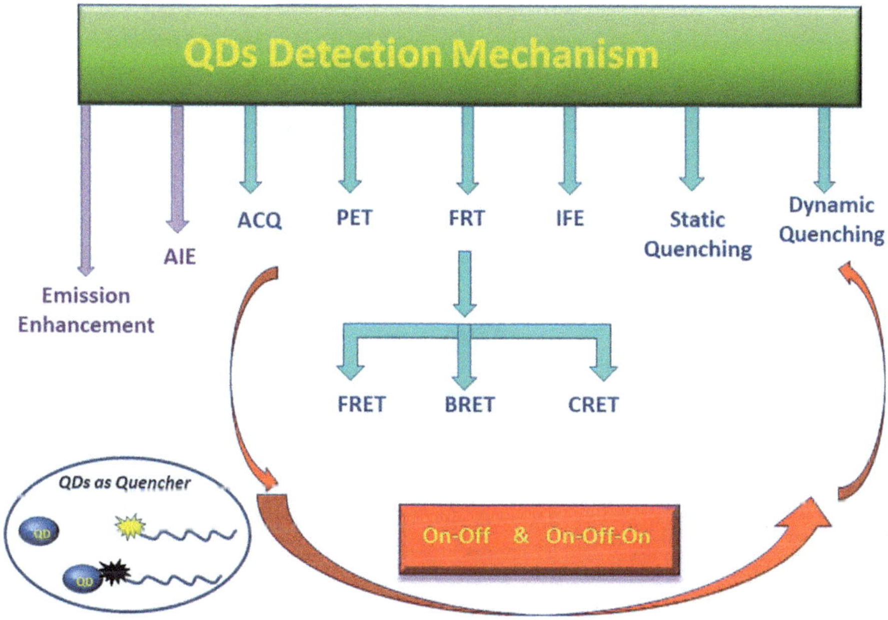

Figure 1.5 Possible detection mechanisms in biosensors-involved fluorescents QDs. Fluorescence resonance energy transfer (FRET), photo-induced electron transfer (PET), inner filter effect (IFE), aggregation-caused quenching (ACQ), aggregation-induced emission (AIE).

ments.[41] Figure 1.5 illustrates possible detection mechanisms when QDs are used as fluorophores in detection science.

1.4.1 Static and Dynamic Quenching

Both dynamic quenching and static quenching are the result of interactions between QDs and the quenching agent, but their effects are different. Static quenching occurs when the interaction between the QDs and the quencher forms a non-fluorescent ground-state complex. In static quenching, there is no emission of a photon when the complex returns to the ground state without delay.

In dynamic quenching, the excited fluorescent QDs collide with the quencher and return to the ground state with energy transfer or charge transfer. This mechanism is represented by the linear Stern–Volmer equation[42]:

$$\frac{I_f^0}{I_f} = 1 + k_q \tau_0 [Q] \tag{1.2}$$

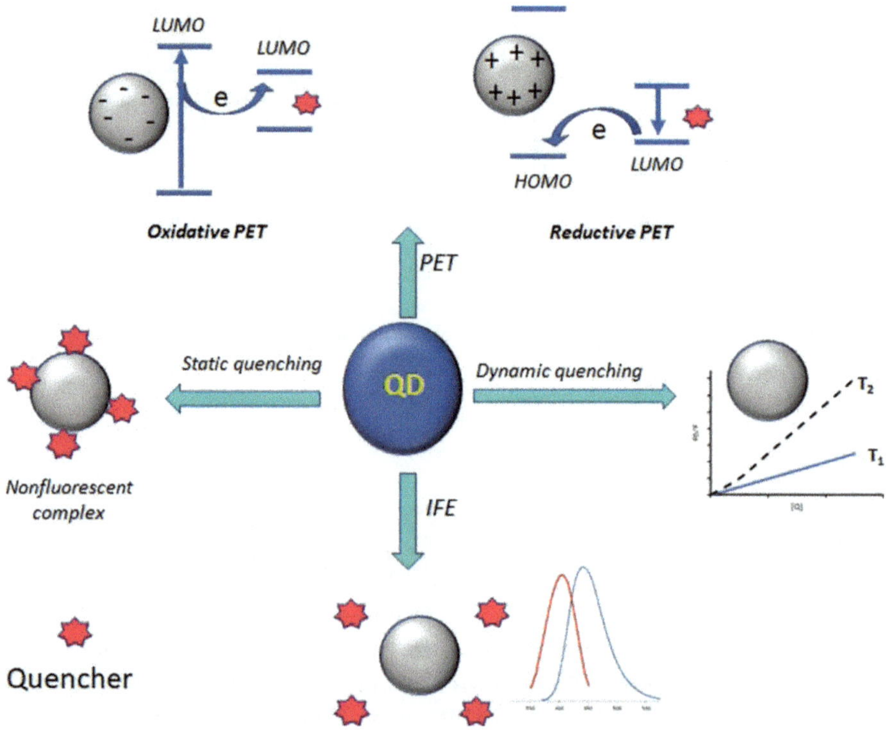

Figure 1.6 Schematic representation of PET, IFE, static, and dynamic quenching of QDs in the presence of a quencher.

where I^0 and I are the intensities of fluorescence without and with a quencher, respectively, k_q is the quencher rate coefficient, τ is the lifetime of the emissive excited state of chemical species, and $[Q]$ is the concentration of the quencher. Several features distinguish dynamic quenching from static quenching: (1) in dynamic quenching, changes in the lifetime of QDs would have occurred in the presence or absence of a quencher; (2) only the excited states of the QDs are affected in dynamic quenching and the absorption spectra of QDs is not observed; and (3) an increase in temperature can cause an increase in the dynamic quenching effect (see Figure 1.6).

1.4.2 Resonance Energy Transfer

RET, as a dipole–dipole interaction, happens when non-radiative energy gets transferred between a donor and an acceptor molecule while the distance between them is less than 10 nm. Therefore, the distance between the QD and the quencher is of vital importance for RET to happen. Distance has an inverse relationship with the emission intensity; if the distance between QDs and the quencher decreases, the emission intensity increases. The term RET may refer to three separate phenomena: (a) chemiluminescence RET (CRET), (b) bioluminescence RET (BRET), and (c) fluorescence RET (FRET), in which the donor molecules are a chemiluminescent structure, a bioluminescent protein, and a fluorophore, respectively, while the acceptor is a fluorophore.[43] This technique is extensively used in sensing and imaging, and in drug tracking both *in vivo* and *in vitro*.[44–51]

1.4.2.1 Fluorescence Resonance Energy Transfer

FRET is regarded as a highly sensitive spectroscopic technique with a wide range of applications for fluorescent probes, such as biomedical and chemical detection (see Table 1.2). FRET is measured not just in bulk fluorescence experiments, but also in fluorescence microscopy techniques such as single molecule fluorescence spectroscopy. FRET occurs when a donor QD in an excited state transfers energy to a nearby acceptor molecule, which in turn decreases the intensity of the donor's (see Figure 1.7A). In the fluorescence process, the rate of energy transfer varies with some parameters such as the distance between the donor and the acceptor and their relative orientations, and also their spectral overlaps.[79,80] There are three points that can demonstrate that FRET is the current mechanism for quenching QDs: (1) overlap between the absorbance spectrum of the quencher and the fluorescence spectrum of QDs; (2) decrease in the fluorescence lifetime of QDs; and (3) the distance between QDs and quencher would be in the range of 10 Å–100 Å.[81]

In some studies, two or more emitting molecules with different emission wavelengths were used to prepare FRET-based dual emission nanoprobes. In these nanoprobes, one of the fluorophores served as the donor, while the other accepted the energy as the receptor. These studies indicate that a FRET system can also be used in ratiometric fluorescence sensing probes. Many sensitive and selective FRET-based ratiometric nanoprobes have been introduced.[44,46,49] QDs co-assembled with different organic fluorophores, other QDs, drug molecules, metals, biomolecules, nanomaterials, and graphene oxide have been used in highly sensitive and selective detection devices. For example, drug molecules as fluorophores attached to CQDs have promising applications for drug delivery to cells and tissues, as well as for bioimaging and biomedical assays. Fluorescent drug molecules used extensively in FRET probe designs include doxorubicin,[82] porphyrins,[83,84] vitamins,[85] curcumin,[86] rhodizonate,[51] and cisplatin (IV) (a prodrug).[87] FRET occurred when doxorubicin (as an acceptor) was coupled with the surfaces of CQDs (as a donor). The FRETs terminated when the drug was released from the CQDs' surfaces in the cell.[88] CQDs were also used for the detection of biologically significant vitamin B_{12}; this was achieved through the development of a FRET system pairing CQDs as donors and vitamin B_{12} as a receptor.[89]

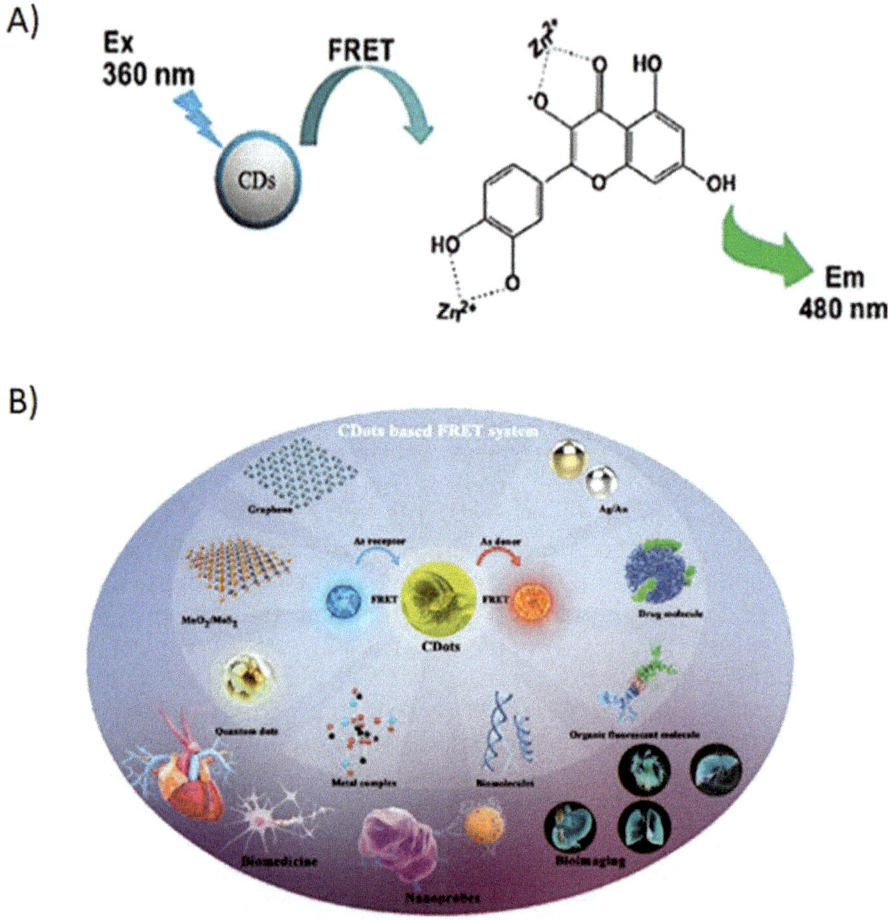

Figure 1.7 FRET in QDs nanoprobes. (A) FRET mechanism, (B) perspective for potential future application of CDs-based FRET system. Reproduced from ref. 48 with permission from the Royal Society of Chemistry.

Considering that CQDs can act as both donors and acceptors in FRET systems, an enormous variety of nanoprobes can be made for different targeted applications (see Figure 1.7B). Recently, a nanoprobe was designed for sensing heavy metal ions based on a FRET system involving GQDs and CQDs.[48] In this probe, GQDs serve as energy donors; they transfer energy to CQDs that act as electron acceptors, resulting in an increase in the intensity of the CQDs' fluorescence. With the addition of Hg^{2+} and As^{3+} ions to the nanoprobe, the FRET signal significantly diminished, and the reduction in FRET intensity was linearly proportional to the heavy metal ions' concentration.

A new nanoprobe for the detection of volatile compounds was developed based on a FRET system using Mn-doped CQDs (Mn-CQDs) as

Table 1.2 Some examples of FRET systems based on fluorescent molecules and CQDs along with their applications. Reproduced from ref. 48 with permission from the Royal Society of Chemistry.

Coupling molecule	Application	Ref.	Coupling molecule	Application	Ref.
Chlorin e6	Bioimaging, biomedicine	52	Naphthalimide	H_2S nanoprobe	53
Naphthalimide	NO nanoprobe	54	Isothiocyanate	pH nanoprobe	55
NBD-PE; BODIPY-PH	Bioimaging	56	Rhodamine B	Fe^{3+} nanoprobe	57
Boronate-protected fluorescein (PFI)	H_2O_2 nanoprobe, bioimaging	58	Trinitrophenol	Bioimaging	
Nitroaromatics	Nanosensor	59	NBD-PE	Nanoprobe, bioimaging	60
Rhodamine B	Up-conversion properties	61	Chlorine e6	Bioimaging	62
Rhodamine 6G	Fe3+ nanoprobe	63	TNB-	DDVP nanoprobe	64
Naphtalimide	Nanoprobe, bioimaging	65	Ethidium chloride	DNA nanosensor	66
Glyphosate	Gly-nanoprobe	67	2,3 Diaminiphenazine	I^- nanoprobe	68
Fluorescein	NH_3 nanoprobe	69	Isothiocyanate	pH nanoprobe	70
Acridon derivate	RNA nanoprobe	71	Naphthalimide	Tyrosinase nanoprobe	72
Rhodamine123	Bioimaging	73	Nitrophenol	Nanoprobe, bioimaging	74
Acriflavine	SEB nanoprobe	75	Fluorescein (FAM)	Zn nanoprobe, bioimaging	76
Boron dipyrromethene	Bioimaging	77	Naphthalimide	Cys nanoprobe, bioimaging	78

both electron donors and acceptors.[90] Intra-particle FRET occurred in the Mn-CQDs from surface-related energy donor groups to metal-related energy acceptor groups. This showed that FRET is also sensitive to polarity due to the inherently fixed positions of the two fluorophores on the Mn-CQD's backbone.

1.4.2.2 Bioluminescence Resonance Energy Transfer

BRET occurs when energy is transferred between a light emitting donor biomolecule (typically a luciferase) and a light-sensitive acceptor molecule (typically a fluorescent protein). In this phenomenon, the energy derived from the luciferase reaction can be used to excite the fluorescent acceptor protein when it is in close proximity to the donor luciferase enzyme.

BRET is very similar to FRET, which makes it appealing as a means of circumventing some practical issues associated with FRET, such as light scattering, autofluorescence, and photobleaching. In FRET, the donor is a fluorescent molecule that must be excited by an external light source, whereas BRET does not require a light source, and natural luminescence from a biomolecule is transferred to an acceptor.

Yao *et al.*[91] designed a BRET system by fusing Luc8 with an MMP-2 substrate peptide and conjugating a six-histidine tag (6His) to the carboxylated CQDs through Ni^{2+}-mediated coordination-chelation interactions. They used the prepared system to detect the activity of protease MMP-2 with improved accuracy and high sensitivity. In fact, when the emission of cCQDs at 655 nm was increased, the QDs quenched the bioluminescent emission of Luc8 at 480 nm. Bioluminescence was recovered when the peptide was cleaved and, accordingly, the BRET process was disrupted.

1.4.2.3 Chemiluminescence Resonance Energy Transfer

CRET is a form of resonance energy transfer in which a non-radiative transfer of energy occurs between a chemiluminescence (CL) donor and an energy acceptor. Recently, QDs have been used as an appropriate type of CRET acceptor due to its good photostability and high fluorescence quantum yield.[47]

Chemiluminescence between luminol and hydrogen peroxide is catalyzed by the hemin/G-quadruplex HRP, the horseradish peroxidase-mimicking DNAzyme. Emitted light (the chemiluminescence) was used as a light source to trigger the emission of CdSe/ZnS QDs through CRET.[92] Moreover, the quantum dots were modified with nucleic acids and aptamers for sensing DNA and ATP *via* CRET. Dual switchable induced CRET based on the catalytic action of hemin/G-quadruplex on the oxidation of luminol by H_2O_2 was designed using nucleic acid functionalized CdSe/ZnS QDs with two different sizes that emit at 610 and 540 nm.[93] An on–off dual CRET-based sensor was designed through the cyclic aggregation and redispersion of QDs coupled with a hemin/G-quadruplex *via* potassium ion and 18-crown-6-ether.[93] The QDs conjugated with nucleic acid aggregated by the hemin/G-quadruplex as a cross linker in the presence of potassium ions; the chemiluminescent energy of luminol oxidation caused the QDs to emit at 610 and 540 nm because of the CRET. When 18-crown-6-ether was added, the bridge was separated, CdSe/ZnS QDs were redispersed, and a weak emission due to uncatalyzed chemiluminescence was observed (emission off).

Wang *et al.*[47] used hapten-functionalized core/multi-shell QDs for sulfamethazine detection based on homologous CRET with improved efficiency. Liu *et al.*[45] used a microwave irradiation technique to synthesize europium ion-modified BP QDs (Eu-BPQDs). They observed efficient CRET, which is attributed to the chelation of organic amines with the surface-functionalized europium.

1.4.3 Photo-induced Electron Transfer

In the PET mechanism, electron transfer occurs between a QD as the excited fluorophore (electron donor or electron receptor) and a quencher (electron receptor or electron donor). There are two types of PET: oxidative PET and reductive PET (see Figure 1.6). Oxidative PET occurs when the electron in the lower unoccupied molecular orbitals (LUMO) energy level of the QD (the electron donor) transfers to the LUMO of the quencher (the electron acceptor). This transfer happens when the energy level of the QD's LUMO is higher than that of the quencher and the distance between them is not far apart. In reductive PET, the electron transfer occurs between the LUMO of the quencher (the electron donor) and the higher occupied molecular orbital (HOMO) energy level of the QD (the electron acceptor).[81] In this process, a complex forms between the electron donor and the electron receptor that can return to the ground state without emitting a photon.

Callan *et al.*[94] have adopted the fluorophore spacer-receptor PET system format to generate a PET-operated QD probe for anions. They have shown that 1-(2-mercapto-ethyl)-3- phenyl-thiourea is a useful receptor for anions when bound to the surface of green-emitting CdSe–ZnS QDs. The observed quench was attributed to an increase in the reduction potential of the receptor upon anion binding, resulting in an increase in PET from the excited QD to the receptor and an attendant decrease in fluorescence intensity. The QD–spacer–receptor conjugate was used to sense the tetrabutylammonium salts of anions; the selectivity and sensitivity were comparable to organic dyes.

1.4.4 Inner Filter Effect

IFE is the result of an overlap between the excitation or emission spectra of QDs and the absorption spectra of a quencher. In fact, the IFE process cannot be called quenching because it occurs as a result of an attenuation of the excitation beam and/or absorption of emitted radiation in the presence of additional absorbing species or additional fluorophores in the solution. It is, therefore, usually called apparent quenching. IFE possibility is evaluated by the correction factor, which is calculated by eqn (1.3), in which F_{obs} is the measured maximum fluorescence intensity in the presence of the quencher and F_{cor} is the corrected fluorescence intensity in the absence of the quenching agent, A_{ex} and A_{em} are the absorbance and emission of the fluorescent molecule with the addition of the quencher, *g* is the distance between the edge of the excitation beam and the edge of the cuvette, *d* is the width of the cuvette, and *S* is the thickness of the excitation beam (slit width).

$$\frac{F_{cor}}{F_{obs}} = \frac{2.3dA_{ex}}{1-10^{dA_{ex}}} 10^{gA_{em}} \frac{2.3SA_{em}}{1-10^{SA_{em}}} \tag{1.3}$$

If the correction factor (F_{cor}/F_{obs}) changes with variations in quencher concentration, the mechanism is IFE; otherwise, other mechanisms may be involved, even if high spectral overlap exists between the absorption and emission of the fluorophore (*e.g.*, QDs) and the quencher.

In IFE, no electron or energy is transferred, and no chemical modification of the fluorophore occurs, in contrast with FRET or PET. IFE has nothing to do with distance and happens even when the distance between the emitter and the absorber exceeds 10 nm (see Figure 1.6).

Zhang *et al.* fabricated cobalt (II)-doped CQDs (Co-CQDs) and used them for the rapid detection of Cr (VI) ions in fish samples and tap water. They showed that Co-CQDs emission was quenched by Cr(VI) *via* an IFE mechanism.[95]

An IFE for radiometric fluorometric assay for uranyl ions[96] was designed using Au nanoparticles (AuNPs), Au nanoclusters (AuNCs), and blue emission CQDs. The ratio of the AuNCs' fluorescence intensity relative to the blue emission intensity of the CQDs was used as the detection signal. The AuNCs' red emissions were quenched when UO_2^{2+} was added to the mixture. IFE was determined to be the quenching mechanism because the absorption spectra of the AuNPs overlapped with the emission wavelengths of the AuNCs.

1.4.5 Aggregation-caused Signal Change

ACSC is used extensively in environmental chemistry, biochemistry, and life sciences. It consists of two opposing strategies: on the one hand, aggregation-induced emission (AIE) and aggregation-induced emission enhancement (AIEE)[97]; on the other hand, aggregation-induced emission quenching (ACQ). Each of these aggregation mechanisms may be observed in fluorescence probes such as organic fluorophores, QDs, or emitting biomolecules. Disaggregation recovered the emission of QDs and other fluorophores.

The functional group on the fluorophore and other environmental conditions such as polarity, pH, and ionic strength affect both the aggregation and the disaggregation of the luminescent molecule or species. AIE was first discovered by Tang *et al.*[98,99] and after that, AIE and AIEE were used extensively in sensing probes for a variety of applications. Many on–off fluorescence probes have been made based on AIE and ACQ strategies. ACQ possesses the advantage of high sensitivity with low-probe fluorophore concentrations compared with AIEE.

The ACQ of fluorescent QDs is due to non-radiative transitions such as energy transfer, surface electron transition, and π–π interaction between the QD's nucleus and the quencher. Most of QDs' fluorescence characteristics in the solution phase can be quenched in the cluster state. Therefore, QDs produce very strong fluorescence when they disperse. If QDs aggregate, energy transfer occurs between the excited and the unexcited QDs. Hence, energy is consumed without radiative transition, and, as a result, QDs' fluorescence fades and disappears.

Wang and co-workers[100] prepared amin-terminated nitrogen GQDs (NGQDs) using polyetherimide (PEI) as the functionalization material and lysine (Lys) as the carbon source. The amin group on the surface of NGQDs was able to coordinate with metal cations like Co^{2+}, leading to the aggregation of the NGQDs (fluorescence off). They determined the quenching mechanism through a set of experiments. Electron transfer quenching was rejected because the energy levels of the LUMO and HOMO of the GQD were −4.11 and −8.63 eV, respectively; this is far from the electric potential of Co^{2+} (−0.28 eV) meaning that the transfer of electrons from the LUMO of the GQDs to the LUMO of Co is not possible (see Figure 1.8a). UV-Vis spectral analysis of Co^{2+} and GQDs showed a large overlap, suggesting that IFE was the most probable mechanism. To check IFE, the correction factor (eqn 1.3) was determined for the system at different Co^{2+} concentrations. The results showed that the correction factors were the same for all Co^{2+} concentrations. Next, they calculated the percent suppression efficiency [(1 F/F_0) × 100] for both observed and corrected GQD fluorescence before and after the addition of Co^{2+} and showed that the $E\%$ *vs.* Co^{2+} concentration (see Figure 1.8b) curves for corrected and observed fluorescence coincided. Despite the large spectral overlap between Co^{2+} and GQDs, IFE was also eliminated as a quenching mechanism. AGQ was monitored by transmission electron microscopy (TEM). TEM images showed some degree of aggregation in the Co^{2+}-GQDs system. Additionally, the FTIR spectra of GQDs-Co^{2+} shifted to a longer wavenumber, which was accompanied by a decrease in the strength of the C–N stretching vibration, indicating that Co^{2+} was inducing the aggregation of GQDs because the Co^{2+} ions had high affinity toward nitrogen-donor atoms from the NGQDs. They used NGQDs (on–off probes) for the intracellular detection of Co^{2+}, while at the same time monitoring cell apoptosis caused by Co^{2+}.

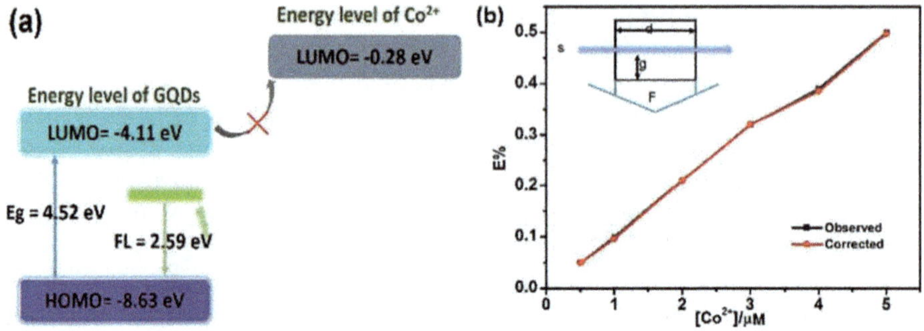

Figure 1.8 (a) The E_{LUMO} and E_{HOMO} of GQDs and Co^{2+}. λ_{EX}: 365 nm, λ_{EM}: 475 nm. (b) Suppressed efficiency (E%) of observed (black curve) and corrected (red curve) measurements for GQDs after addition of Co^{2+} at different concentrations, $E = 1{-}F/F_0$. F_0 and F are the steady-state fluorescence intensities of GQDs in the absence and presence of Co^{2+}, respectively. Reproduced from ref. 100 with permission from the Royal Society of Chemistry.

1.4.6 QDs as Quencher

Recently, nanomaterials have been used as nano-quenchers to quench the emissions of various fluorophores with different emission wavelengths and improve detection signals. These nano-quenchers were used successfully in a fluorescent probe for the detection of nucleic acids. QDs are also reliable candidates as quenchers in biosensing systems. They interact with fluroprobes through different forces and reduce emissions' intensity. LOO *et al.*[41] recently developed a fluorescent-sensing platform for DNA detection using carboxylated CQDs (cCQDs) as quenchers. A fluorescently labeled single-stranded DNA (ssDNA) probe (FAM-L probe) was adsorbed on the surface of a cCQD, and the emission of the FAM was quenched. When target DNA coupled with the FAM and double-stranded DNA (dsDNA) was formed, the FAM-L probe desorbed from the cCQD, and its fluorescence recovered. The adsorption and desorption were explained by competition between electrostatic repulsion and π–π interactions. Electrostatic repulsion was caused by repulsion between the carboxyl group on the cCQD's surface and the negative phosphodiester backbone of the DNA, while π–π interaction was due to π–π stacking between the hydrophobic region of the cCQD and the DNA's nucleobase. Therefore, when electrostatic repulsion was lower than hydrophobic interaction, DNA was adsorbed on the surface of cCQDs (as in ssDNA); it was desorbed if repulsive forces dominated (as in dsDNA).

1.5 Conclusion

Conventional SQDs were first prepared from elements in group IV and III–IV on the periodic table; development subsequently extended to elements in other groups. Innovations in SQD synthesis, modification, and passivation methods improved their photoluminescent properties, water solubility, and photo and chemical stability. Research efforts have also reduced their toxicity to make them more suitable for applications in biological systems. The toxicity of SQDs is mostly due to the leakage of heavy metals into the surrounding environment; this issue has been reduced *via* surface passivation and *via* core/shell preparation with more than one level of coating. SQDs have attracted much scientific attention, and their development is still an active field in biomedical research because they have a very stable and bright fluorescence in the UV to NIR regions with a quantum yield over 90%. Their research applications are limited by their high preparation costs as well as their high toxicity to the environment, humans, and living cells. Research has, therefore, focused on the development of low-cost, environmentally friendly, non-toxic, and high fluorescence efficiency QDs for various applications.

Recently, CQDs have attracted scientific attention due to their extraordinary properties. CQDs can be prepared from inexpensive carbon sources (natural, chemical, or biological) with simple synthetic procedures, and they exhibit highly stable and bright fluorescence with low toxicity. Overall, they are a reliable alternative for organic dyes in biomedical, environmental, and optoelectronic applications.

Detection signals in QD-based biosensors/sensors are produced mostly through the suppression or depression of emissions *via* common mechanisms, including static and dynamic quenching, RET, PET, IFE, and ACSC. Three RET strategies are FRET, BRET, and CRET; FRET is the most commonly used quenching mechanism in a variety of QD-based sensors. Oxidative and reductive PET depends on the direction of electron transfer between LUMO and HOMO energy levels. Future experimental results and observations should support the suggested mechanism. The ACSC mechanism, referring to the aggregation of QDs resulting in emission increase/enhancement (AIE/AIEE) or quenching (ACQ), and the TEM and FTIR spectra of the mixture, distinguish the aggregation. If the extent to which emissions are quenched changes with temperature, dynamic quenching is the mechanism of emission depression, and changes in emissions exhibit a linear relationship (derived from the Stern–Volmer equation) with the concentration of the quencher. If the absorption or emission spectrum of the quencher overlaps with the emission spectrum of the fluorophore, the quenching may be due to IFE. Sometimes more than one mechanism may be involved in fluorescent signal changes.

Abbreviations

ACSC	Aggregation-caused signal change
AGQ	Aggregation-caused quenching
AIE	Aggregation-induced emission
AIEE	Aggregation-induced emission enhancement
AuNCs	Au nanoclusters
AuNPs	Au nanoparticles
BP	Black phosphorus
cCQDs	Carboxylated carbon-based quantum dots
CQDs	Carbon-based QDs
CNQDs	Graphitic carbon nitride quantum dots
CSQDs	Core–shell quantum dots
dsDNA	Double stranded DNA
Eu-BPQDs	Eu-black phosphorus quantum dots
FRET	Fluorescence resonance energy transfer
FWHM	Full width at half emission maximum
g-C_3N_4	Graphitic carbon nitride
GQDs	Graphene quantum dots
h-BN	Hexagonal boron nitride
h-BN QDs	Hexagonal boron nitride quantum dots
HOMO	Higher occupied molecular orbital
IFE	Inner filter effect
LUMO	Lower unoccupied molecular orbital
Lys	Lysine
NGQDs	Nitrogen GQDs
PEI	Polyetherimide
PET	Photoinduced electron transfer
PL	Photoluminescence
QDs	Quantum dots
QY%	Quantum yield percent
RET	Resonance energy transfer
SQDs	Semiconductor quantum dots
ssDNA	Single stranded DNA
TEM	Transmission electron microscopy
TMD	Transition metal dichalcogenide

References

1. A. Sukhanova, S. Bozrova, E. Gerasimovich, M. Baryshnikova, Z. Sokolova, P. Samokhvalov, C. Guhrenz, N. Gaponik, A. Karaulov and I. Nabiev, Dependence of Quantum Dot Toxicity In Vitro on Their Size, Chemical Composition, and Surface Charge, *Nanomaterials*, 2022, **12**(16), 2734.

2. A. Sahu and D. Kumar, Core-shell quantum dots: A review on classification, materials, application, and theoretical modeling, *J. Alloys Compd.*, 2022, **924**, 166508.
3. A. Topcu, E. Bağda, T. Oymak and M. Durmuş, Development of quantum dot-phthalocyanine integrated G-quadruplex/double-stranded DNA biosensor, *Anal. Biochem.*, 2022, **654**, 114777.
4. A. L. Efros and L. E. Brus, Nanocrystal quantum dots: from discovery to modern development, *ACS Nano*, 2021, **15**(4), 6192–6210.
5. X. Dai, Z. Zhang, Y. Jin and Y. Niu, *Nature*, 2014, **515**, 96–99.
6. M. n. A. Cotta, Quantum Dots and Their Applications: What Lies Ahead?, *ACS Appl. Nano Mater.*, 2020, **3**, 4920–4924.
7. J. Kabel, S. Sharma, A. Acharya, D. Zhang and Y. K. Yap, Molybdenum Disulfide Quantum Dots: Properties, Synthesis, and Applications, *C*, 2021, **7**(2), 45.
8. R. Priyadarshi, P. Ezati and J.-W. Rhim, Synthesis, properties and food packaging applications of sulfur quantum dots: a review, *Environ. Chem. Lett.*, 2022, 3993–4008.
9. X. Huang and M. Tang, Research advance on cell imaging and cytotoxicity of different types of quantum Dots, *J. Appl. Toxicol.*, 2021, **41**(3), 342–361.
10. K. Hagiwara, S. Horikoshi and N. Serpone, Photoluminescent Carbon Quantum Dots: Synthetic Approaches and Photophysical Properties, *Chem. Eur. J.*, 2021, **27**(37), 9466–9481.
11. N. Azam, M. Najabat Ali and T. Javaid Khan, Carbon Quantum Dots for Biomedical Applications: Review and Analysis, *Front. Mater.*, 2021, **8**, 700403.
12. Y. R. Kumar, K. Deshmukh, K. K. Sadasivuni and S. K. K. Pasha, Graphene quantum dot based materials for sensing, bio-imaging and energy storage applications: a review, *RSC Adv.*, 2020, **10**(40), 23861–23898.
13. M. Fang, C.-W. Peng, D.-W. Pang and Y. Li, Quantum dots for cancer research: current status, remaining issues, and future perspectives, *Cancer Biol. Med.*, 2012, **9**(3), 151–163.
14. W. Zhang, G. Dang, J. Dong, Y. Li, P. Jiao, M. Yang, X. Zou, Y. Cao, H. Ji and L. Dong, A multifunctional nanoplatform based on graphitic carbon nitride quantum dots for imaging-guided and tumor-targeted chemo photodynamic combination therapy, *Colloids Surf. B Biointerfaces*, 2021, **199**, 111549.
15. A. J. Parker and A. S. Barnard, Unsupervised structure classes vs. supervised property classes of silicon quantum dots using neural networks, *Nanoscale Horiz.*, 2021, **6**(3), 277–282.
16. N. F. Johnson, Quantum dots: few-body, low-dimensional systems, *J. Phys. Condens. Matter*, 1995, **7**(6), 965–989.

17. L. Wang, D. Xu, J. Gao, X. Chen, Y. Duo and H. Zhang, Semiconducting quantum dots: Modification and applications in biomedical science, *Sci. China Mater.*, 2020, **63**(9), 1631–1650.
18. H. J. Park, D. J. Shin and J. Yu, Categorization of quantum dots, clusters, nanoclusters, and nanodots, *J. Chem. Educ.*, 2021, **98**, 703–709.
19. S. Chinnathambi and N. Shirahata, Recent advances on fluorescent biomarkers of near-infrared quantum dots for in vitro and in vivo imaging, *Sci. Technol. Adv. Mater.*, 2019, **20**(1), 337–355.
20. P. Zhao, Q. Xu, J. Tao, Z. Jin, Y. Pan, C. Yu and Z. Yu, Near infrared quantum dots in biomedical applications: current status and future perspective, *Wiley Interdiscip. Rev. Nanomed. Nanobiotechnol.*, 2018, **10**(3), e1483.
21. L.-W. Wang, C.-W. Peng, C. Chen and Y. Li, Quantum dots-based tissue and in vivo imaging in breast cancer researches: current status and future perspectives, *Breast Cancer Res. Treat.*, 2015, **151**(1), 7–17.
22. F. Bian, L. Sun, L. Cai, Y. Wang and Y. Zhao, Quantum dots from microfluidics for nanomedical application, *Wiley Interdiscip. Rev. Nanomed. Nanobiotechnol.*, 2019, **11**(5), e1567.
23. T. Torchynska and Y. Vorobiev, Semiconductor II-VI Quantum Dots with Interface States and Their Biomedical Applications, in *Advanced Biomedical Engineering*, ed. D. G. Gargiulo, InTech, 2011.
24. K. Kalyanasundaram, E. Borgarello, D. Duonghong and M. Grätzel, Cleavage of Water by Visible-Light Irradiation of Colloidal CdS Solutions; Inhibition of Photocorrosion by RuO2, *Angew. Chem. Int. Ed. Engl.*, 1981, **20**, 987–988.
25. R. Rossetti, S. Nakahara and L. E. Brus, Quantum size effects in the redox potentials, resonance Raman spectra, and electronic spectra of CdS crystallites in aqueous solution, *J. Chem. Phys.*, 1983, **79**(2), 1086–1088.
26. A. M. Wagner, J. M. Knipe, G. Orive and N. A. Peppas, Quantum dots in biomedical applications, *Acta Biomater.*, 2019, **94**, 44–63.
27. X. Cheng, S. B. Lowe, P. J. Reece and J. J. Gooding, Colloidal silicon quantum dots: from preparation to the modification of self-assembled monolayers (SAMs) for bio-applications, *Chem. Soc. Rev.*, 2014, **43**(8), 2680–2700.
28. S. Tao, T. Feng, C. Zheng, S. Zhu and B. Yang, Carbonized Polymer Dots: A Brand New Perspective to Recognize Luminescent Carbon-Based Nanomaterials, *J. Phys. Chem. Lett.*, 2019, **10**(17), 5182–5188.
29. Y. Chong, Y. Ma, H. Shen, X. Tu, X. Zhou, J. Xu, J. Dai, S. Fan and Z. Zhang, The in vitro and in vivo toxicity of graphene quantum dots, *Biomaterials*, 2014, **35**(19), 5041–5048.

30. Z. Zhang, J. Zhang, N. Chen and L. Qu, Graphene quantum dots: an emerging material for energy-related applications and beyond, *Energy Environ. Sci.*, 2012, **5**(10), 8869–8890.

31. L. Li, G. Wu, G. Yang, J. Peng, J. Zhao and J.-J. Zhu, Focusing on luminescent graphene quantum dots: current status and future perspectives, *Nanoscale*, 2013, **5**(10), 4015–4039.

32. B. Huo, B. Liu, T. Chen, L. Cui, G. Xu, M. Liu and J. Liu, One-Step Synthesis of Fluorescent Boron Nitride Quantum Dots via a Hydrothermal Strategy Using Melamine as Nitrogen Source for the Detection of Ferric Ions, *Langmuir*, 2017, **33**(40), 10673–10678.

33. P. Mohammad-Andashti, Z. Ramezani, V. Zare-Shahabadi and P. Torabi, Rapid and green synthesis of highly luminescent MoS2 quantum dots via microwave exfoliation of MoS2 powder and its application as a fluorescence probe for cortisol detection in human saliva, *Colloids Surf. A: Physicochem. Eng. Asp.*, 2022, **647**, 129048.

34. Y. Guo and J. Li, MoS_2 quantum dots: synthesis, properties and biological applications, *Mater. Sci. Eng. C Mater. Biol. Appl.*, 2020, **109**, 110511.

35. X. Ding, F. Peng, J. Zhou, W. Gong, G. Slaven, K. P. Loh, C. T. Lim and D. T. Leong, Defect engineered bioactive transition metals dichalcogenides quantum dots, *Nat. Commun.*, 2019, **10**(1), 41.

36. B. Ma, W. Dang, Z. Yang, J. Chang and C. Wu, MoS2 Nanoclusters-based biomaterials for disease- impaired wound therapy, *Appl. Mater. Today*, 2020, **20**, 100735.

37. X. Zhang, H. Xie, Z. Liu, C. Tan, Z. Luo, H. Li, J. Lin, L. Sun, W. Chen, Z. Xu, L. Xie, W. Huang and H. Zhang, Black Phosphorus Quantum Dots, *Angew. Chem. Int. Ed. Engl.*, 2015, **54**(12), 3653–3657.

38. Z. Sun, H. Xie, S. Tang, X.-F. Yu, Z. Guo, J. Shao, H. Zhang, H. Huang, H. Wang and P. K. Chu, Ultrasmall Black Phosphorus Quantum Dots: Synthesis and Use as Photothermal Agents, *Angew. Chem. Int. Ed. Engl.*, 2015, **54**(39), 11526–11530.

39. D. M. Wang, K. L. Lin and C. Z. Huang, Carbon dots-involved chemiluminescence: Recent advances and developments, *Luminescence*, 2019, **34**(1), 4–23.

40. M. J. Molaei, Principles, mechanisms, and application of carbon quantum dots in sensors: a review, *Anal. Methods*, 2020, **12**(10), 1266–1287.

41. A. H. Loo, Z. Sofer, D. Bouša, P. Ulbrich, A. Bonanni and M. Pumera, Carboxylic carbon quantum dots as a fluorescent sensing platform for DNA detection, *ACS Appl. Mater. Interfaces*, 2016, **8**(3), 1951–1957.

42. T. R. Shojaei, M. A. M. Salleh, K. Sijam, R. A. Rahim, A. Mohsenifar, R. Safarnejad and M. Tabatabaei, Fluorometric immunoassay for detecting the plant virus Citrus tristeza using carbon nanoparticles acting as

quenchers and antibodies labeled with CdTe quantum dots, *Microchim. Acta*, 2016, **183**(7), 2277–2287.
43. M. Iranifam, Analytical applications of chemiluminescence systems assisted by carbon nanostructures, *TrAC Trends Anal. Chem.*, 2016, **80**, 387–415.
44. Q. Zhang, H. Xue, J. Niu, M. Tian and X. Yu, Construction of fluorescent probes for ratiometric visualization of mitochondrial depolarization based on the FRET mechanism, *New J. Chem.*, 2022, **46**(21), 10067–10072.
45. H. Liu, Y. Zhou, Y. Zhou and J. Gou, Europium functionalized black phosphorus quantum dots as a CRET platform for synergistically enhanced chemiluminescence, *Chem. Commun.*, 2022, **58**(33), 5168–5171.
46. L.-K. Li, Y.-M. Hou, X.-C. Liu, M.-J. Tian, Q.-J. Ma, N.-N. Zhu and S.-Z. Liu, An ICT-FRET-based fluorescent probe for the ratiometric sensing of hypochlorous acid based on a coumarin–naphthalimide derivative, *New J. Chem.*, 2022, **46**(14), 6596–6602.
47. Y. Yan, X. -y. Wang, X. Hai, W. Song, C. Ding, J. Cao and S. Bi, Chemiluminescence resonance energy transfer: From mechanisms to analytical applications, *TrAC Trends Anal. Chem.*, 2020, **123**, 115755.
48. S. Miao, K. Liang and B. Kong, Förster resonance energy transfer (FRET) paired carbon dot-based complex nanoprobes: versatile platforms for sensing and imaging applications, *Mater. Chem. Front.*, 2020, **4**(1), 128–139.
49. Y. Huang, F. Qiu, R. Chen, D. Yan and X. Zhu, Fluorescence resonance energy transfer-based drug delivery systems for enhanced photodynamic therapy, *J. Mater. Chem. B*, 2020, **8**(17), 3772–3788.
50. W. Gao, Y. Ma and W. Lin, Design of a FRET-based fluorescent probe for the reversible detection of SO_2 and formaldehyde in living cells and mice, *New J. Chem.*, 2020, **44**(32), 13654–13658.
51. M. Ganiga and J. Cyriac, FRET based ammonia sensor using carbon dots, *Sens. Actuators B Chem.*, 2016, **225**, 522–528.
52. P. Huang, J. Lin, X. Wang, Z. Wang, C. Zhang, M. He, K. Wang, F. Chen, Z. Li and G. Shen, Light-triggered theranostics based on photosensitizer-conjugated carbon dots for simultaneous enhanced-fluorescence imaging and photodynamic therapy, *Adv. Mater.*, 2012, **24**(37), 5104–5110.
53. C. Yu, X. Li, F. Zeng, F. Zheng and S. Wu, Carbon-dot-based ratiometric fluorescent sensor for detecting hydrogen sulfide in aqueous media and inside live cells, *Chem. Commun.*, 2013, **49**(4), 403–405.
54. C. Yu, Y. Wu, F. Zeng and S. Wu, A fluorescent ratiometric nanosensor for detecting NO in aqueous media and imaging exogenous and endogenous NO in live cells, *J. Mater. Chem. B*, 2013, **1**(33), 4152–4159.

55. F. Du, Y. Ming, F. Zeng, C. Yu and S. Wu, A low cytotoxic and ratiometric fluorescent nanosensor based on carbon-dots for intracellular pH sensing and mapping, *Nanotechnology*, 2013, **24**(36), 365101.
56. S. Nandi, R. Malishev, K. P. Kootery, Y. Mirsky, S. Kolusheva and R. Jelinek, Membrane analysis with amphiphilic carbon dots, *Chem. Comm.*, 2014, **50**(71), 10299–10302.
57. S. Hu, Q. Zhao, Q. Chang, J. Yang and J. Liu, Enhanced performance of Fe 3+ detection via fluorescence resonance energy transfer between carbon quantum dots and Rhodamine B, *RSC Adv.*, 2014, **4**(77), 41069–41075.
58. F. Du, Y. Min, F. Zeng, C. Yu and S. Wu, A targeted and FRET-based ratiometric fluorescent nanoprobe for imaging mitochondrial hydrogen peroxide in living cells, *Small*, 2014, **10**(5), 964–972.
59. Z. Liang, M. Kang, G. F. Payne, X. Wang and R. Sun, Probing energy and electron transfer mechanisms in fluorescence quenching of biomass carbon quantum dots, *ACS Appl. Mater. Interfaces*, 2016, **8**(27), 17478–17488.
60. S. Nandi, R. Malishev, S. K. Bhunia, S. Kolusheva, J. Jopp and R. Jelinek, Lipid-Bilayer Dynamics Probed by a Carbon Dot-Phospholipid Conjugate, *Biophys. J.*, 2016, **110**(9), 2016–2025.
61. S. Yu, S. Y. Lee, H. N. Umh and J. Yi, Energy conversion of sub-band-gap light using naked carbon nanodots and rhodamine B, *Nano Energy*, 2016, **26**, 479–487.
62. K. Yang, F. Li, W. Che, X. Hu, C. Liu and F. Tian, Increment of the FRET efficiency between carbon dots and photosensitizers for enhanced photodynamic therapy, *RSC Adv.*, 2016, **6**(103), 101447–101451.
63. M. Deng, S. Wang, C. Liang, H. Shang and S. Jiang, A FRET fluorescent nanosensor based on carbon dots for ratiometric detection of Fe3+ in aqueous solution, *RSC Adv.*, 2016, **6**(32), 26936–26940.
64. J. Hou, Z. Tian, H. Xie, Q. Tian and S. Ai, A fluorescence resonance energy transfer sensor based on quaternized carbon dots and Ellman's test for ultrasensitive detection of dichlorvos, *Sens. Actuators B Chem.*, 2016, **232**, 477–483.
65. J. S. Sidhu, A. Singh, N. Garg and N. Singh, Carbon dot based, naphthalimide coupled FRET pair for highly selective ratiometric detection of thioredoxin reductase and cancer screening, *ACS Appl. Mater. Interfaces*, 2017, **9**(31), 25847–25856.
66. J. Kudr, L. Richtera, K. Xhaxhiu, D. Hynek, Z. Heger, O. Zitka and V. Adam, Carbon dots based FRET for the detection of DNA damage, *Biosens. Bioelectron.*, 2017, **92**, 133–139.
67. Y. Yuan, J. Jiang, S. Liu, J. Yang, J. Yan and X. Hu, Fluorescent carbon dots for glyphosate determination based on fluorescence resonance

energy transfer and logic gate operation, *Sens. Actuators B Chem.*, 2017, **242**, 545–553.

68. H. Wang, Q. Lu, Y. Liu, H. Li, Y. Zhang and S. Yao, A dual-signal readout sensor for highly sensitive detection of iodide ions in urine based on catalase-like reaction of iodide ions and N-doped C-dots, *Sens. Actuators B Chem.*, 2017, **250**, 429–435.
69. C.-P. Hsu, Z. Hejazi, E. Armagan, S. Zhao, M. Schmid, H. Zhang, H. Guo, L. Weidenbacher, R. M. Rossi and M. M. Koebel, Carbon dots and fluorescein: The ideal FRET pair for the fabrication of a precise and fully reversible ammonia sensor, *Sens. Actuators B Chem.*, 2017, **253**, 714–722.
70. H. Liu, C. Xu, Y. Bai, L. Liu, D. Liao, J. Liang, L. Liu and H. Han, Interaction between fluorescein isothiocyanate and carbon dots: Inner filter effect and fluorescence resonance energy transfer, *Spectrochim. Acta A Mol. Biomol. Spectrosc.*, 2017, **171**, 311–316.
71. Y. Xia, L. Wang, J. Li, X. Chen, J. Lan, A. Yan, Y. Lei, S. Yang, H. Yang and J. Chen, A ratiometric fluorescent bioprobe based on carbon dots and acridone derivate for signal amplification detection exosomal microRNA, *Anal. Chem.*, 2018, **90**(15), 8969–8976.
72. J. S. Sidhu and N. Singh, FRET and PET paired dual mechanistic carbon dots approach for tyrosinase sensing, *J. Mater. Chem. B*, 2018, **6**(24), 4139–4145.
73. A. Das, D. Roy, M. Mandal, C. Jaiswal, M. Ta and P. K. Mandal, Carbon dot with pH independent near-unity photoluminescence quantum yield in an aqueous medium: electrostatics-induced Förster resonance energy transfer at submicromolar concentration, *J. Phys. Chem. Lett.*, 2018, **9**(17), 5092–5099.
74. A. Chatzimarkou, T. G. Chatzimitakos, A. Kasouni, L. Sygellou, A. Avgeropoulos and C. D. Stalikas, Selective FRET-based sensing of 4-nitrophenol and cell imaging capitalizing on the fluorescent properties of carbon nanodots from apple seeds, *Sens. Actuators B Chem.*, 2018, **258**, 1152–1160.
75. S. Kumari, M. Tiwari and P. Das, Multi format compatible visual and fluorometric detection of SEB toxin in nanogram range by carbon dot-DNA and acriflavine nano-assembly, *Sens. Actuators B Chem.*, 2019, **279**, 393–399.
76. Y. Shu, N. Zheng, A.-Q. Zheng, T.-T. Guo, Y.-L. Yu and J.-H. Wang, Intracellular zinc quantification by fluorescence imaging with a FRET system, *Anal. Chem.*, 2019, **91**(6), 4157–4163.
77. Y. Su, S. Lu, P. Gao, M. Zheng and Z. Xie, BODIPY@ carbon dot nanocomposites for enhanced photodynamic activity, *Mater. Chem. Front.*, 2019, **3**(9), 1747–1753.

78. J. S. Sidhu, A. Singh, N. Garg, N. Kaur and N. Singh, Gold conjugated carbon dots nano assembly: FRET paired fluorescence probe for cysteine recognition, *Sens. Actuators B Chem.*, 2019, **282**, 515–522.
79. J. Ren, F. Weber, F. Weigert, Y. Wang, S. Choudhury, J. Xiao, I. Lauermann, U. Resch-Genger, A. Bande and T. Petit, Influence of surface chemistry on optical, chemical and electronic properties of blue luminescent carbon dots, *Nanoscale*, 2019, **11**(4), 2056–2064.
80. M. F. Frasco and N. Chaniotakis, Semiconductor quantum dots in chemical sensors and biosensors, *Sensors*, 2009, **9**(9), 7266–7286.
81. F. Zu, F. Yan, Z. Bai, J. Xu, Y. Wang, Y. Huang and X. Zhou, The quenching of the fluorescence of carbon dots: a review on mechanisms and applications, *Microchim. Acta*, 2017, **184**(7), 1899–1914.
82. H. Chen, Z. Wang, S. Zong, P. Chen, D. Zhu, L. Wu and Y. Cui, A graphene quantum dot-based FRET system for nuclear-targeted and real-time monitoring of drug delivery, *Nanoscale*, 2015, **7**(37), 15477–15486.
83. C. Fowley, N. Nomikou, A. P. McHale, B. McCaughan and J. F. Callan, Extending the tissue penetration capability of conventional photosensitisers: a carbon quantum dot-protoporphyrin IX conjugate for use in two-photon excited photodynamic therapy, *Chem. Commun.*, 2013, **49**(79), 8934–8936.
84. J. Wang, Z. Zhang, S. Zha, Y. Zhu, P. Wu, B. Ehrenberg and J.-Y. Chen, Carbon nanodots featuring efficient FRET for two-photon photodynamic cancer therapy with a low fs laser power density, *Biomaterials*, 2014, **35**(34), 9372–9381.
85. L. Lin, Y. Wang, Y. Xiao and X. Chen, Ratiometric fluorescence detection of riboflavin based on fluorescence resonance energy transfer from nitrogen and phosphorus co-doped carbon dots to riboflavin, *Anal. Bioanal. Chem.*, 2019, **411**(13), 2803–2808.
86. Y. Shi, C. Li, S. Liu, Z. Liu, J. Zhu, J. Yang and X. Hu, Facile synthesis of fluorescent carbon dots for determination of curcumin based on fluorescence resonance energy transfer, *RSC Adv.*, 2015, **5**(79), 64790–64796.
87. T. Feng, H. J. Chua and Y. Zhao, Reduction-Responsive Carbon Dots for Real-Time Ratiometric Monitoring of Anticancer Prodrug Activation in Living Cells, *ACS Biomater. Sci. Eng.*, 2017, **3**(8), 1535–1541.
88. T. Jing, K. Biao, W. Hao, X. Ming, W. Yongcheng, W. Yanli, Z. Dongyuan and Z. Gengfeng, *Adv. Mater.*, 2013, **25**, 6569–6574.
89. J. Wang, J. Wei, S. Su and J. Qiu, Novel fluorescence resonance energy transfer optical sensors for vitamin B12 detection using thermally reduced carbon dots, *New J. Chem.*, 2015, **39**(1), 501–507.
90. Y. Wangn, H. Meng, M. Jia, Y. Zhang, H. Li and L. Feng, *Nanoscale*, 2016, **8**, 17190–17195.

91. H. Yao, Y. Zhang, F. Xiao, Z. Xia and J. Rao, Quantum dot/bioluminescence resonance energy transfer based highly sensitive detection of proteases, *Angew. Chem. Int. Ed. Engl.*, 2007, **119**(23), 4424–4427.
92. R. Freeman, X. Liu and I. Willner, Chemiluminescent and chemiluminescence resonance energy transfer (CRET) detection of DNA, metal ions, and aptamer-substrate complexes using hemin/G-quadruplexes and CdSe/ZnS quantum dots, *J. Am. Chem. Soc.*, 2011, **133**(30), 11597–11604.
93. L. Hu, X. Liu, A. Cecconello and I. Willner, Dual Switchable CRET-Induced Luminescence of CdSe/ZnS Quantum Dots (QDs) by the Hemin/G-Quadruplex-Bridged Aggregation and Deaggregation of Two-Sized QDs, *Nano Lett.*, 2014, **14**(10), 6030–6035.
94. J. F. Callan, R. C. Mulrooney, S. Kamila and B. McCaughan, Anion sensing with luminescent quantum dots--a modular approach based on the Photoinduced Electron Transfer (PET) mechanism, *J. Fluoresc.*, 2008, **18**(2), 527–532.
95. H.-Y. Zhang, Y. Wang, S. Xiao, H. Wang, J.-H. Wang and L. Feng, Rapid detection of Cr (VI) ions based on cobalt(II)-doped carbon dots, *Biosens. Bioelectron.*, 2017, **87**, 46–52.
96. H.-R. Nan, Y.-H. Liu, W.-J. Gong, H.-B. Peng, Y.-Q. Wang, Z.-B. Zhang and X.-H. Cao, An inner-filter-effect based ratiometric fluorescent sensor for the detection of uranyl ions in real samples, *Anal. Methods*, 2022, **14**(5), 532–540.
97. Q. Zeng, Z. Li, Y. Dong, A. Qin, Y. Hong, L. Ji, Z. Zhu, C. K. W. Jim, G. Yu and Q. Li, Fluorescence enhancements of benzene-cored luminophors by restricted intramolecular rotations: AIE and AIEE effects, *Chem. Commun.*, 2007, (1), 70–72.
98. D. Ding, K. Li, B. Liu and B. Z. Tang, Bioprobes based on AIE fluorogens, *Acc. Chem. Res.*, 2013, **46**(11), 2441–2453.
99. A. Chae, Y. Choi, S. Jo, P. Paoprasert, S. Y. Park and I. In, Microwave-assisted synthesis of fluorescent carbon quantum dots from an A 2/B 3 monomer set, *RSC Adv.*, 2017, **7**(21), 12663–12669.
100. N. Wang, Z. X. Liu, R. S. Li, H. Z. Zhang, C. Z. Huang and J. Wang, The aggregation induced emission quenching of graphene quantum dots for visualizing the dynamic invasions of cobalt(ii) into living cells, *J. Mater. Chem. B*, 2017, **5**(31), 6394–6399.

CHAPTER 2

How Functionalization Affects the Detection Ability of Quantum Dots

ZAHRA RAMEZANI[a,b] AND MICHAEL THOMPSON*[c]

[a]Nanotechnology Research Center, Faculty of Pharmacy, Ahvaz Jundishapur University of Medical sciences, Ahvaz, Iran; [b]Medicinal Chemistry Department, Faculty of Pharmacy, Ahvaz Jundishapur University of Medical sciences, Ahvaz, Iran; [c]Chemistry Department, University of Toronto, Toronto, Canada
*E-mail: mikethom@chem.utoronto.ca; m.thompson@utoronto.ca

2.1 Introduction

Quantum dots (QDs) have significant photoelectric properties that make them a promising material in photovoltaic cells and in light-emitting devices, and they act as catalysts in different industries, in environmental remediation and detection, and in biomedical engineering.[1–3] They have unique properties compared with organic dyes, including high fluorescence intensity, tunable emission, high fluorescence stability and lifetime, high quantum yield, high resistance to photobleaching, and surface modifiability.[4] Surface modification of QDs reduces their toxicity, improves their photoluminescent properties, and makes them an ideal material for certain purposes, such as selective sensing and molecule detection, cell imaging, and targeted drug delivery.[5]

Detection Science Series No. 22
Quantum Dots in Bioanalytical Chemistry and Medicine
Edited by Michael Thompson and Zahra Ramezani

Published by the Royal Society of Chemistry, www.rsc.org

Traditional semiconductor QDs (SQDs) such as CdSe and CdS have excellent properties and are used extensively for many purposes, including detection and various biomedical applications. The release of Cd as a toxic element limited their applications until scientists introduced core–shell SQDs such as CdSe/ZnS to overcome this problem. ZnS as a shell improved the QDs' luminescent properties but could not completely prevent heavy metal leakage in acidic media. Thereafter, researchers considered a different approach for improving QDs' properties and lowering their toxicity. First, the surface of the core–shell QDs was modified and functionalized with different materials such as silica polymers, amphiphilic lipids, polyethylene glycol (PEG), and many other functional groups that were specific to particular targets or events and that could also be used in the parallel detection of different targets. All these changes affected the QDs' lipophilicity, photostability, intensity, and quantum yields. QDs' quantum yield is significantly influenced by their surface properties. The photoluminescent properties of SQDs can be increased with the addition of different surface functional groups; however, in solution, QDs' quantum yield is reduced due to partial heavy metal ligand removal, which in turn releases heavy metals (such as Cd) into the solution.[6] Therefore, Cd-free (*i.e.*, heavy metal-free) SQDs such as Ag_2S, ZnO, Si-based QDs, and carbon-based QDs have been introduced and modified for various uses.

Carbon-based QDs have attracted many researchers' attention because of their enhanced performance in electronics, sensing, drug delivery, and cell imaging. They can be functionalized easily, they have stable photoluminescence, and they are more biocompatible. Furthermore, they can be prepared from cheap and readily available sources. Carbon QDs (CQDs) are a class of carbon-based QDs that can be prepared from natural carbon sources, such as quince fruit[7] ground coffee,[8] and grass,[9] and small chemical carbon precursors, such as citric acid and urea,[10] through simple procedures. CQDs are usually composed of discrete, quasi-spherical nanoparticles with sizes below 10 nm. They have many functional groups on their surfaces, such as amino, epoxy, ether, carboxy, hydroxy, and carboxylic acid.[11,12] The high concentration of functional groups on QDs' surfaces results in high hydrophilicity and readiness for functionalization with various species; these qualities in turn enable their use in target-specific probes and events.

Herein, we will discuss several common functionalization and conjugation methods. We will also explain how the attachment of different functional groups to QDs' surfaces changes their luminescence as well as their photo-physical and chemical properties. QDs' biochemical and chemical sensing and detection applications, their toxicity, and their therapeutic usage are also discussed.

2.2 Functionalization of QDs

Many years have passed since SQDs were first introduced, and QDs are composed of various materials. QDs are commercially available with different structures and surface states. Their extraordinary optical and electronic characteristics make them useful in a wide variety of contexts.[13] Yet, research on QDs' synthesis and applications is active and ongoing. Progress in QD research has expanded their elemental composition, and QDs have been synthesized with various surface states and functional groups. In addition to traditional SQDs, perovskites, transition metal dichalcogenides, and carbon QDs have thus been used in the preparation of detection probes in chemical and biological systems. QDs are chosen for targeted applications based on their toxicity, selectivity or specificity, and stable fluorescence. These specifications are affected by the type of QDs used as well as by the QDs' surface functional groups. All QDs, from traditional QDs to the recently developed carbon-based QDs, can be functionalized with different chemicals and biomolecules for targeted applications (Figure 2.1). Cd-based SQDs covered with different shells prior to functionalization with miscellaneous biomolecules (Figure 2.1A) are used for a variety of biochemical and chemical applications. QDs' hydrophilicity must be increased to improve their availability to biological media and make them suitable for the preparation of targeted probes. This is achieved by attaching different ligands such as alkylthios, PEG, silica polymers, and lipids to the surfaces of QDs. QDs are functionalized and conjugated using different strategies depending on their desired applications.

Conjugated carbon-based materials such as CQDs are easily prepared, as different functional groups can be attached to their surfaces during their synthesis. Depending on the carbon precursors, carboxyl, amine, amide, hydroxyl, or organic molecules are anchored to QDs' surfaces after one-pot synthesis (Figure 2.1B, Figure 2.2A). The water solubility of carbon-based QDs is due to these functionalities.[14] Compared with SQDs, the functional groups on carbon-based QDs make bioconjugation and the addition of other recognition elements easier. Antibody- or aptamer-conjugated QDs have a variety of applications, from single cell detection to drug delivery (Figure 2.1C).

2.2.1 Bioconjugation of QDs

Pairing biomolecules with QDs through covalent or non-covalent bonds is known as bioconjugation. The resulting particles are used as probes for sensing, detection, imaging, and other *in vivo* or *in vitro* purposes (Figure 2.1C).[15] These molecules include fatty acids, carbohydrates, lipids, peptides (glutathione, enzyme inhibitors, cell penetrating peptides, *etc.*), nucleic acids (DNA, RNA, ribozymes, *etc.*), proteins (antibodies, enzymes, *etc.*), and any other entities found in biological environments. They also include monomeric forms and macromolecules. Biomolecules may be

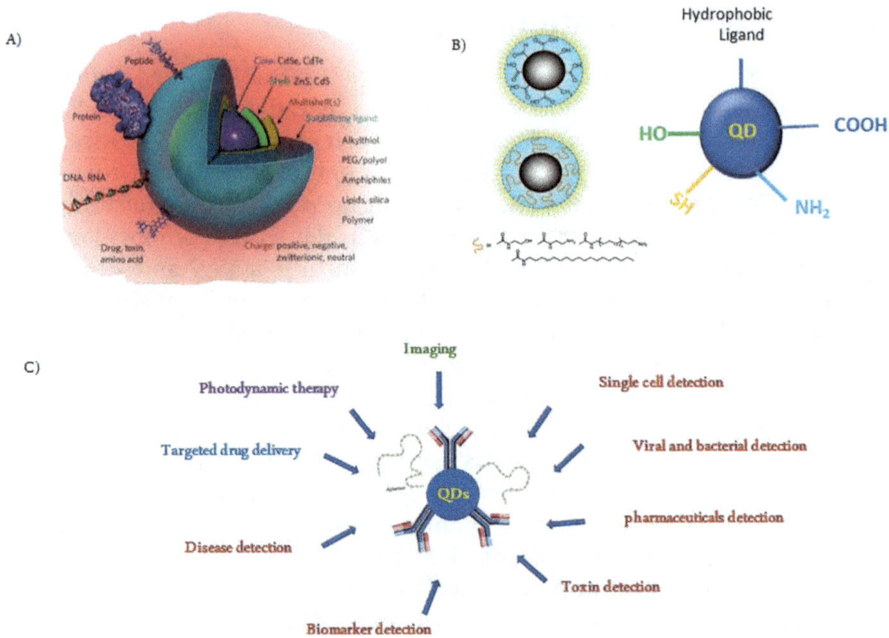

Figure 2.1 (A) Some possible biomolecule attached to the semiconductor core–shell QDs, (B) most common functional groups on the surface of CQDs after one-pot synthesis using different carbon sources, and (C) various biomedical and chemical applications of antibody- or aptamer-loaded QDs.

active or passive when attached to the surface of a QD. They are referred to as active if their intrinsic functionality is retained when conjugated with QDs; otherwise, they are referred to as passive or inert and are used simply as coating or structural intermediate materials. Diagnostic contrast agents or fluorophores (used in MRI and cell imaging), enzymatic cofactors, and drugs exhibit their desired activities in biological environments, and therefore, they are considered to be active biomolecules. There are two general types of bioconjugates: homogeneous and heterogeneous. In homogeneous particles, the QD's surface is decorated with one type of molecule, while in heterogeneous particles, more than one biomolecule is placed on the surface.[16] Heterogeneous particles are used for more than one application at the same time and are thus described as multipurpose QDs. The type of bioconjugate selected for use depends on the targeted applications.

QDs should have a bright PL with a high quantum yield. They should also be colloidally stable in biological buffers and media for conjugation with biomolecules in biological applications. SQDs with suitable PL specifications are usually prepared with hydrophobic surface ligands and are insoluble in aqueous media. Therefore, surface modifications are

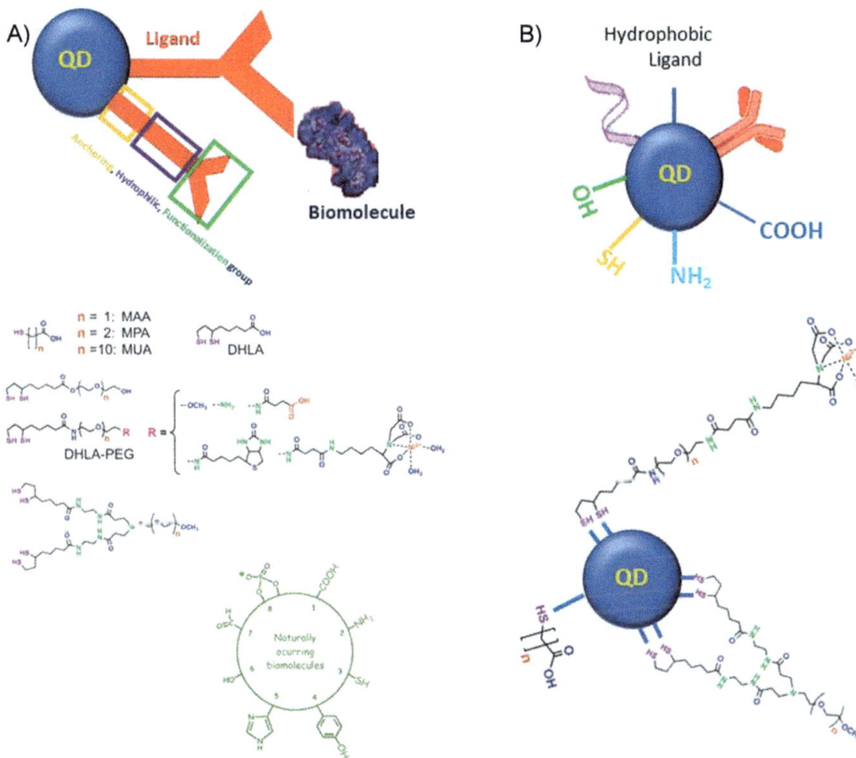

Figure 2.2 Schematic representation of surface modifications and functionalization. (A) Ligand design around QDs (anchoring group attached to the surface and two other groups for conjugation and/or stabilization), and (B) different thiols functional ligands for conjugation with QDs and the functional groups on naturally occurring biomolecules.

mandatory to ensure that QDs *are uniformly dispersed in biological media and to increase their biocompatibility*.[1,17,18] Furthermore, the nature of the functionalized ligands is of utmost importance for bioconjugation and its consequent bio-application. Therefore, the surface chemistry of the QDs determines their solubility in biological and aqueous media and protects the biological environment from leakages of QDs' core material. These are the minimum requirements for *in vitro* and *in vivo* assays, as well as for cellular imaging applications.

As demonstrated in Figure 2.2, most QDs' surface ligands have two basic parts: an anchoring group and a hydrophilic group. The anchoring group interacts with the QD's surface to provide a stable connection between the QD, the ligand, and the hydrophilic part of the ligand.[1] The hydrophilic group impairs the QD's solubility in aqueous media. The ligand may have a third functional group for conjugation with biomolecules. The selection of the cross-linker (surface ligand) depends on many criteria, such as the ionic strength and pH of the medium, as well as the intended use. There are seven qualities of ideal ligands as QD surface modifiers prior to functionalization with biomolecules: (1) strong affinity for the QD's surface; (2) minimal toxicity; (3) minimal non-specific adsorption in the medium; (4) small final hydrodynamic diameter (D_h); (5) colloidal stability across a wide pH range and in high salt concentrations; (6) inexpensive and commercially available or easily synthesized in large quantities; and (7) availability of a functional group for conjugation with the target biomolecule. All these concerns should be considered when designing ligands for specific applications. For example, one of the requirements for *in vivo* sensing and cell imaging is the stability of QDs in acidic media for cellular uptake.[19] Biological conditions differ across a variety of applications; for example, some coupling reactions happen in acidic conditions, while others occur at the basic pH limit. Therefore, representative QDs must be stable over a wide pH range. Besides pH stability, salt stability and small D_h are both critical for biological applications. D_h is crucial, especially when FRET is the sensing mechanism. FRET sensing efficiency directly depends on the distance between the energy donor and the acceptor. Minimizing the thickness of the surface coating of QDs reduces donor–acceptor separation and increases FRET intensity by maximizing energy transfer.[20,21] Peptides, proteins, and electrolytes with high concentrations are present in biological media and are adsorbed on QDs' surfaces (high non-specific adsorption), preventing normal functions in certain applications. Therefore, it is crucial to minimize non-specific adsorption when developing QDs for biomedical applications.

Having considered the importance of specific adsorption, we will now consider three general approaches to the surface modification of hydrophobic QDs: (1) ligand exchange; (2) silica coating; and (3) encapsulation. In ligand exchange, a hydrophilic ligand takes the place of the hydrophobic group originally present on the surface of the QD. Thiol-containing ligands such as mercaptoacetic acid (MAA), 3-mercaptopropionic acid (MPA), and 11-mercaptoundecanoic acid (Figure 2.2B) are frequently utilized as ligands in ligand exchange protocols.[22] The ligands are attached to the surface *via* the sulfur group because heavy metals such as Zn, Pb, and Cd have a strong affinity toward sulfur. The aqueous solubility of as-prepared functionalized QDs is due to the deprotonation of their carboxylic groups (Figure 2.2), which limits the QDs' solubility to a basic pH and moderate ionic strength.[23] Monolithic ligands modified with terminal PEG oligomers (Figure 2.2B) can expand the ionic strength and pH range,[24] but the dynamic thiol on/off rate from the surface limits QDs' colloidal stability. However, multidentate thiol ligands have been used to resolve this issue.[25] Dihydrolipoic acid (DHLA) has become one of the most important mooring ligands to impart aqueous colloidal stability to QDs, but it still cannot be used in a wide pH range or with high ionic strength; accordingly, these QDs are not suitable for many biosensing and imaging applications. To address this issue, PEG can be attached to DHLA (Figure 2.2B). The thiol groups of DHLA attach strongly to the QDs' surfaces, and the PEG segment stabilizes the QDs in biological fluid in a wide pH range, from weakly acidic to strongly basic. PEG-appended DHLA has a series of functional groups that are easily utilized in conjugation with biomolecules.[26] These include methoxy, carboxylic, hydroxy, biotin, amino, and several others. In one study, PEG was appended to tetradentate thiol groups, and remarkable stability from pH 1.1 to pH 13.9 at 2M NaCl (high salt concentration) was observed.[26] The observations showed the importance of multidentate ligands to QDs' long-term colloidal stability in biological fluids and solids. The main advantage of the ligand exchange method is the small D_h of the as-prepared QDs; however, this method significantly decreases the brightness of the QDs. The reduction in fluorescence quantum yield is due to the increase in surface defects during modification, which consequently limits these QDs' use as fluorescence probes.

In the silica coating method, a silica network covers QDs' surfaces through the hydrolysis of silica molecules. In this method, the surface ligand is replaced with (3-mercaptopropyl) trimethoxysilane, which provides nucleation sites on the surface for further shell growth with silane-containing molecules such as tetraethoxysilane *via* hydrolysis and condensation.[27] Reverse microemulsion methods are sometimes used for coating the surface with silica.[28]

Encapsulation is the process of overcoating QDs' original surfaces with amphiphilic ligands. Amphiphilic ligands naturally have both hydrophilic and hydrophobic ends. The hydrophobic end interacts with the hydrophobic ligand that is originally available on a QD's surface, and the hydrophilic end imparts the QD's aqueous solubility. Surfactants such as phospholipids are the most commonly used amphiphilic ligands.[29] The long-term stability of cetyltrimethylammonium bromide (CTAB)-stabilized QDs has been improved through silica mesoporous encapsulation and PEG modification.[30] The stability of these QDs has also been increased through the removal of excess surfactant (CTAB).

Other strategies for modifying and functionalizing QDs' surfaces are available, and most of the strategies occur during QDs' synthesis. One of these miscellaneous methods is polymerization, which covers the surface almost completely and includes the required functional group for further conjugation. For example, MoS_2 as a transition-metal dichalcogenide quantum dot is prepared through a simple process: MoS_2 bulk is exfoliated to an MoS_2 nanosheet through sonication, and the QDs are formed by heating the nanosheets for 6 h at 140 °C (Figure 2.3). Bare MoS_2 QDs cannot be used for imaging and therapy purposes since they have no functional molecules and biomolecules that would enable easy uptake by target cells, and they exhibit non-specific adsorption in *in vitro* assays. The conjugation of as-prepared QDs with PEG provides a multifunctional nanotheranostic agent.[31] COOH functional groups are required for the

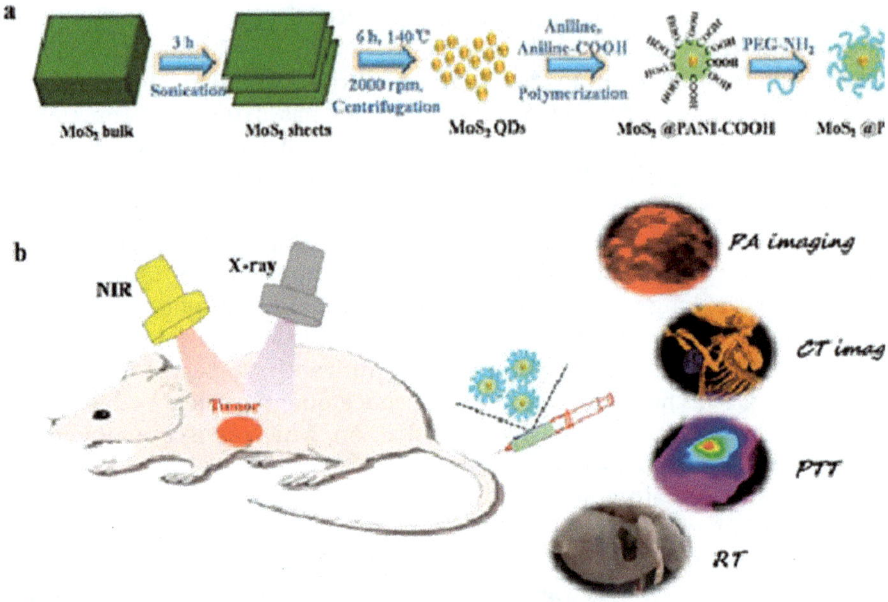

Figure 2.3 MoS_2 from synthesis to multiple applications. Reproduced from ref. 31 with permission from American Chemical Society, Copyright 2016.

attachment of PEG. However, the surfaces of bare MoS_2 QDs can be covered with polyaniline (PANI) through polymerization, which makes plenty of carboxylic groups available for bioconjugation to PEG (Figure 2.3). As-prepared MoS_2@PANI-PEG has been used in dual-purpose probes, namely, simultaneous imaging and therapy. These probes enhance photoacoustic (PA) imaging and X-ray computed tomography (CT) signals, and they perform efficient radiotherapy (RT) and photothermal therapy (PTT) in cancer treatment.

Some QDs, apart from SQDs, have a variety of functional groups on their surfaces after synthesis; the type of group(s) present will depend on the precursors and the synthesis medium. These QDs require no further modifications for conjugation with biomolecules. In some synthetic protocols, bioconjugation is not even necessary. Water-soluble QDs are usually prepared by a simple one-pot method and are economical.[11,32] Carbon-based QDs form a large subgroup of these QDs; among them, carbon QDs (CQDs) are the most recent innovation. CQDs are prepared from various chemical and natural carbon sources,[7–10,33] and they have a variety of surface functional groups and doping elements. Some do not require modification or bioconjugation for applications in biological media. For example, CQDs were prepared by mixing citric acid with manganese tetraphenyl porphyrin complex (MnTTP) as the carbon source (Figure 2.4); these CQDs were utilized in cell imaging and as contrast agents in magnetic resonance imaging (MRI).[34] The preparation methods

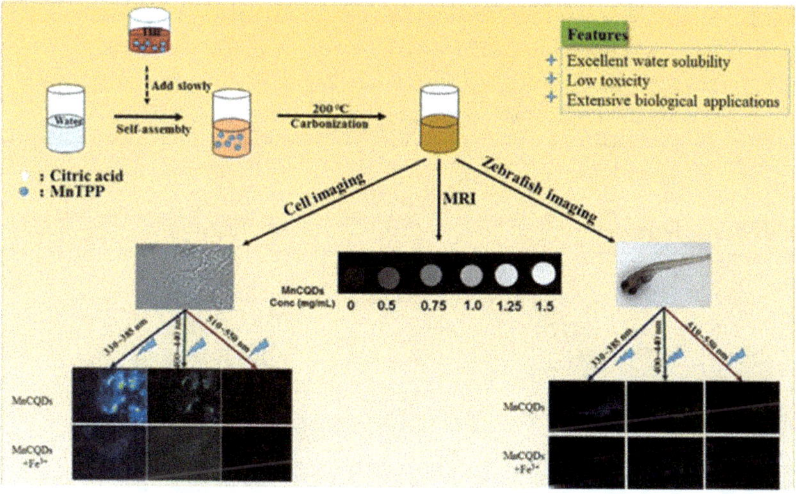

Figure 2.4 Mn-doped carbon quantum dots (MnCQDs) using citric acid and MnTPP as carbon precursors. Reproduced from ref. 34 with permission from Elsevier, Copyright 1966.

were simple, and no further surface modification or functionalization was required (Figure 2.4).

Surface modification affects QDs' physiochemical and photoluminescent properties. Table 2.1 lists various QD types and outlines changes in emission wavelength that occur with changes to their surface modification and functionalization. The table also lists these QDs' corresponding applications.

Protein and peptide labeling is currently the most common process used in the bioconjugation of biomolecules,[15] so that the carboxyl, amine, and, to a lesser extent, thiol groups on these biomolecules attach directly to the QDs' surface (Figure 2.2B). A good bioconjugation process should allow researchers to control the ratio of biomolecules attached per QD, the orientation of the biomolecule on the QDs' surfaces, the distance between the biomolecules and the QDs, and the strength and stability of QDs–biomolecules. The process should also preserve QDs' and biomolecules' normal functions.[29] The chemistry commonly used to attach biomolecules to QDs' surfaces has some limitations, and the resulting conjugates apparently differ from these ideal behaviors.[13,17,29] The most common conjugation protocols are carbodiimide, avidin biotin, dative bonding, and functional group interconversion. Polyhistidine-metal-affinity coordination to QDs is another bioconjugation process that is useful for specific biological applications.[29] Having considered the tremendous changes to QDs' functions and sensing abilities that occur with bioconjugation, we will next describe several strategies for bioconjugation.

Table 2.1 Summary of representative QDs and their surface modifications, emission maxima, and bio-applications. Reproduced from ref. 3, https://doi.org/10.1080/14686996.2019.1590731, under the terms of the CC BY 4.0 license, http://creativecommons.org/licenses/by/4.0/.

Types of QDs	Surface coating	Emission maxima (nm)	Applications	Ref.
II-VI Group				
CdTe	Mercaptoundecanoic acid (MPA)	700–800	*In vivo* tumor targeting	35
CdTe/CdSe	PMAS-PEG	801	Multiplex imaging in live animals	36
CdTe/ZnSe	Phospholipids	730	Imaging of panc-1 tumor-bearing mice	37
CdTe1−xSex/CdS	PEG-grafted phospholipid	845	Cancer imaging and therapy	38
CdHgTe/CdS/ZnS	*N*-acetyl-L-cysteine	650–800	*In vivo* tumor targeting	39
IV-VI Group				
PbS	Recombinant protein (GST-EGFP-GB!)	1150	Breast tumor imaging	40
PbS	Dihydrolipoic acid-polyethylene glycol	1300	Mice *in vivo*	41
PbS	MPA	1100	Imaging of the thrombotic state in septic mouse brain	42
PbS	BS-functionalized AuNCs	813	Bioimaging of ascorbic acid	43
PbS/CdS/ZnS	MPA	1270	*In vivo* fluorescence imaging agents	44
I-VI Group				
Ag_2Se	Polyethylene glycol (PEG)	1300	Blood clearance and bio-distribution	45
Ag_2Se	Multidentate polymer	950–1250	*In vivo* fluorescence images	46
Ag_2Se	C_{18}-PMH-PEG	1300	*In vivo* imaging of live mice	47
Ag_2S	Glutathiose	1015	Nitric oxide delivery and fluorescence imaging	48
Ag_2S	Dihydrolipoic acid	1000–1400	Early-stage tumor diagnosis	49
Ag_2S	Carboxylic acid group	825	EGFR-targeted tumor imaging	50

(*continued*)

Table 2.1 (*continued*)

Types of QDs	Surface coating	Emission maxima (nm)	Applications	Ref.
I-III-VI Group				
$CuInS_2$	RNase A-capped $CuInS_2$	700–800	*In vivo* imaging	51
$AgInSe_2$-ZnSe	Multidentate polymer	625–940	Cell imaging	52
$AgInS_2$	Multidentate polymer	820	*In vivo* imaging	53
III-V Group				
InAs core–shell	Phospholipid Nanosomes composite particles	1280 1300 1280	Multicolor imaging Quantitative metabolic imaging Blood flow map	54
InAs/InP/ZnSe	MPA	800	Tumor detection in living mice	55
IV Group				
Carbon	Ethylenediamine	750	Two-photon fluorescence imaging	56
Carbon	S, Se-co-doped	731 & 820	Fluorescence imaging and photothermal therapy	57
Carbon	Poly(vinylpyrrolidone)	750	*In vivo* NIR fluorescence imaging	58
Carbon	PEG-800	600–900	Photothermal cancer therapy	59
Carbon	Fluorescein isothiocyanate	683	Two-photon fluorescence bioimaging	60
Si	Pluronic F127	650	Two-photon fluorescence cell imaging	61
Si	4,7-Di(2-thienyl)-2,1,3-benzothiadiazole	900	Two-photon absorbing antenna	62
Si	B, P-co-doped	600–1200	Time-gated fluorescence imaging in the second near-infrared window	63

2.2.1.1 *Carbodiimide Chemistry*

Carboxyls and amines are condensed by *N*-(3-dimethylaminopropyl)-*N*'-ethylcarbodiimide or carbodiimide (EDC) to form an amide bond, which

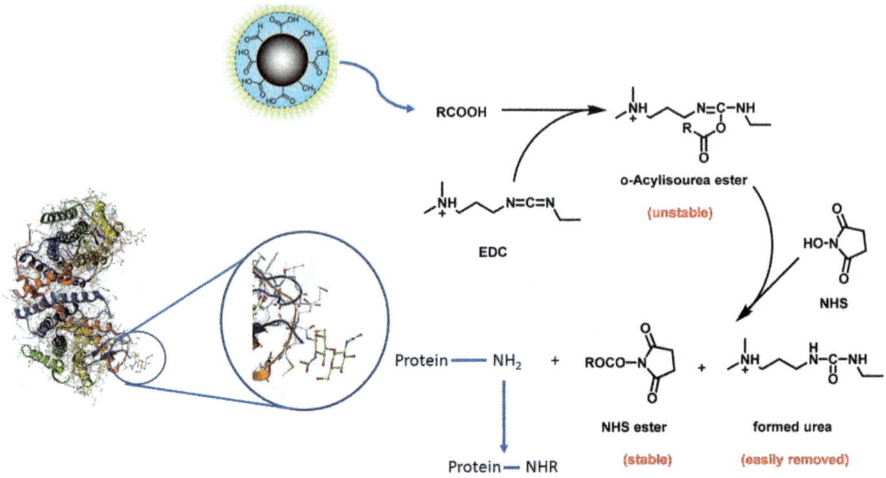

Figure 2.5 Carbodiimide-mediated linkage of an amine (on the protein) to the carboxyl group (on the QDs surface) by amid bond formation.

can attach antibodies and other proteins to QDs' surfaces.[64] EDC activates the coupling reaction. This is a popular conjugation method because proteins have plenty of amines and carboxyls on their N and C terminals as well as on their amine side chains. For this bioconjugation process to work, the QDs must have amines or carboxyls on their surfaces (Figure 2.5); carboxyl-functionalized QDs are the easiest type of QD to use in this case. The proteins, EDC, and functionalized QDs are mixed in a suitable buffer and allowed to sit for a period of time before the conjugate is purified.[65,66] *N*-Hydroxysulfosuccinimide (Sulfo-NHS) is added to increase the coupling efficiency and improve the solubility of the intermediate.[25] This coupling reaction is pH-dependent, and an excessive amount of EDC must be added; purification is subsequently required to remove excess reagents. Additionally, side reactions such as protein cross-linking and QD-protein–QD cross-linking are also probable. To resolve this cross-linking issue, the reaction conditions must be optimized. In ideal conditions, one species has a carboxyl group and the other has only an amine group available to react. Although EDC is inexpensive and easily performed, the application of this method is limited due to its insoluble intermediate and its issues of competing hydrolysis, cross-linking, and lack of control over the final orientation of the biomolecule.[66] Figure 2.5 shows the conjugation of proteins to QDs *via* EDC/NHS chemistry.

2.2.1.2 Dative Bonding

Thiol-containing molecules can be attached to metallic QDs' surfaces through dative coordination. Sulfur has a strong affinity for metal atoms (such as Zn, Cd, Au, Ag, and sulfur) that are available on the surface of SQDs.[29] The main issue of this conjugation procedure is the nature of QDs' surface ligands and how they are attached to the QDs, and equally important, the conformation and nature of biomolecules. The proteins, as biomolecules, should have thiol at their accessible surface, and this thiol should not be part of an active structure. This limitation is even more significant when small thiolate peptides or DNAs are used to cap the QDs.[67] Another problem is the dynamic on–off rate of single thiol interaction, which shortens the reagent's lifetime.[68] This limitation has been overcome through the use of multiple thiols on the protein, as is seen in metallothionein.[69] However, this thiol-bonding bioconjugation procedure is rarely used in current research due to the advent of more versatile chemistry.

Ligands and biomolecules containing N and O may also attach to QDs' surfaces *via* dative interaction to a lesser extent since their bonding strength is usually weaker than that of sulfur.

2.2.1.3 Avidin–Biotin Chemistry

Supramolecular conjugation is based on the molecular recognition between ligands and receptors without the formation of covalent bonds, and it is reversible. The main advantage of this protocol is that the biomolecule's orientation can be maintained; indeed, even the conformation of the enzyme is not heavily altered.[70,71] Avidin–biotin interactions constitute a supramolecular immobilization method that is used extensively for bioconjugation on QDs' surfaces.[70] Avidin is a glycoprotein, a protein found in egg whites. It has four identical subunits with high affinity and specificity for biotin. Streptavidin, a non-glycosylated tetramer protein, is the most important avidin analogue for biomolecule immobilization. Antibodies, single-stranded DNA, oligonucleotides, enzymes, and biotinylated proteins can be attached to QDs' surfaces *via* avidin–biotin interaction. The resulting conjugated particles can be employed in biosensors for a variety of biological and chemical purposes. This technique is used extensively for selective probe design.

2.2.1.4 Functional Group Interconversion

Given the requirements of biocompatibility and stability, only a limited number of functional groups can be attached to QDs' surfaces for bioconjugation purposes. These groups are mostly amines, carboxyls, and, in rare situations, thiols, alkenes, alkynes, and azides (Figure 2.2). Likewise, biomolecules have a limited number of functional groups (Figure 2.2B). It is therefore imperative that researchers select specific QDs and biomolecules with suitable functional groups based on the desired bioconjugation outcome.

The best solution is to use commercial QD conjugates or QDs with functionalized ligands to design specific biosensors or detection systems

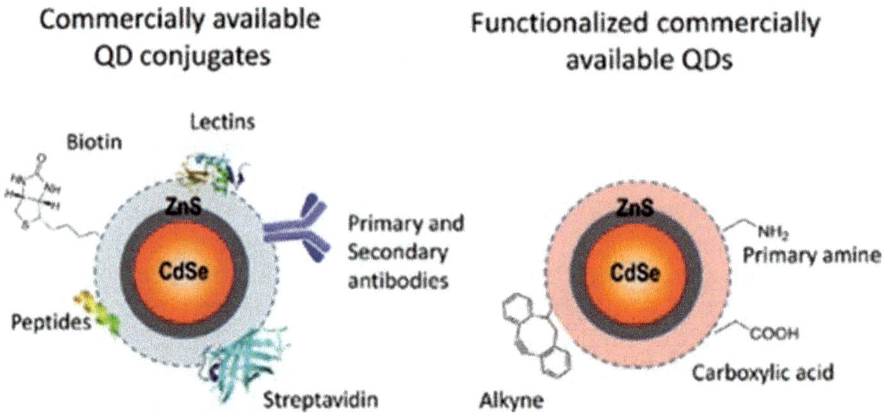

Figure 2.6 Commercially available bioconjugate and functionalized QDs. Reproduced from ref. 13 with permission from Elsevier, Copyright 2020.

(Figure 2.6). Another approach is to block undesirable functional groups on both QDs and biomolecules[72] using various chemical approaches. For example, if QDs with functionalized sulfhydryl are required, the amine should be converted to sulfhydryl by a precisely designed experiment. Generally, several procedures are required to arrive at optimum conditions for this kind of interconversion. Sometimes, multistep chemistry is required to block the functional groups on QDs and biomolecules. Multistep purification chemistry is required to block the functional groups on QDs and biomolecules, which limits the overall conjugate yield.

2.2.1.5 Other Bioconjugation Techniques

Physical adsorption, direct antibody/antigen (Ab/Ag) or aptamer attachment, molecularly imprinted polymer formation of the targeted analyte, host–guest protocol, and polyhistidine–metal-affinity coordination are among the most frequently used protocols for attaching biological molecules to QDs.[4,70] Figure 2.7 provides an overview of these techniques.

Biomolecules are physically adsorbed on QDs' surfaces *via* hydrogen bonding, electrostatic attraction, van der Waals forces, and hydrophobic interactions. The specific type of adsorption depends on the coating of the QDs and the traits of the biomolecules.[70] These physical interactions are used in the preparation of composites for drug delivery and controlled drug release because of their reversible nature. Variations in media conditions such as pH, temperature, and polarity promote the desorption of the immobilized drug.[70] This strategy again suffers from a lack of control over the orientation of the immobilized biomolecules, which in turn reduces their biological activity.

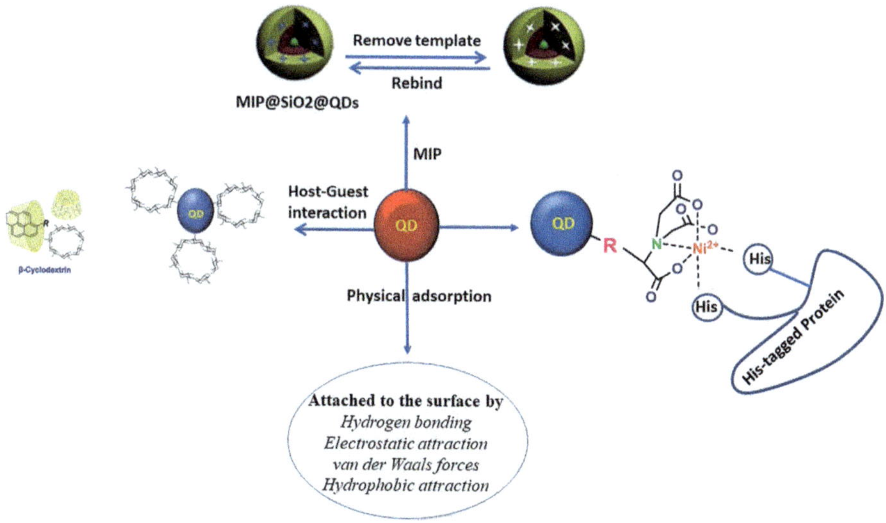

Figure 2.7 Conjugation on QDs by MIP, physical adsorption, host–guest interaction, and polyhistidine–metal affinity coordination.

Histidine (His) coordinates with metal ions such as Cu^{2+}, Ni^{2+}, Zn^{2+}, and Co^{2+} *via* the electron-donor group on imidazole side chains. Immobilized metal affinity chromatography (IMAC) is a protein purification technique that is widely used in biological laboratories[73] and that takes advantage of His–metal binding to form ligand–metal–His complexes. In IMAC, a ligand is immobilized on a solid support that can bind to Cu^{2+}, Ni^{2+}, or Zn^{2+}. During 1987–1988, Hochuli introduced the use of a Ni^{2+}-charged nitrotriacetic acid (NTA)-modified support for the efficient and selective purification of proteins with a His tag in their C and N terminals. The high affinity between His and nickel, as well as the possibility of tagging biomolecules with His, attracted scientists' attention for use in the preparation of QD bioconjugates. Three approaches to this type of binding are possible: (1) direct binding of polyhistidine to ligand-coated QDs, (2) binding of polyhistidine to carboxylated polymer-coated QDs, and (3) binding of polyhistidine to Ni^{2+}-NTA-modified QDs. Figure 2.7 schematically represents His–Ni–L bioconjugation.

Multiple His-tagged biomolecules are directly attached to CdSe/ZnS QDs because the imidazole chain of the His tag has a strong affinity for Zn sites at the surface of the ZnS shell. This enables the direct conjugation of QDs with biomolecules. The affinity is higher if the proteins are tagged with more than one His biomolecule due to multidentate dative interactions. Even in ligand-coated CdSe/ZnS, the His tag has no attraction toward the ligands, and the protein has strong affinity for the QDs surface with zero distance.[74]

As noted earlier, PEG-QDs are an ideal particle for many biological applications, but His-tagged proteins prefer to attach to the shell of PEG-SQDs with zero distance, which reduces PEG's availability in the bioenvironment. To resolve this problem, NTA is attached to PEG and the resulting group is used to cover the QD's surface. When PEG-NTA-QDs are charged with Ni^{2+} (Figure 2.2), a stable attachment site for binding to His-tagged proteins is provided.

QDs encapsulated with amphiphilic ligands have high water solubility and stability and are better passivated. They also have a high quantum yield and photostable PL. The amphiphilic ligand impairs carboxylic groups, and the shell metal is not available for His–metal affinity coordination. Meanwhile, electrostatic interactions between the surfaces of negatively charged QDs (the COO^- group) and cationic His are probable. This interaction is non-specific and sensitive to the pH and ionic strength of the medium. Scientists, therefore, used the ability of the carboxylic group to chelate with Ni^{2+} and added His-tagged protein (carboxylated-coated His–metal affinity coordination). This technique was originally used for the conjugation of a His6-tagged Renilla luciferase mutant (Luc8) to carboxylated polymer-coated QDs.[75] In this technique, free Ni^{2+} should be separated from the conjugate by dialysis.

Molecularly imprinted polymers (MIPs) have emerged as an approach to the preparation of polymers with tailor-made receptor cavities for molecules and cells.[76,77] The analyte of interest is referred to as the template; the MIP's cavity has the exact same geometry and size as the template molecule. Polymerization occurs around the template molecule. After the reaction is complete, the template is removed, leaving a cavity with the same size and topography as the analyte of interest (*e.g.*, a specific biomolecule). MIPs have been used to coat various QDs to prepare optical sensors based on a "key and lock" model. Many procedures for MIP synthesis have been introduced to date with the goal of improving assays' sensitivity and selectivity.[78–81] Figure 2.7 represents the MIP production process graphically.

QDs functionalized with cyclodextrin (CD) act as hosts and enable conjugation with biomolecules via inclusion complex formation (Figure 2.7). This type of conjugation is usually used for targeted drug delivery[81] and cell imaging.

2.2.2 Toxicity Effect of Functionalized QDs

QDs have been used for both *in vitro* and *in vivo* applications such as cell imaging, physiochemical change tracking in cells, and *in vitro* sensing. They have several advantages that make them useful in these applications, including: (1) they are more fluorescent compared to proteins and organic dyes; (2) their fluorescence can be adjusted by altering their size and functional groups; (3) their fluorescence is more stable in cells (420-fold higher than the Alexa 488 fluorophore); and (4) they possess extraordinary photochemical stability, which makes long-term optical tracking possible. However, many problems restrict their medical applications. QDs' main limiting factor is their toxicity,[5,82,83] but the temperature-dependent internalization of SQDs is also a vital concern, as it prevents QDs from entering cells at temperatures lower than 4 °C.

The mechanisms behind QDs' toxicity have been studied in detail. Research has shown that they may cause DNA damage, cytoskeletal alterations, deactivation of some enzymes, and oxidative damage due to the formation of reactive oxygen species (ROS).[82,83] Scientists have evaluated QDs' interactions with the immune system, their potential for cell membrane perturbation, and their possible transportation of toxic parts to cells, all with the goal of designing safer and more stable QDs. In addition to studying QDs' cytotoxicity, researchers have investigated the environmental toxicity, genotoxicity, and carcinogenic effects of these particles. Most studies on *in vitro* toxicity are performed on mammalian cells, namely, cancer and normal cells.[84]

The cytotoxicity of QDs depends on their chemical composition, size, and shape, as well as on the physiochemical properties of their inorganic and organic outer shells, such as surface charge, hydrophilicity/hydrophobicity, and resistance to environmental factors. These properties are different for metallic QDs compared to other QD types.[82–84]

The cytotoxicity of most metallic SQDs is due to heavy metal leakage. Among metallic SQDs, CdSe is most often used in biological media because it is easily prepared, it has outstanding PL properties, and it is easily coated with ZnS. This coating protects the core from oxidation and other degradation processes, which consequently reduce Cd leakage into the working environment, but these SQDs can still release Cd into the medium. Cd-free QDs were developed as an alternative to eliminate Cd contamination. Adding an organic ligand to the surface of the QD strongly prevents heavy metal leakage, while at the same time improving the QD's solubility and photoluminescence.[17]

In core–shell QDs, toxicity depends on the chemical composition of the core and the inorganic shell, the size of the QD, and the physiochemical specifications of the organic outer shell. A recent study that compared the impact of composition, size, and charge on QDs' toxicity determined that of these three parameters, the charge of the particles had the greatest effect, the type of functional groups had a lesser effect, and the size of the QDs least affected their toxicity (charge > functionalization > size).[82]

Recently, Sukhanova *et al.*[82] synthesized different kinds of SQDs with different functional groups on the outer shell. They analysed as-prepared SQDs' cytotoxicity to compare the effects of functionalization, composition, and size effect on QDs' toxicity. To the best of our knowledge, this is the first systematic investigation that has shown the dependence of cytotoxicity on QDs' size, chemical composition, and surface functional groups.

The tested QDs were CdSe/CdS/ZnS (6 monolayer [ML] CdS and 3 ML Zns)-PEG-OH, CdSe/ZnS(8 ML)-PEG-OH, CdSe/ZnS(3 ML)-PEG-OH, CdSe/ZnS(3 ML)-PEG-COOH, and CdSe/ZnS(3 ML)-PEG-NH_2. Pegylated QDs were selected because of the high biocompatibility of PEG and the possibility to impair different electrical charges to the final QDs since PEG derivative has amine, carboxyl, and hydroxyl at one end, SH at the other end, and 11 carbon atoms in its aliphatic chain. SK-BR-3 human breast cancer cells and WI38 normal cells were used in the cytotoxicity evaluations because the intracellular transport rate is higher in tumor cells. As-prepared QDs were in contact with the cells for 24, 48, and 72 h, and their corresponding cell viability was recorded. PEG-OH and PEG-COOH derivatives imparted a low negative charge and a high negative charge to the QDs' surfaces, respectively, while PEG-NH_2 imparted a positive charge. The toxicity of the particles is due to their transportation to the cell *via* the plasma membrane and their interaction rate with the cell's constituents. The toxicity test results showed that the QDs with a low negative charge on their surfaces were the most toxic to both cancer and normal cells, while QDs with a positive charge (PEG-NH_2) exhibited no strong toxicity. Fluorescence evaluations indicated that both low negative and positive QDs accumulated in cells. The IC50 of the QDs was reduced after 48 h of incubation with both cell lines. The data showed that even strongly positive QDs penetrated cells after 24 h of incubation because the negative charge of the cells' membranes facilitated their uptake. The QDs were also able to penetrate the nuclear membrane and interact with the sugar-phosphate backbone of DNA. The QDs with a small positive charge (-PEG-NH_2) were more effective at penetrating cells' nuclei after incubation for 48 and 72 h, which indicates that the weakly positive QDs were the most toxic to the SK-BR-3 human breast cancer cells and WI38 normal cells *in vitro*.

CQDs' toxicity mechanisms have not been fully investigated, but their toxicity has been evaluated by MTT assay. They are supposed to be safe and environmentally friendly, and they exhibit low cytotoxicity compared with traditional semiconductors.[8,10,11,85,86] For example, CQDs were prepared using soot as the carbon source and were passivated by PEG1500; these CQDs were compared with semiconductor CdSe/ZnS QDs for bioimaging and cytotoxicity.[11] The CQDs were smaller in size (below 10 nm), brighter, more biocompatible, and less toxic compared with CdSe/ZnS QDs (size below 20 nm).

2.2.3 Sensing Ability of Conjugated QDs

To date, a variety of QD bioconjugates have been developed for different biological applications.[1,3,15,16,87] In some cases, photoluminescence intensity is used for signaling purposes, with the biomolecules themselves playing a passive role. For example, antibodies in QD–antibody conjugates are passive in flow cytometry and immunoassays because the biomolecules are used only for targeting and biorecognition.[88,89]

The biomolecule on the QD's surface is active when the QD's PL changes upon conjugation. Therefore, QDs are used both as platforms for biomolecules and as dynamic signal transducers. In these situations, changes in QDs' PL occur due to various mechanisms (such as active Forster resonance energy transfer [FRET] or electron transfer [ET], previously described in Chapter 1). Scientists can take advantage of the unique photophysical properties of these QDs for active biosensing. For example, the binding or enzymatic cleavage of target molecules with biomolecules on QDs' surfaces (peptides or receptors) alters the conformation or position of biomolecules relative to QDs; this change alters the rate of energy transfer, which in turn changes the QDs' PL signals. Researchers can therefore monitor these changes to detect the target analyte. Table 2.1 presents several conjugated QDs from different categories, their corresponding PLs, and their applications. QDs can also be coated with multiple antibodies or biomolecules for multiple diagnostic and sensing purposes.[3]

2.2.3.1 Passive Role of Detection-based QDs Conjugates

Biomolecules are said to play a passive role in detection when they are attached to QDs' surfaces for use in recognition and specification but do not undergo a specific biological interaction. Next, we discuss several recent examples in which the functionalization of QDs enables recognition in specific applications.

Functionalized QDs are hydrodynamic and small in size (10–20 nm) and are suitable for neuron and glial cell labeling and tracking. These QDs can cross the brain–blood barrier and reach brain tissues; as such, they are used extensively in brain-targeting imaging and are also employed as vehicles for drug delivery to the brain. They are also important from a clinical point of view. Zhang *et al.* synthesized functionalized core–shell CdSe/CdS QDs and evaluated their use[90] as a stable, highly bright PL fluorescence material for cellular uptake *in vitro* in cultured brain slices (*ex vivo*); they also evaluated their utility for *in vivo* studies.[90] The QDs were functionalized with 3-mercaptopropionic acid (MPA), poly(ethylene glycol) (PEG)-5000k-methoxy (PEG-OMe), and (PEG)-5000k-hydroxyl (PEG-OH). The core–shell CdSe/CdS QD architecture maximized the luminescence efficiencies. The stability and cellular uptake of as-prepared QDs depend strongly on their chemistry and model. Zhang *et al.* observed that surface chemistry strongly influenced QD behavior in the brain. PEG coatings improved QDs' stability in complex neurophysiologically relevant fluids and led to stable, diffuse cellular uptake. All QDs were entered into microglia, mostly in the corpus callosum, which was further confirmed *in vivo*. PEGylated QDs induced an increase in Mt expression in the organotypic whole hemisphere (OWH) slices as a function of time and surface chemistry. Non-PEGylated QDs destabilized in brain fluids, which prevented their transfer to brain tissue. Figure 2.8 shows the Iba1+ microglia and lysosome uptake of QDs in *ex vivo* OWH slices.

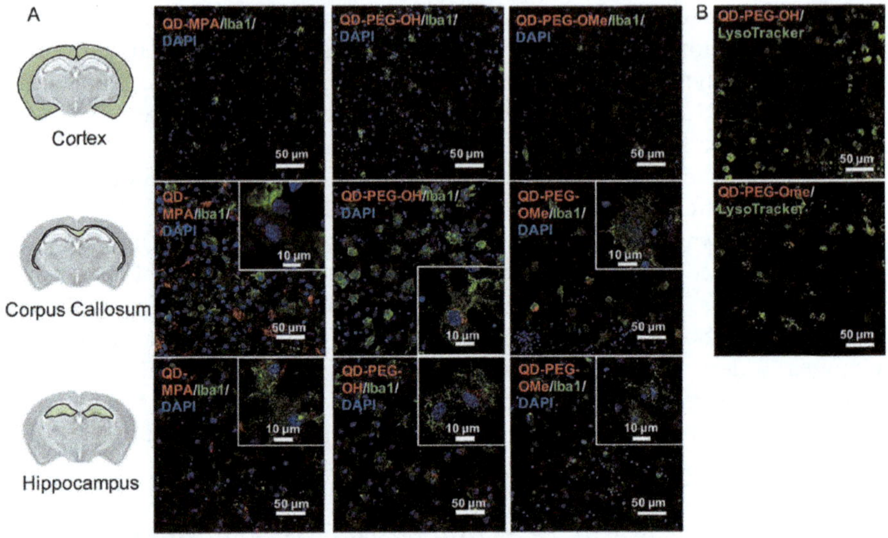

Figure 2.8 Iba1+ microglia and lysosome uptake of QDs in *ex vivo* OWH slices. (A) Representative 40× magnification images of QD-MPA, QD-PEGOH, and QD-PEG-OMe distribution and Iba1+ cellular uptake in the cortex, corpus callosum, and hippocampus in P14 Sprague-Dawley (SD) rat OWH slices (300 mm thickness). QDs (red, all images) were found internalized into Iba1+ microglia (green, all images), especially in the corpus callosum and hippocampus regions. Representative high-magnification images of QD-MPA, QD-PEG-OH, and QD-PEG-OMe localization inside Iba1+ cells are shown as insets. Cell nuclei were stained with DAPI and displayed blue luminescence. (B) Representative 40× magnification images of QD-PEG-OH (red) and QD-PEG-OMe (red) internalization into LysoTracker-stained lysosomes (green) in the corpus callosum region in a live OWH slice after 2 h of QD exposure. Reproduced from ref. 90 with permission from the Royal Society of Chemistry.

CQDs have bright multiwavelength emissions with low toxicity and high photostability and chemical stability; they are good candidates for bioimaging.[10] Poly(propionylethylenimine-co-ethylenimine) (PPEI-EI) passivated CQDs were incubated with MCF-7 human cancer cells at 37 °C for 2 h. A bright PL in the cytoplasmic area and cell membrane was observed when excited at 800 nm.[91] The surfaces of CQDs derived from natural sources possess different functional groups, which can be used in cell imaging. CQDs prepared from coffee grounds were localized in the cell membrane and cytoplasm of LLC-PK1 cells through endocytosis.[92] CQDs synthesized with green tea as a carbon source have been used for both imaging and inhibiting cancer cell growth.[93] Multi-emission CQDs successfully penetrated MCF-10A and LLC-PK1 normal cells as well as MCF-7 and

MDA-MB-231 cancer cells. They exhibited PL in blue, green, and red while excited at wavelengths of 360–380, 460–480, and 510–530 nm, respectively.

Conjugated QDs have recently been used for contrast in MRI[94–96] and as fluorescent tags in flow cytometry. Table 2.1 shows different kinds of QDs, their surface modifications, and their biological applications. NIR QDs are used for cell imaging, targeted drug delivery, and cell imaging.

2.2.3.2 Active Roles of Detection Based on QDs' Conjugates

Biomolecules maintain their active functionality as well as their recognition abilities when conjugated with QDs. These molecules are used for simultaneous biorecognition and assay, both *in vitro* and *in vivo*.[97] Next, we summarize several new investigations of biomolecules that play active roles within bioconjugated or functionalized QDs for detection and quantitation purposes.

Peptide-coated CdSe/ZnS QDs were labeled with rhodamine and used as FRET probes for tracking metalloproteinases (MMPs) in cancerous and normal cell cultures.[98] Each CdSe/ZnS QD's capping ligand was replaced with a tetrapeptide molecule to increase its solubility in biological media. Rhodamine was attached as a fluorescence tag to the end of the peptide chain and acted as an electron acceptor. While excited at 445 nm, rhodamine-labeled peptide QDs showed two emission wavelengths: one at 545 nm (QD emissions) and another at 590 nm (rhodamine emissions). When rhodamine concentration increased, emissions at 590 nm increased and emissions at 545 nm decreased because FRET occurred between the QDs as donors and the rhodamine tags as acceptors. It was also shown that rhodamine had 10-fold lower emission (with the same concentration) when attached to the peptide-coated CdSe/ZnS QDs. This increase in emissions is due to the FRET that occurs between the QDs and the rhodamine. This FRET probe successfully determined the proteolytic activity of MMPs in the cell cultures in less than 15 min. The results indicated that the MMPs activity in breast cancer cell cultures was significantly higher than that of normal cells. Therefore, it became a fast method to distinguish between normal and cancerous cells. The fast FRET response was due to the high monodispersity of the rhodamine-labeled peptide-coated CdSe/ZnS QDs and the small distance between QDs and acceptor molecule.

Carboxylated CQDs (cCQDs) have been used as nanoquenchers in a fluorometric nucleic acid detection probe, which constitutes a new detection application.[99] A fluorescently labeled single-stranded DNA (ssDNA) probe (FAM-Lprobe) was adsorbed on the surface of a cCQD, which quenched its fluorescence. When the FAM-Lprobe underwent hybridization with its complementary DNA target, a double-stranded DNA (dsDNA) was formed, and the fluorescence of the FAM-Lprobe recovered due to the weak interaction between the dsDNA and the cCQD. Two forces were involved in this interaction between cCQDs and DNA: electrostatic repulsion and hydrophobic π–π interaction. Electrostatic repulsion occurred between the negatively charged carboxylic group on the cCQD and the negatively charged phosphodiester backbone of the DNA, while the hydrophobic π–π interaction occurred due to π–π stacking between the DNA's nucleobase and the CQD's hydrophobic surface. However, the electrostatic repulsion was stronger than the hydrophobic interaction when the FAM-Lprobe was hybridized with dsDNA, and as a result, the FAM-Lprobe's fluorescence recovered.

Multifunctional fluorescent magnetic CQDs have been used to isolate methicillin-resistant *Staphylococcus* (MRSA) and the *Salmonella* DT104 superbug from whole blood samples; they have also been used to identify these bacteria *via* multicolor fluorescence imaging.[100] The magnetic properties of the particles easily enriched superbug from the blood sample for its determination at a lower amount, a key step in the early-stage diagnosis of an infection. Pardaxin antimicrobial peptide-conjugated anti-MRSA-antibody magnetic-CQDs (red emission) and pardaxin antimicrobial peptide-conjugated anti-*Salmonella*-antibody-attached magnetic-CQDs (blue emission) were designed to target, capture, and kill *Salmonella* DT104 and MRSA. To determine the percentage of live superbugs, the functionalized magnetic CQDs were separated from blood by a magnet and washed with distilled water, and then, the number of colonies was counted. The results implied that the superbugs were separated from blood samples with high selectivity and accuracy. MRSA was distinguished from *Salmonella* DT104 because of the multicolor nature of the functionalized CQDs. Because the recognition units are the CQDs' corresponding antibodies, this detection and isolation method can be extended to other superbugs by conjugating their antibodies with magnetic CQDs. The as-prepared particles can also be used in MRSA and *Salmonella* DT104 imaging. Moreover, these functionalized particles can be used to clean the blood in circulation.

Nitrogen-doped GQDs (N-GQDs) were prepared by the nanospace-confined method and then coordinated with Fe^{3+} (N-GQD-Fe^{3+}) for use in the highly selective fluorescence determination of ascorbic acid in fish blood.[101] This detection is based on a fluorescence on–off–on strategy. The addition of Fe^{3+} to N-GQDs quenched the fluorescence of the QDs; the emission was recovered with the addition of ascorbic acid. N-GQDs were synthesized by steaming N-doped graphene layers in the confined space of ordered mesoporous SBA-15 with nitric acid vapor. This nanospace-confined strategy enables an exceptional degree of control over the product's monodispersity and provides a faster separation and purification process.

Many other conjugated QDs with various applications in bioassays have been reported in recent years. The above examples constitute just a few illustrations of how functionalization alters the particles' properties and opens up potential applications.

2.2.4 QDs Conjugation in Electrochemical Detection

Electrochemical sensing is a very sensitive, inexpensive, rapid, and easy-to-handle method of determination. It is widely applied in clinical diagnosis, food analysis, detection of biological processes, and environmental monitoring. The working electrode, which is the electrode at which electron transfer occurs, consists of a bare or modified solid electrode. The electrode's surface can be modified to improve electron transfer kinetics or to accumulate electroactive molecules on the electrode surface and accordingly increase the electrochemical sensor's sensitivity and stability. Recently, QDs have been proven to be suitable electrode modifiers. Functionalized and conjugated QDs are attached to the surfaces of solid electrodes through simple procedures; the modified electrodes can be used in electrochemical sensors or biosensors depending on the type of conjugated molecules used.[102] Electrodes' surfaces are coated with QDs *via* drop casting or by *in situ* electrochemical simultaneous synthesis and/or deposition.

The APO *e4* allele is a lipid transport protein. It is one of the three common alleles of apolipoprotein E, and it may be a major risk factor for diseases such as Alzheimer's disease, Parkinson's disease, coronary artery disease, and Huntington's disease. Functionalized graphene has been used to modify an Indium-Tin oxide (ITO) electrode to detect and determine this protein as a potential biomarker for Alzheimer's disease.[103] A GQD-Curcumin-DNA sensor was prepared through step-by-step deposition and functionalization (Figure 2.9). GQDs were dropped onto a pre-treated and clean ITO electrode surface. Curcumin was coated on the surface of the electrode *via* an electropolymerization process. In brief, the modified GQDs-ITO electrode was incubated in a curcumin solution in phosphate buffer pH 8, and 20 consecutive voltametric cycles were performed in the potential range of 0.1–0.6 V at a scan rate of 50 mV/s. Adsorbed curcumin (CM) was conjugated to amino-terminated DNA by EDC/NHS chemistry (DNA-CM-GQDs-ITO) with malonic acid as a spacer. The non-specific adsorption of DNA was prevented through the incubation of the DNA-CM-GQDs-ITO in 1% BSA solution for 30 min.

A CQD-modified glassy carbon (GC) electrode (CQD/GCE) was prepared simply by adding Nafion oil to a clean GCE surface, then adding CQDs, and finally air-drying the resulting modified electrode for 15 min.[104] This CQD/GCE was used for the electrochemical assay of metanil yellow and curcumin simultaneously or individually.

Another approach to the preparation of QCD-modified electrodes is to first prepare a QD composite and then cast it on an electrode's surface. Tufa *et al.*[105] modified GCEs' surfaces with GQD-coated Fe_3O_4@Ag core–shell nanostructures (Fe_3O_4@Ag/GQDs) by first preparing the nanomaterials and then attaching them to the electrodes' surfaces. GQDs were functionalized with thiol ligands (Thio-GQDs), cast on GCEs, and then attached to amine-modified aptamers using silver as the linker.[106] They

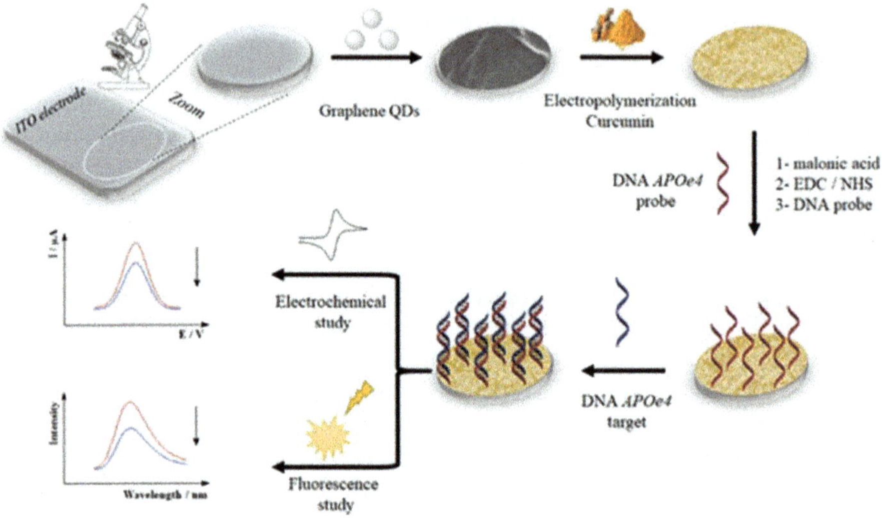

Figure 2.9 Schematic representation of stepwise formation of GQDs-curcumin-DNA APO e4 DNA sensor and electrochemical assay. Reproduced from ref. 103 with permission from Elsevier, Copyright 2018.

designed an aptasensor for the determination of trinitrotoluene (TNT). Overall, conjugation chemistry enables the easy attachment of QDs to electrode surfaces in electrochemical cells and to a variety of other solid substrates.

2.2.5 Functionalized QDs for Theranostic Purposes

Due to their small size, QDs can easily penetrate cells or move through the body *via* capillaries. They are biocompatible, and by choosing specific QDs with the right functional groups, researchers can target them to specific areas of the body. They have therefore attracted much scientific attention recently. Next, we discuss several theranostic applications that demonstrate how QDs' functional groups can be tailored for the treatment of various diseases.

Phenol-like group functionalized graphene QDs (h-GQDs) as antioxidants can efficiently treat acute kidney injury (AKI).[107] AKI is a syndrome in which the kidneys rapidly lose their excretory functions. It is a serious condition with a high in-hospital death risk. The only current clinical treatment is renal replacement therapy (RRT), which has a high mortality rate (50–60%). Nearly 20% of surviving patients require dialysis after hospital discharge. The major cause of renal failure in AKI is excessive ROS production. However, antioxidant therapy through ROS scavenging is considered as an effective procedure to protect renal function in AKI. *N*-acetylcysteine (NAC) as an antioxidant has poor bioavailability and specificity, which hampers its clinical utility in various types of AKI therapy. Recently, a series of carbon-based QDs, such as nitrogen-doped CQDs, selenium-doped CQDs, chlorine-doped CQDs, nitrogen/sulfur-doped CQDs, and GQDs, were developed and used as antioxidants to scavenge ROS.[108] Among these QDs, GQDs proved the most effective at targeting the kidneys: they were retained the longest in the body and exhibited the most efficient renal clearance.[109]

Normally, GQDs exhibit low antioxidant activity; however, this activity can be increased through the functionalization of their surfaces with enriched oxygenated groups. To design an antioxidant GQD, they looked to the functional groups of polyphenols, which are important natural antioxidants. The phenolic group of polyphenols acts as H atom donors for ROS scavenging:[110] when a free radical captures a phenolic H atom, it forms a phenoxy radical intermediate that is resonance stabilized by p–π orbital overlap and that can couple rapidly with another free radical. Many oxygenated functional groups such as hydroxyl (C-OH), carbonyl (C = O), and carboxyl (COOH) can be anchored on the surface of GQDs. C-OH on graphene can attach to the sp^2 or the sp^3 carbon center. The resulting C-OH sp^2 carbon centers on the graphene structure are called phenol-like groups[111] and perform the same ROS scavenging function as the H atoms in polyphenols. As-prepared h-GQDs highly accommodated in the kidney and show high-efficacy antioxidant activity to treat AKI. These functionalized GQDs exhibited high *in vivo* and *in vitro* biocompatibility, even at high concentrations.[112] The ROS trapping ability of h-GQDs could be suppressed extensively by adjacent carbonyl groups and also significantly improved by other phenol-like moieties. Therefore, h-GQDs must be well tailored in order to achieve the highest possible antioxidant activity for effective AKI treatment. Figure 2.10 schematically illustrates the antioxidant activity of designed GQDs and characterization of as-prepared GQDs for AKI therapy.

CQDs prepared from ginger, a natural carbon source, strongly inhibited the growth of human hepatocellular carcinoma cells (HepG2) and exhibited low toxicity to normal mammary epithelial cells (MCF-10A) and normal liver cells (FL83B).[113] Testing against a human lung cancer cell line (A549), a human breast cancer cell line (MDA-MB-231), and a human cervical cancer cell line (Hela) indicated that the CQDs were selective for HepG2. The carcinoma cells exhibited greater QCD uptake (1.1 mg mL^{-1}) as well as an 18.2 fold increase in intercellular [or intracellular?] ROS production compared to the other cell lines. The CQDs also significantly reduced the weight of HepG2-induced tumors in mice after 14 days of treatment.

CQDs synthesized using green tea as a carbon source have been used to inhibit cancer cell growth.[93] As-prepared dots inhibited the growth of MCF-7 and MDA-MB-231 cancer cells by 80 and 82%, respectively, while only inhibiting the growth of MCF-10A and LLC-PK1 normal cells by 5%.

Functionalized QDs are also used in photothermal and photodynamic therapy[31] and in the treatment of bacterial infections.[100]

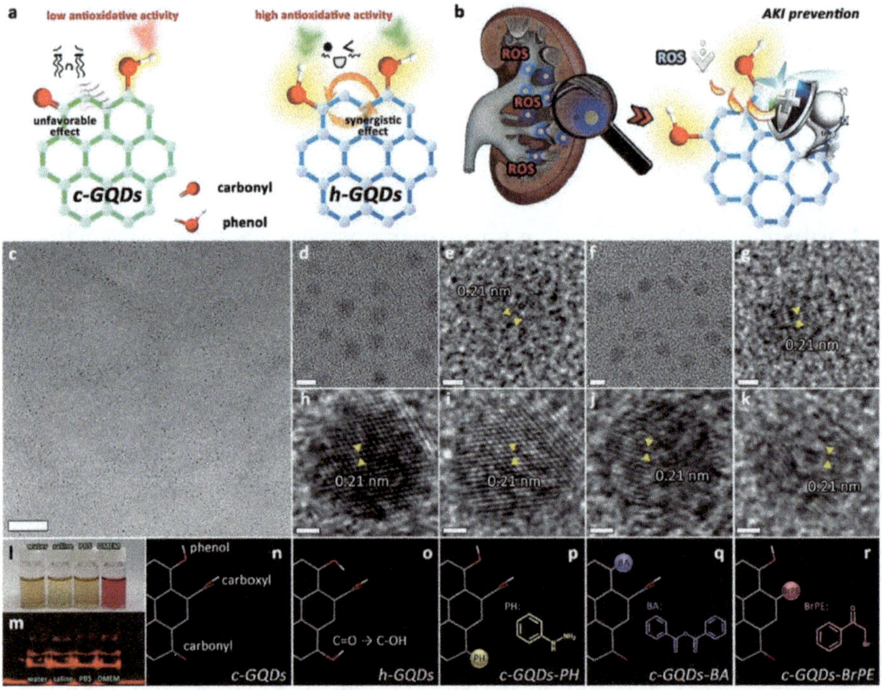

Figure 2.10 Design and characterization of h-GQDs for AKI treatment.[107] Schematic illustration of antioxidative activity of h-GQDs (a) and their usage as ROS scavengers for treating AKI (b). TEM images of h-GQDs (c–e), p-GQDs (f, g), c-GQDs (h), c-GQDs–PH (i), c-GQDs–BA (j), and c-GQDs–BrPE (k). Scale bars are equal to 100 nm (c), 5 nm (d, f), and 1 nm (e, g–k). Photograph of h-GQDs dissolved in different solutions including water, saline, PBS, and DMEM containing 10% FBS (l). Photograph of the Tyndall effect of h-GQDs dissolved in different solutions (m). Schematic illustrations of the chemical structures of c-GQDs (n), h-GQDs (o), c-GQDs–PH (p), c-GQDs–BA (q), and c-GQDs–BrPE (r). Red bands indicate O atoms, white bands indicate H atoms, and grey bands indicate C atoms. Insets: the molecular structure of PH (p), BA (q), and BrPE (r). Reproduced from ref. 107 with permission from the Royal Society of Chemistry.

2.3 Conclusion and Future Prospects

QDs have been developed for use in a vast array of biomedical applications owing to their wide variety of specifications. SQDs are the most commonly used, commercially available QDs. Although SQDs have extraordinary and bright PL, size-dependent tunable emission, and excellent physiochemical properties, they are insoluble in biological media and toxic due to leakage of the heavy metal to the bio-environment. The modification and functionalization of QDs' surfaces can improve their photoluminescence and physiochemical properties, and it can also decrease their toxicity. QDs' water solubility can also be increased through various procedures, rendering them bioavailable for *in vitro* and *in vivo* detection and assays. Functionalization also makes bioconjugation more straightforward and improves QDs' sensing and detection applications.

CQDs, a newer generation of QDs, have attracted recent scientific attention for several reasons: functionalized CQDs can be simply prepared from low-cost carbon sources, their fluorescence is highly stable, and they exhibit low toxicity. They have, therefore, seen extensive use in many fields, especially biomedicine. CQDs such as SQDs generate ROS and can be utilized as biocompatible therapeutic and diagnostic agents. However, their low quantum yield and poor size distribution limit their applications. CQD research is ongoing; scientists continue to modify them, improve their properties and functionalities, and investigate new avenues for biomedical applications. Although CQDs are a reliable alternative to SQDs, there are few reports on their use as tags or labels in biomedicine. Scientists are still working to improve CQDs' sensitivity and selectivity as elements in biosensor and chemosensor design. Last but not least, CQDs' cytotoxicity mechanisms have not yet been fully investigated. The toxicity of various CQDs is determined mostly by MTT assays using *in vitro* experiments.

Today, enormous progress has been made in the field of QD bioconjugation. Each bioconjugation chemistry has its own benefits and disadvantages. The best bioconjugation system is the one that needs low purification of the conjugate and that has been used for a more specific application.

Abbreviations

AKI	Acute kidney injuring
BRET	Biological fluorescence resonance energy transfer
cCQDs	Carboxylated CQDs
CD	Cyclodextrin
CDs	Carbon based nanodots
CM	Curcumin
CQD/GCE	CQD-modified glassy carbon electrode
CQDs	Carbon quantum dots
CT	X-ray computed tomography
CTAB	Cethyltrimethylammonium bromide
DHLA	Dihydrolipoic acid
dsDNA	Double-stranded DNA
EDC	*N*-(3-dimethylaminopropyl)-*N*′-ethylcarbodiimide or carbodiimide
ET	Electron transfer

FL83B	Normal liver cell
FRET	Forster resonance energy transfer
h-GQDs	Phenol-like group functionalized graphene QDs
HepG2	Human hepatocellular carcinoma cells
HIS	Histidine
IMAC	Immobilized metal affinity chromatography
ITO	Indium–tin oxide
MAA	Mercaptoacetic acid
MCF-10A	Normal mammary epithelial cells
MIP	Molecularly imprinted polymers
ML	Monolayer
MMPs	Metalloproteinases
MnTTP	Manganese tetraphenyl porphyrin complex
MPA	3-Mercaptopropionic acid
MRI	Magnetic resonance imaging
MRSA	Methicillin-resistant *Staphylococcus*
N-GQDs	Nitrogen-doped GQD
NAC	*N*-acetylcysteine
NTA	Nitrotriacetic acid
PA	Photoacoustic
PANI	Polyaniline
PEG	Polyethylene glycol
PL	Photoluminescence
PPEI-EI	Poly(propionylethylenimine-co-ethylenimine)
PTT	Photothermal therapy
QDs	Quantum dots
ROS	Reactive oxygen species
RT	Radiotherapy
SQDs	Semiconductor quantum dots
ssDNA	Single-stranded DNA
Sulfo-NHS	*N*-hydroxysulfosuccinimide
TNT	Trinitrotoluene

References

1. A. M. Wagner, J. M. Knipe, G. Orive and N. A. Peppas, Quantum dots in biomedical applications, *Acta Biomater*, 2019, **94**, 44–63.
2. Z. Liu, C. Chang, W. Zhang, M. Yang and Q. Zhang, Research on Ligand Properties of CdSe Quantum Dots, *IOP Conf. Ser.: Mater. Sci. Eng.*, 2019, **562**(1), 012067.
3. S. Chinnathambi and N. Shirahata, Recent advances on fluorescent biomarkers of near-infrared quantum dots for in vitro and in vivo imaging, *Sci. Technol. Adv. Mater*, 2019, **20**(1), 337–355.

4. L. Wang, D. Xu, J. Gao, X. Chen, Y. Duo and H. Zhang, Semiconducting quantum dots: Modification and applications in biomedical science, *Sci. China Mater.*, 2020, **63**(9), 1631–1650.
5. X. Huang and M. Tang, Research advance on cell imaging and cytotoxicity of different types of quantum Dots, *J. Appl. Toxicol.*, 2021, **41**(3), 342–361.
6. M. Grabolle, M. Spieles, V. Lesnyak, N. Gaponik, A. Eychmüller and U. Resch-Genger, Determination of the Fluorescence Quantum Yield of Quantum Dots: Suitable Procedures and Achievable Uncertainties, *Anal. Chem.*, 2009, **81**(15), 6285–6294.
7. Z. Ramezani, M. Qorbanpour and N. Rahbar, Green synthesis of carbon quantum dots using quince fruit (Cydonia oblonga) powder as carbon precursor: Application in cell imaging and As3+ determination, *Colloids Surf. A Physicochem. Eng. Asp.*, 2018, **549**, 58–66.
8. A. I. Costa, P. D. Barata, B. Moraes and J. V. Prata, Carbon Dots from Coffee Grounds: Synthesis, Characterization, and Detection of Noxious Nitroanilines, *Chemosensors*, 2022, **10**(3), 113.
9. H. Khalid, M. F. Khan, B. Ahmad, M. Ismail, M. Zahid and A. Ismail, Grass-derived carbon nanodots as a fluorescent-sensing platform for label-free detection of Cu (II) ions, *J. Mater. Sci: Mater. Electron.*, 2022, **33**(8), 5626–5634.
10. Z. Ji, Z. Yin, Z. Jia and J. Wei, Carbon Nanodots Derived from Urea and Citric Acid in Living Cells: Cellular Uptake and Antioxidation Effect, *Langmuir*, 2020, **36**(29), 8632–8640.
11. N. Azam, M. Najabat Ali and T. Javaid Khan, Carbon Quantum Dots for Biomedical Applications: Review and Analysis, *Front. Mater.*, 2021, **8**, 8.
12. P. Teng, J. Xie, Y. Long, X. Huang, R. Zhu, X. Wang, L. Liang, Y. Huang and H. Zheng, Chemiluminescence behavior of the carbon dots and the reduced state carbon dots, *J. Lumin.*, 2014, **146**, 464–469.
13. C. Tosat-Bitrián and V. Palomo, CdSe quantum dots evaluation in primary cellular models or tissues derived from patients, *Nanomedicine*, 2020, **30**, 102299.
14. M. Zhang, Y. Jia, J. Cao, G. Li, H. Ren, H. Li and H. Yao, Carbon dots-enhanced luminol chemiluminescence and its application to 2-methoxyestradiol determination, *Green Chem. Lett. Rev.*, 2018, **11**(4), 379–386.
15. A. Foubert, N. V. Beloglazova, A. Rajkovic, B. Sas, A. Madder, I. Y. Goryacheva and S. De Saeger, Bioconjugation of quantum dots: Review & impact on future application, *TrAC, Trends Anal. Chem.*, 2016, **83**, 31–48.

16. L.-W. Wang, C.-W. Peng, C. Chen and Y. Li, Quantum dots-based tissue and in vivo imaging in breast cancer researches: current status and future perspectives, *Breast Cancer Res. Treat.*, 2015, **151**(1), 7–17.
17. A. Sahu and D. Kumar, Core-shell quantum dots: A review on classification, materials, application, and theoretical modeling, *J. Alloys Compd.*, 2022, **924**, 166508.
18. Y. Wang and L. Chen, Quantum dots, lighting up the research and development of nanomedicine, *Nanomedicine*, 2011, **7**(4), 385–402.
19. L. W. Zhang, W. Bäumer and N. A. Monteiro-Riviere, Cellular uptake mechanisms and toxicity of quantum dots in dendritic cells, *Nanomedicine*, 2011, **6**(5), 777–791.
20. I. L. Medintz and H. Mattoussi, Quantum dot-based resonance energy transfer and its growing application in biology, *Phys. Chem. Chem. Phys.*, 2009, **11**(1), 17–45.
21. Q. Zhang, H. Xue, J. Niu, M. Tian and X. Yu, Construction of fluorescent probes for ratiometric visualization of mitochondrial depolarization based on the FRET mechanism, *New J. Chem.*, 2022, **46**(21), 10067–10072.
22. W. C. W. Chan and S. Nie, Quantum Dot Bioconjugates for Ultrasensitive Nonisotopic Detection, *Science*, 1998, **281**(5385), 2016–2018.
23. J. K. Oh, Surface modification of colloidal CdX-based quantum dots for biomedical applications, *J. Mater. Chem.*, 2010, **20**(39), 8433–8445.
24. R. Hong, N. O. Fischer, A. Verma, C. M. Goodman, T. Emrick and V. M. Rotello, Control of protein structure and function through surface recognition by tailored nanoparticle scaffolds, *J. Am. Chem. Soc.*, 2004, **126**(3), 739–743.
25. K. Susumu, H. T. Uyeda, I. L. Medintz, T. Pons, J. B. Delehanty and H. Mattoussi, Enhancing the stability and biological functionalities of quantum dots via compact multifunctional ligands, *J. Am. Chem. Soc.*, 2007, **129**(45), 13987–13996.
26. M. H. Stewart, K. Susumu, B. C. Mei, I. L. Medintz, J. B. Delehanty, J. B. Blanco-Canosa, P. E. Dawson and H. Mattoussi, Multidentate Poly(ethylene glycol) Ligands Provide Colloidal Stability to Semiconductor and Metallic Nanocrystals in Extreme Conditions, *J. Am. Chem. Soc.*, 2010, **132**(28), 9804–9813.
27. D. Gerion, F. Pinaud, S. C. Williams, W. J. Parak, D. Zanchet, S. Weiss and A. P. Alivisatos, Synthesis and Properties of Biocompatible Water-Soluble Silica-Coated CdSe/ZnS Semiconductor Quantum Dots, *J. Phys. Chem. B*, 2001, **105**(37), 8861–8871.
28. S. T. Selvan, P. K. Patra, C. Y. Ang and J. Y. Ying, Synthesis of silica-coated semiconductor and magnetic quantum dots and their use in the imaging of live cells, *Angew. Chem.*, 2007, **119**(14), 2500–2504.

29. W. R. Algar, K. Susumu, J. B. Delehanty and I. L. Medintz, Semiconductor quantum dots in bioanalysis: crossing the valley of death, *Anal. Chem.*, 2011, **83**(23), 8826–8837.
30. X. Hu, P. Zrazhevskiy and X. Gao, Encapsulation of single quantum dots with mesoporous silica, *Ann. Biomed. Eng.*, 2009, **37**(10), 1960–1966.
31. J. Wang, X. Tan, X. Pang, L. Liu, F. Tan and N. Li, MoS_2 Quantum Dot@Polyaniline Inorganic–Organic Nanohybrids for in Vivo Dual-Modal Imaging Guided Synergistic Photothermal/Radiation Therapy, *ACS Appl. Mater. Interfaces*, 2016, **8**(37), 24331–24338.
32. K. Hagiwara, S. Horikoshi and N. Serpone, Photoluminescent Carbon Quantum Dots: Synthetic Approaches and Photophysical Properties, *Chem. Eur. J.*, 2021, **27**(37), 9466–9481.
33. M. J. Molaei, Principles, mechanisms, and application of carbon quantum dots in sensors: a review, *Anal. Methods*, 2020, **12**(10), 1266–1287.
34. L. Yue, H. Li, Q. Liu, D. Guo, J. Chen, Q. Sun, Y. Xu and F. Wu, Manganese-doped carbon quantum dots for fluorometric and magnetic resonance (dual mode) bioimaging and biosensing, *Microchim. Acta*, 2019, **186**(5), 315.
35. Y. He, Y. Zhong, Y. Su, Y. Lu, Z. Jiang, F. Peng, T. Xu, S. Su, Q. Huang, C. Fan and S.-T. Lee, Water-dispersed near-infrared-emitting quantum dots of ultrasmall sizes for in vitro and in vivo imaging, *Angew. Chem.*, 2011, **123**(25), 5813–5816.
36. D. Hu, P. Zhang, P. Gong, S. Lian, Y. Lu, D. Gao and L. Cai, A fast synthesis of near-infrared emitting CdTe/CdSe quantum dots with small hydrodynamic diameter for in vivo imaging probes, *Nanoscale*, 2011, **3**(11), 4724–4732.
37. K.-T. Yong, I. Roy, W.-C. Law and R. Hu, Synthesis of cRGD-peptide conjugated near-infrared CdTe/ZnSe core-shell quantum dots for in vivo cancer targeting and imaging, *Chem. Commun.*, 2010, **46**(38), 7136–7138.
38. R. Hu, K.-T. Yong, I. Roy, H. Ding, W.-C. Law, H. Cai, X. Zhang, L. A. Vathy, E. J. Bergey and P. N. Prasad, Functionalized near-infrared quantum dots for in vivo tumor vasculature imaging, *Nanotechnology*, 2010, **21**(14), 145105.
39. X. Liu, P. Zhou, H. Liu, H. Zhan, Q. Zhang, Y. Zhao and Y. Chen, Design of bright near infrared-emitting quantum dots capped with different stabilizing ligands for tumor targeting, *RSC Adv.*, 2018, **8**(8), 4221–4229.
40. A. Sasaki, Y. Tsukasaki, A. Komatsuzaki, T. Sakata, H. Yasuda and T. Jin, Recombinant protein (EGFP-Protein G)-coated PbS quantum dots

for in vitro and in vivo dual fluorescence (visible and second-NIR) imaging of breast tumors, *Nanoscale*, 2015, **7**(12), 5115–5119.

41. F. Zamberlan, L. Turyanska, A. Patanè, Z. Liu, H. E. L. Williams, M. W. Fay, P. A. Clarke, Y. Imamura, T. Jin, T. D. Bradshaw, N. R. Thomas and A. M. Grabowska, Stable DHLA-PEG capped PbS quantum dots: from synthesis to near-infrared biomedical imaging, *J. Mater. Chem. B*, 2018, **6**(4), 550–555.
42. Y. Imamura, S. Yamada, S. Tsuboi, Y. Nakane, Y. Tsukasaki, A. Komatsuzaki and T. Jin, Near-infrared emitting PbS quantum dots for in vivo fluorescence imaging of the thrombotic state in septic mouse brain, *Molecules*, 2016, **21**(8), 1080.
43. P. Zhao, K. He, Y. Han, Z. Zhang, M. Yu, H. Wang, Y. Huang, Z. Nie and S. Yao, Near-infrared dual-emission quantum dots-gold nanoclusters nanohybrid via co-template synthesis for ratiometric fluorescent detection and bioimaging of ascorbic acid in vitro and in vivo, *Anal. Chem.*, 2015, **87**(19), 9998–10005.
44. A. Benayas, F. Ren, E. Carrasco, V. Marzal, B. del Rosal, B. A. Gonfa, Á. Juarranz, F. Sanz-Rodríguez, D. Jaque, J. García-Solé, D. Ma and F. Vetrone, PbS/CdS/ZnS quantum dots: a multifunctional platform for in vivo near-infrared low-ddose fluorescence imaging, *Adv. Funct. Mater.*, 2015, **25**(42), 6650–6659.
45. X.-L. Ge, Z.-L. Zhang, Z.-X. Xie, R. Cui and D.-W. Pang, Revealing the biodistribution and clearance of Ag_2Se near-infrared quantum dots in mice, *New J. Chem.*, 2017, **41**(21), 12721–12725.
46. L. Tan, A. Wan, T. Zhao, R. Huang and H. Li, Aqueous synthesis of multidentate-polymer-capping Ag2Se quantum dots with bright photoluminescence tunable in a second near-infrared biological window, *ACS Appl. Mater. Interfaces*, 2014, **6**(9), 6217–6222.
47. B. Dong, C. Li, G. Chen, Y. Zhang, Y. Zhang, M. Deng and Q. Wang, Facile synthesis of highly photoluminescent Ag_2Se quantum dots as a new fluorescent probe in the second near-infrared window for in vivo imaging, *Chem. Mater.*, 2013, **25**(12), 2503–2509.
48. L. Tan, A. Wan and H. Li, Conjugating S-nitrosothiols with glutathiose stabilized silver sulfide quantum dots for controlled nitric oxide release and near-infrared fluorescence imaging, *ACS Appl. Mater. Interfaces*, 2013, **5**(21), 11163–11171.
49. Y.-P. Gu, R. Cui, Z.-L. Zhang, Z.-X. Xie and D.-W. Pang, Ultrasmall near-infrared Ag2Se quantum dots with tunable fluorescence for in vivo imaging, *J. Am. Chem. Soc.*, 2012, **134**(1), 79–82.
50. T. Torimoto, T. Kameyama and S. Kuwabata, Photofunctional materials fabricated with chalcopyrite-type semiconductor nanoparticles composed of AgInS2 and its solid solutions, *J. Phys. Chem. Lett.*, 2014, **5**(2), 336–347.

51. H. Tang, S.-T. Yang, Y.-F. Yang, D.-M. Ke, J.-H. Liu, X. Chen, H. Wang and Y. Liu, Blood clearance, distribution, transformation, excretion, and toxicity of near-infrared quantum dots Ag2Se in mice, *ACS Appl. Mater. Interfaces*, 2016, **8**(28), 17859–17869.
52. D. Che, X. Zhu, H. Wang, Y. Duan, Q. Zhang and Y. Li, Aqueous synthesis of high bright and tunable near-infrared AgInSe2-ZnSe quantum dots for bioimaging, *J. Colloid Interface Sci.*, 2016, **463**, 1–7.
53. L. Tan, S. Liu, X. Li, I. S. Chronakis and Y. Shen, A new strategy for synthesizing $AgInS_2$ quantum dots emitting brightly in near-infrared window for in vivo imaging, *Colloids Surf., B Biointerfaces*, 2015, **125**, 222–229.
54. O. T. Bruns, T. S. Bischof, D. K. Harris, D. Franke, Y. Shi, L. Riedemann, A. Bartelt, F. B. Jaworski, J. A. Carr and C. J. Rowlands, Next-generation in vivo optical imaging with short-wave infrared quantum dots, *Nat. Biomed. Eng.*, 2017, **1**(4), 1–11.
55. J. Gao, K. Chen, R. Xie, J. Xie, S. Lee, Z. Cheng, X. Peng and X. Chen, Ultrasmall near-infrared non-cadmium quantum dots for in vivo tumor imaging, *Small*, 2010, **6**(2), 256–261.
56. D. Li, D. Wang, X. Zhao, W. Xi, A. Zebibula, N. Alifu, J.-F. Chen and J. Qian, Short-wave infrared emitted/excited fluorescence from carbon dots and preliminary applications in bioimaging, *Mater. Chem. Front.*, 2018, **2**(7), 1343–1350.
57. M. Lan, S. Zhao, Z. Zhang, L. Yan, L. Guo, G. Niu, J. Zhang, J. Zhao, H. Zhang and P. Wang, Two-photon-excited near-infrared emissive carbon dots as multifunctional agents for fluorescence imaging and photothermal therapy, *Nano Res*, 2017, **10**(9), 3113–3123.
58. D. Li, P. Jing, L. Sun, Y. An, X. Shan, X. Lu, D. Zhou, D. Han, D. Shen and Y. Zhai, Near-infrared excitation/emission and multiphoton-induced fluorescence of carbon dots, *Adv. Mater.*, 2018, **30**(13), 1705913.
59. M. Zheng, Y. Li, S. Liu, W. Wang, Z. Xie and X. Jing, One-pot to synthesize multifunctional carbon dots for near infrared fluorescence imaging and photothermal cancer therapy, *ACS Appl. Mater. Interfaces*, 2016, **8**(36), 23533–23541.
60. L. Pan, S. Sun, L. Zhang, K. Jiang and H. Lin, Near-infrared emissive carbon dots for two-photon fluorescence bioimaging, *Nanoscale*, 2016, **8**(39), 17350–17356.
61. S. Chandra, B. Ghosh, G. Beaune, U. Nagarajan, T. Yasui, J. Nakamura, T. Tsuruoka, Y. Baba, N. Shirahata and F. M. Winnik, Functional double-shelled silicon nanocrystals for two-photon fluorescence cell imaging: spectral evolution and tuning, *Nanoscale*, 2016, **8**(16), 9009–9019.

62. L. Ravotto, Q. Chen, Y. Ma, S. A. Vinogradov, M. Locritani, G. Bergamini, F. Negri, Y. Yu, B. A. Korgel and P. Ceroni, Bright long-lived luminescence of silicon nanocrystals sensitized by two-photon absorbing antenna, *Chem.*, 2017, **2**(4), 550–560.
63. M. Sakiyama, H. Sugimoto and M. Fujii, Long-lived luminescence of colloidal silicon quantum dots for time-gated fluorescence imaging in the second near infrared window in biological tissue, *Nanoscale*, 2018, **10**(29), 13902–13907.
64. K. E. Sapsford, W. R. Algar, L. Berti, K. B. Gemmill, B. J. Casey, E. Oh, M. H. Stewart and I. L. Medintz, Functionalizing nanoparticles with biological molecules: developing chemistries that facilitate nanotechnology, *Chem. Rev.*, 2013, **113**(3), 1904–2074.
65. J. Kim, Y. Seo, Y. Jeong, M. P. Hwang, J. Hwang, J. Choo, J. W. Hong, J. H. Jeon, G. Rhie and J. Choi, A novel nanoprobe for the sensitive detection of Francisella tularensis, *J. Hazard. Mater.*, 2015, **298**, 188–194.
66. M. P. Hwang, J.-W. Lee, K. E. Lee and K. H. Lee, Think modular: a simple apoferritin-based platform for the multifaceted detection of pancreatic cancer, *ACS Nano*, 2013, **7**(9), 8167–8174.
67. T. Torchynska and Y. Vorobiev, Semiconductor II-VI Quantum Dots with Interface States and Their Biomedical Applications, in *Advanced Biomedical Engineering*, ed. D. G. Gargiulo, InTech, 2011.
68. W. C. W. Chan, T. L. Prendergast, M. Jain and S. Nie, One-step conjugation of biomolecules to luminescent nanocrystals, in *Molecular Imaging: Reporters, Dyes, Markers, and Instrumentation*, SPIE, 2000, pp. 2–9.
69. E. Guszpit, L. Krejčová, S. Křížková, M. Kepinska, L. Richtera, P. Kopel, V. Adam and H. Milnerowicz, Kinetic analysis of human metallothionein and CdTe quantum dot complexes using fluorescence and voltammetry techniques, *Colloids Surf., B*, 2017, **160**, 381–389.
70. M. Díaz-González, A. de la Escosura-Muñiz, M. T. Fernandez-Argüelles, F. J. García Alonso and J. M. Costa-Fernandez, Quantum Dot Bioconjugates for Diagnostic Applications, *Top. Curr. Chem.*, 2020, **378**(2), 35.
71. C. Iriarte-Mesa, Y. C. López, Y. Matos-Peralta, K. de la Vega-Hernández and M. Antuch, Gold, Silver and Iron Oxide Nanoparticles: Synthesis and Bionanoconjugation Strategies Aimed at Electrochemical Applications, *Top. Curr. Chem.*, 2020, **378**(1), 12.
72. G. Hermanson, Bioconjugate Techniques, Academic Press, London, UK, 2nd edn, 2008.
73. M. Rowinska-Zyrek, D. Witkowska, S. Potocki, M. Remelli and H. Kozlowski, His-rich sequences–is plagiarism from nature a good idea?, *New J. Chem.*, 2013, **37**(1), 58–70.

74. K. E. Sapsford, T. Pons, I. L. Medintz, S. Higashiya, F. M. Brunel, P. E. Dawson and H. Mattoussi, Kinetics of metal-affinity driven self-assembly between proteins or peptides and CdSe–ZnS quantum dots, *J. Phys. Chem. C*, 2007, **111**(31), 11528–11538.
75. H. Yao, Y. Zhang, F. Xiao, Z. Xia and J. Rao, Quantum dot/bioluminescence resonance energy transfer based highly sensitive detection of proteases, *Angew. Chem.*, 2007, **119**(23), 4424–4427.
76. J. Hou, H. Li, L. Wang, P. Zhang, T. Zhou, H. Ding and L. Ding, Rapid microwave-assisted synthesis of molecularly imprinted polymers on carbon quantum dots for fluorescent sensing of tetracycline in milk, *Talanta*, 2016, **146**, 34–40.
77. R. I. Boysen, Advances in the development of molecularly imprinted polymers for the separation and analysis of proteins with liquid chromatography, *J. Sep. Sci.*, 2019, **42**(1), 51–71.
78. J. Wackerlig and P. A. Lieberzeit, Molecularly imprinted polymer nanoparticles in chemical sensing–Synthesis, characterisation and application, *Sens. Actuators, B*, 2015, **207**, 144–157.
79. M. Marć and P. P. Wieczorek, Introduction to MIP synthesis, characteristics and analytical application, in *Comprehensive Analytical Chemistry*, Elsevier, 2019, vol. **86**, pp. 1–15.
80. A. Poma, A. P. F. Turner and S. A. Piletsky, Advances in the manufacture of MIP nanoparticles, *Trends Biotechnol.*, 2010, **28**(12), 629–637.
81. N. S. Kulkarni, Y. Guererro, N. Gupta, A. Muth and V. Gupta, Exploring potential of quantum dots as dual modality for cancer therapy and diagnosis, *J. Drug Delivery Sci. Technol.*, 2019, **49**, 352–364.
82. A. Sukhanova, S. Bozrova, E. Gerasimovich, M. Baryshnikova, Z. Sokolova, P. Samokhvalov, C. Guhrenz, N. Gaponik, A. Karaulov and I. Nabiev, Dependence of Quantum Dot Toxicity In Vitro on Their Size, Chemical Composition, and Surface Charge, *Nanomaterials*, 2022, **12**(16).
83. G. Libralato, E. Galdiero, A. Falanga, R. Carotenuto, E. de Alteriis and M. Guida, Toxicity Effects of Functionalized Quantum Dots, Gold and Polystyrene Nanoparticles on Target Aquatic Biological Models: A Review, *Molecules*, 2017, **22**(9).
84. J. Kabel, S. Sharma, A. Acharya, D. Zhang and Y. K. Yap, Molybdenum Disulfide Quantum Dots: Properties, Synthesis, and Applications, *C*, 2021, **7**(2).
85. Y. Li and S. Han, Carbon dots enhanced chemiluminescence method for the sensitive determination of iodide, *Microchem. J.*, 2020, **154**, 104638.
86. D. M. Wang, K. L. Lin and C. Z. Huang, Carbon dots-involved chemiluminescence: Recent advances and developments, *Luminescence*, 2019, **34**(1), 4–22.

87. P. Zhao, Q. Xu, J. Tao, Z. Jin, Y. Pan, C. Yu and Z. Yu, Near infrared quantum dots in biomedical applications: current status and future perspective, *WIREs Nanomedicine and Nanobiotechnology*, 2018, **10**(3), e1483.
88. T. L. Jennings, S. G. Becker-Catania, R. C. Triulzi, G. Tao, B. Scott, K. E. Sapsford, S. Spindel, E. Oh, V. Jain and J. B. Delehanty, Reactive semiconductor nanocrystals for chemoselective biolabeling and multiplexed analysis, *ACS Nano*, 2011, **5**(7), 5579–5593.
89. K. E. Sapsford, S. Spindel, T. Jennings, G. Tao, R. C. Triulzi, W. R. Algar and I. L. Medintz, Optimizing Two-Color Semiconductor Nanocrystal Immunoassays in Single Well Microtiter Plate Formats, *Sensors*, 2011, **11**(8), 7879–7891.
90. M. Zhang, B. P. Bishop, N. L. Thompson, K. Hildahl, B. Dang, O. Mironchuk, N. Chen, R. Aoki, V. C. Holmberg and E. Nance, Quantum Dot Cellular Uptake and Toxicity in the Developing Brain: Implications for Use as Imaging Probes, *Nanoscale Adv.*, 2019, **1**(9), 3424–3442.
91. L. Cao, X. Wang, M. J. Meziani, F. Lu, H. Wang, P. G. Luo, Y. Lin, B. A. Harruff, L. M. Veca and D. Murray, Carbon dots for multiphoton bioimaging, *J. Am. Chem. Soc.*, 2007, **129**(37), 11318–11319.
92. P.-C. Hsu, Z.-Y. Shih, C.-H. Lee and H.-T. Chang, Synthesis and analytical applications of photoluminescent carbon nanodots, *Green Chem.*, 2012, **14**(4), 917–920.
93. P.-C. Hsu, P.-C. Chen, C.-M. Ou, H.-Y. Chang and H.-T. Chang, Extremely high inhibition activity of photoluminescent carbon nanodots toward cancer cells, *J. Mater. Chem. B*, 2013, **1**(13), 1774–1781.
94. S. Pandey, R. Ghosh and A. Ghosh, Preparation of Hydrothermal Carbon Quantum Dots as a Contrast Amplifying Technique for the diaCEST MRI Contrast Agents, *ACS Omega*, 2022, **7**(38), 33934–33941.
95. J. Zhang, Y. Yuan, M. Gao, Z. Han, C. Chu, Y. Li, P. C. M. van Zijl, M. Ying, J. W. M. Bulte and G. Liu, Carbon Dots as a New Class of Diamagnetic Chemical Exchange Saturation Transfer (diaCEST) MRI Contrast Agents, *Angew. Chem. Int. Ed.*, 2019, **58**(29), 9871–9875.
96. S. Radhakrishnan, A. Samanta, P. M. Sudeep, K. L. Maldonado, S. A. Mani, G. Acharya, C. S. Tiwary, A. K. Singh and P. M. Ajayan, Metal-Free Dual Modal Contrast Agents Based on Fluorographene Quantum Dots, *Part. Part. Syst. Charact.*, 2017, **34**(1), 1600221.
97. J. L. Santos, J. X. Soares, S. S. M. Rodrigues and D. S. Ribeiro, Quantum Dots: Light Emitters for Diagnostics and Therapeutics, in *Nanoparticles in Life Sciences and Biomedicine*, Jenny Stanford Publishing, 2018, pp. 467–501.

98. L. Shi, V. De Paoli, N. Rosenzweig and Z. Rosenzweig, Synthesis and application of quantum dots FRET-based protease sensors, *J. Am. Chem. Soc.*, 2006, **128**(32), 10378–10379.
99. A. H. Loo, Z. Sofer, D. Bouša, P. Ulbrich, A. Bonanni and M. Pumera, Carboxylic carbon quantum dots as a fluorescent sensing platform for DNA detection, *ACS Appl. Mater. Interfaces*, 2016, **8**(3), 1951–1957.
100. A. Pramanik, S. Jones, F. Pedraza, A. Vangara, C. Sweet, M. S. Williams, V. Ruppa-Kasani, S. E. Risher, D. Sardar and P. C. Ray, Fluorescent, magnetic multifunctional carbon dots for selective separation, identification, and eradication of drug-resistant superbugs, *ACS Omega*, 2017, **2**(2), 554–562.
101. H. Xu, S. Zhou, J. Liu and Y. Wei, Nanospace-confined preparation of uniform nitrogen-doped graphene quantum dots for highly selective fluorescence dual-function determination of Fe^{3+} and ascorbic acid, *RSC Adv.*, 2018, **8**(10), 5500–5508.
102. S. Campuzano, P. Yáñez-Sedeño and J. M. Pingarrón, Carbon Dots and Graphene Quantum Dots in Electrochemical Biosensing, *Nanomaterials*, 2019, **9**, 634.
103. A. Mars, M. Hamami, L. Bechnak, D. Patra and N. Raouafi, Curcumin-graphene quantum dots for dual mode sensing platform: Electrochemical and fluorescence detection of APOe4, responsible of Alzheimer's disease, *Anal. Chim. Acta*, 2018, **1036**, 141–146.
104. R. M. Shereema, T. P. Rao, V. B. Sameer Kumar, T. V. Sruthi, R. Vishnu, G. R. D. Prabhu and S. Sharath Shankar, Individual and simultaneous electrochemical determination of metanil yellow and curcumin on carbon quantum dots based glassy carbon electrode, *Mater. Sci. Eng. C*, 2018, **93**, 21–27.
105. L. T. Tufa, S. Oh, V. T. Tran, J. Kim, K.-J. Jeong, T. J. Park, H.-J. Kim and J. Lee, Electrochemical immunosensor using nanotriplex of graphene quantum dots, Fe3O4, and Ag nanoparticles for tuberculosis Electrochim, *Electrochim. Acta*, 2018, **290**, 369–377.
106. F. Shahdost-Fard and M. Roushani, Designing an ultra-sensitive aptasensor based on an AgNPs/thiol-GQD nanocomposite for TNT detection at femtomolar levels using the electrochemical oxidation of Rutin as a redox probe, *Biosens. Bioelectron.*, 2017, **87**, 724–731.
107. H. Wang, D. Yu, J. Fang, Y. Zhou, D. Li, Z. Liu, J. Ren and X. Qu, Phenol-like group functionalized graphene quantum dots structurally mimicking natural antioxidants for highly efficient acute kidney injury treatment, *Chem. Sci.*, 2020, **11**(47), 12721–12730.
108. Y. Wang, W. Kong, L. Wang, J. Z. Zhang, Y. Li, X. Liu and Y. Li, Optimizing oxygen functional groups in graphene quantum dots for improved antioxidant mechanism, *Phys. Chem. Chem. Phys.*, 2019, **21**(3), 1336–1343.

109. S. Zhao, M. Lan, X. Zhu, H. Xue, T.-W. Ng, X. Meng, C.-S. Lee, P. Wang and W. Zhang, Green synthesis of bifunctional fluorescent carbon dots from garlic for cellular imaging and free radical scavenging, *ACS Appl. Mater. Interfaces*, 2015, **7**(31), 17054–17060.
110. T. Szabó, O. Berkesi, P. Forgó, K. Josepovits, Y. Sanakis, D. Petridis and I. Dékány, Evolution of surface functional groups in a series of progressively oxidized graphite oxides, *Chem. Mater.*, 2006, **18**(11), 2740–2749.
111. Y. Liu and J. Shi, Antioxidative nanomaterials and biomedical applications, *Nano Today*, 2019, **27**, 146–177.
112. D. Zhang, Z. Zhang, Y. Wu, K. Fu, Y. Chen, W. Li and M. Chu, Systematic evaluation of graphene quantum dot toxicity to male mouse sexual behaviors, reproductive and offspring health, *Biomaterials*, 2019, **194**, 215–232.
113. C.-L. Li, C.-M. Ou, C.-C. Huang, W.-C. Wu, Y.-P. Chen, T.-E. Lin, L.-C. Ho, C.-W. Wang, C.-C. Shih and H.-C. Zhou, Carbon dots prepared from ginger exhibiting efficient inhibition of human hepatocellular carcinoma cells, *J. Mater. Chem. B*, 2014, **2**(28), 4564–4571.

CHAPTER 3

Quantum Dots in Medical Detection/Diagnosis

NAHID REZVANI JALAL[a], FARIBA MOLLARASOULI[b], MOHAMMAD REZA JALALI SARVESTANI[a], SINA KHALILI[a], SEPIDEH ASADI[a], ZAHRA DERAKHSHAN[c,d], TAYYEBEH MADRAKIAN[a], ABBAS AFKHAMI*[a] AND MAZAHER AHMADI[a]

[a]Faculty of Chemistry, Bu-Ali Sina University, Hamedan, Iran; [b]Department of Chemistry, Yasouj University, Yasouj, Iran; [c]Research Center for Health Sciences, Department of Environmental Health, School of Health, Shiraz University of Medical Sciences, Shiraz, Iran; [d]Department of Environmental Health Engineering, School of Health, Shiraz University of Medical Sciences, Shiraz, Iran
*E-mail: abbas.afkhami@gmail.com

3.1 Introduction

One area of interest in nanotechnology, particularly in nanobiotechnology, is the study of optical and electrical phenomena related to nanometer-scale semiconductors.[1,2] Quantum dots (QDs) are semiconductor nanocrystals (NCs) (usually 2–10 nm in diameter) whose electrons and holes are quantum-confined in all three spatial dimensions.[3,4] The exciton, an electron–hole Coulomb attraction, is the most fascinating aspect of QDs. Excitons are formed when external energy (*e.g.*, light input) acts upon materials such as semiconductors or insulators.[2] The extraordinary size-dependent optical properties of semiconductor QDs, such as

Detection Science Series No. 22
Quantum Dots in Bioanalytical Chemistry and Medicine
Edited by Michael Thompson and Zahra Ramezani

Published by the Royal Society of Chemistry, www.rsc.org

absorbance and photoluminescence, were discovered in the early 1980s by the research groups of Brus, Henglein, and Grätzel.[5,6] In the early 1980s, QDs of groups II–VI, III–V, IV–VI, and IV were synthesized specifically to study size quantization effects. Numerous QDs were synthesized in this period, including CdS, CdSe, CdTe, InP, GaAs, GaP, GaN, GaInP, core–shell InP/ZnCdSe2, Ge, and Si QDs.[3] QDs have developed as a novel class of fluorescent contrast agents for biomedical applications since their initial demonstration as optical probes for cell labeling in the late 1990s.[7] To date, most QD-related studies in biology and medicine use Cd-based QDs, sometimes with additional protective shells. The main problem with the use of these QDs is their toxicity due to the presence of heavy metal ions such as cadmium ions, which has led scientists to look at heavy Cd-free QDs.[8]

QDs come in a variety of forms, including spheres, rods, and tetrapods; spherical QDs are most commonly employed in biological applications.[9] Depending on their intended applications, QDs can be synthesized as dispersions in either organic or aqueous solutions. They are employed in biological applications in solution or colloidal form. QDs must be water dispersible in order to be used in biological applications. Surface modification is required to make QDs dispersible in aqueous biological media after they are initially manufactured as colloidal dispersions in nonpolar chemical media. Ligand exchange and encapsulation of QDs in hydrophilic/amphiphilic compounds or polymers are typical methods for converting hydrophobic QDs to water-dispersible QDs.[10,11]

QDs feature a wide absorption band, a narrow emission band, size-tunable emission from visible to near-infrared (NIR), exceptional brightness, and a long fluorescent lifetime. Most importantly, they are very resistant to photobleaching, which allows them to be used for long-term *in vitro* and *in vivo* imaging. Therefore, QDs' unique optical features make them suitable as optical probes or optically trackable biomolecule carriers for *in vitro* and *in vivo* research in biological applications. QDs can be used to target specific areas *in vitro* and *in vivo* through the conjugation of relevant functional biomolecules onto their surfaces.[12–14] Functionalization and surface alteration enable their use in various pharmaceutical applications. QDs have been employed, for example, as nanoplatforms for delivering pharmacological, therapeutic, and diagnostic nanoprobes.[7] Additionally, QDs have shown significant promise as components in sensitive bioanalysis systems; their unique optical and microfluidic capabilities make them useful in point-of-care testing (POCT).[10] In this chapter, we focus our attention on QDs' medical applications. More specifically, we discuss QDs' clinical use and examine their applications in medical detection and diagnosis. We also briefly discuss the practical challenges associated with QDs' use in these contexts.

3.2 QDs: From Concept to Clinics

3.2.1 Types, Synthesis, and Characterization

QDs are semiconductor materials that have attracted much attention due to their unique properties.[15] QDs, often known as artificial atoms, were first discovered in 1981.[16] They are widely used in various fields and scientific applications, such as light-emitting diodes, lasers, solar cells, and even targeted drug delivery and biological applications.[17] There are twelve types of QDs; they are categorized based on their elemental composition.[18] Common types of QDs consist of two separate parts.[19] A core of heavy metals and a surface with a wide band containing various semiconductors covered by shells of SiO_2 or ZnS increase their quantum efficiency.[20] The main problem with the use of these QDs is their toxicity due to the presence of heavy metal ions such as cadmium ions, which has led scientists to look at heavy metal-free QDs. Alternatives to Cd-based QDs include silicon QDs, carbon QDs (CQDs) or carbon dots (C-dot), and graphene QDs (GQDs); they are very suitable for use in biological fields because they do not induce cadmium toxicity.[6,21]

QDs can be classified into three types according to their chemical composition, namely, core-type, shell–core, and alloy QDs. The properties of a QD do not depend solely on its size; a dot's composition and structure also significantly affect its properties.[22] Based on the core–shell combination and the band alignment of the valence and conduction bands of the constituent materials, QDs can be divided into different types.[23] The shell around the QDs protects them from oxidation. It also prevents the leakage of nucleus elements, such as cadmium, which makes QDs more suitable for biological applications.[24] These types of QDs are considered very promising, efficient, and safe in bioimaging applications.[25]

In general, all top-down and bottom-up techniques can be used to synthesize QDs. These techniques are performed by using physical approaches without chemical reagents, chemical methods through chemical reactions, and biological processes through the activities of living organisms.[26] QDs' sizes can vary from a few nanometers to a few micrometers depending on their production path. One of the first methods for synthesizing QDs is organic metal synthesis, which is a bottom-up method. QDs synthesized in this way have excellent optical properties.[27] This method consists of three general steps. The first step is the formation of the nucleus, followed by its growth and development. The final step is the termination step, which is performed in organic solvents at high temperatures in inert media. In this method, the emission size and the wavelength of the QDs can be adjusted by changing the reaction conditions. For example, decreasing the size of the QDs changes the color of their emissions from red to blue. In general, QDs synthesized through this method have a high quantum efficiency and a narrow emission peak. These qualities make them suitable for use in bioimaging.[27,28] QDs synthesized in this method are usually hydrophobic and require further surface modification to increase their solubility in aqueous media. Several methods are used to improve QDs' solubility, including the exchange of hydrophobic ligands with hydrophilic ones, the encapsulation of QDs with layers of amphiphilic copolymers, and the coating of QDs with silica.[28,29] Table 3.1 summarizes the methods of QD synthesis and lists the advantages and disadvantages of each.[30]

3.2.2 Properties: Optical and Electrical Features

QDs are often referred to as semiconductor NCs consisting of elements from groups I–VI, III–V, or IV. The small size of QDs leads to different optical and electronic properties compared to bulk materials.[4] QDs can be divided into core-type QDs, core–shell QDs, and alloy QDs. Core-type QDs are composed of single-component materials with uniform internal compositions, such as S, Se, Te, Zn, or Cd. These QDs have high optical fluorescence properties and are widely used in biosensors and imaging. The optical and electrochemical properties of CdX are excellent, but the cytotoxicity of these QDs has prevented their widespread use in medical and biological fields. Heavy metal-free QDs (such as Group VI QDs, carbon- or graphene-based QDs, and silicon-based QDs) are more biocompatible materials. CQDs and GQDs are widely used due to their lower toxicity, higher solubility in various solvents, good electronic properties, high specific surface area, simple functionalization, *etc.*[31] Silicon QDs are widely used in electronic and optical instruments. Their high biocompatibility has led to their use in the environment and in living organisms, as well as in biological sensors and in biological and imaging applications.[32] CQDs are a group of carbon nanomaterials that act as competitors to conventional QDs. C-dots have unique physicochemical properties such as surface deactivation and functionalization, high chemical stability, environmental compatibility, low emission toxicity, and strong luminescence.[33] The carboxyl groups on their surfaces make them soluble in aqueous and biocompatibility media.[34] A GQD is a small piece of two-dimensional lateral graphene that contains only carbon atoms.[35–37]

Core–shell QDs show better optical properties than simple QDs. For example, QDs with CdSe in their core and ZnS in their shell have a 50% higher quantum efficiency than core-type QDs. This difference is due to the growth of the shell around the core–shell QDs' cores, which increases and improves their optical properties and optical stability. These types of QDs are generally formed from IV–VI, II–VI, or III–V group elements.[38]

Alloy QDs (*e.g.*, CdSeS) have higher luminescence and unique optical properties. The structure and the internal structure are among the parameters that can be used to adjust the optical and electronic properties of this type of QD.[39,40]

In general, QDs have found wide applications in various fields due to their unique optical properties,[41] which originate from their mineral core.[42]

Table 3.1 The respective advantages and disadvantages of different methods.[30]

Technique	Methods	Advantages	Disadvantages
Top-down strategy	Oxidative cleavage	It is the most widely used method that can achieve mass production because it is simple and effective.	Some strong oxidizers, which need to be used, may cause burning or explosion, and the postprocessing method is complicated.
	Hydrothermal/Solvothermal method	It is a simple and rapid method. Additionally, the hydrothermal method is environmentally friendly.	The carbon materials need to be treated through strong oxidation before the reactions happen.
	Microwave-assisted/Ultrasonic-assisted process	It can not only shorten the reaction time but also improve the production yield.	Microwave/Ultrasonic reactor is expensive, and the volume is limited in industrial production.
	Electrochemical oxidation	The GQDs obtained by this method show high levels of stability, and the size distribution is uniform.	Pretreatment of raw materials and the purification of GQD products take a long time. Moreover, it is difficult to realize mass production because of the low product yield.
Bottom-up strategy	Controllable synthesis	The as-prepared GQDs have an accurate number of carbon atoms, uniform size, and shape.	The preparation process includes multistep complicated chemical reactions, which not only consume a long time but also are of low yield.
	Carbonization	It is an environmentally friendly and facile method.	GQDs with polydispersity are obtained because the size and structure are difficult to be controlled precisely.

These properties make them suitable candidates for imaging. Compared to common organic fluorophores, CQDs have non-blinking fluorescence and good light stability. Single-molecule tracking is enabled by single-molecule fluorescence; long-term imaging is enabled by excellent optical stability.[34] QDs with optical properties, such as broad absorption spectra, narrow emission spectra, and high quantum efficiency, are widely used as fluorescents in bio-photography. The distance between an electron and a hole in a bulk semiconductor's electron–hole pair is called the exciton Bohr radius. As QDs get smaller, this radius decreases, energy levels separate, and the distance between gaps increases. It is therefore possible to create QDs with a constant chemical composition but different fluorescence emissions by changing their size.[43,44]

Synthesis conditions significantly influence QDs' fluorescence because their emissions are highly size-dependent. Other properties of QDs include resistance to photobleaching. Extensive absorption bandwidth also enables the simultaneous excitation of multiple colors by a single light source due to the presence of numerous electron modes at higher energy levels.[45,46]

Zinc sulfide QDs can be synthesized homogeneously at a low cost. Modified sulfides are very suitable for bioimaging applications due to their adjustable emission wavelengths.[47–49]

Among QDs, C-dots have gained attention for their outstanding electronic properties. They can act as electron donors and receivers, which gives them luminescence properties.[50] Today, the use of QDs as labels is expanding due to their unique optical properties compared to conventional fluorescents. Conventional fluorescents have several limitations for imaging, including their narrow excitation spectrum, their low stability, and their need for stimulation by light at specific wavelengths. QDs have a comparatively broader absorption spectrum, which allows them to be excited by a wide range of wavelengths. Multicolor QDs can be excited by a monotone wavelength. Narrow light emission has led to the widespread use of QDs in imaging. It is noteworthy that spectrum overlap is lower in QDs. As a result, they can be used as labels in biological fields.[51–53]

3.2.3 Surface Modification/Coating and Its Effect on QDs' Features

One of the limitations of QDs for use in biological fields is their toxicity. Their toxicity is related to the release of cadmium, lead, or arsenic from their structures into biological media. Another limitation is QDs' low solubility. Modifications and appropriate surface coatings are essential to overcome both problems. These modification methods include surface ligand replacement, surface coating using encapsulated polymers, and silanization.

3.2.3.1 Ligand Exchange or Cap Exchange

In this process, hydrophobic ligands in QDs are replaced with various suitable hydrophilic ligands, which improve the QDs' water solubility. This method is commonly used for organometallic QDs. Molecules such as peptides,[54] mercaptopropionic acid,[55] and thiol polyethylene glycol[56] are used in this method. These molecules for binding to the QD and other molecules, including antibodies, proteins, and drug molecules, as well as improving solubility have two suitable functional groups.[57]

3.2.3.2 Silanization

One of the most suitable methods of improving QDs' properties is silanization, in which QDs are covered by silica shells. First, surface ligands with a thiol silane are used to improve water solubility. Next, tetraethyl orthosilicate undergoing hydrolysis is used to organize silica nanospheres and cover the QDs' surfaces. The final product is highly stable and biocompatible; however, the complexity of the preparation method and the need to dilute the product limit silanized QDs' production in large quantities.[58–60]

3.2.3.3 Polymeric Coatings

Polymer coatings are used to determine the solubility of QDs, provide functional groups, and improve the dots' performance. In this method, QDs' surface ligands are replaced with suitable polymeric ligands. Two types of amphiphilic polymers are used in this approach, namely, high molecular weight polymers and low molecular weight phospholipids. The type of polymer used depends on the final product's intended use. The use of hydrophilic block polymers to regulate size, composition, and chemical properties has also been reported. One issue with this method is that polymeric ligands increase the diameter of QDs compared to simple organic ligands.[61–63]

3.2.4 QD Features Used for Medical Detection/Diagnosis

Nanobiotechnology is one of the most promising areas in modern nanotechnology that has focused on the optical and electrical phenomena associated with nanomaterials at the nanometer scale.[64] Several nanomaterials have been introduced, such as metal–organic frameworks, natural polymers, metal nanoparticles (NPs), and QDs. Nanostructures have attracted more interest in medical and biological detection in recent years because of their fascinating electronic and optical properties, as well as their potential utility as biocompatible systems through conjugation with various biomaterials.[65,66] The use of QDs improves several immunological and molecular assays for various pathogens and biomarkers. Semiconductor nanostructures, generally referred to as QDs, exhibit unique properties (such as extinction coefficients and photostability), and they can serve as useful synthetic tools for various biotechnological applications. QDs are available in several dimensions, including QDs with zero dimensions, quantum wires with one dimension, and quantum wells with two dimensions. QDs have engendered considerable interest in the field of nanomedicine, owing to properties such as their size-dependent emission wavelength, high stability against photobleaching, narrow emission peak, high fluorescent quantum yield, size-tunable photoluminescence, high extinction coefficient, and broad excitation range. The optical properties of the QDs are the critical features that play a significant role in the classification of materials, and these properties are strongly dependent on particle size.[67] The following section presents various features of QDs that have been extensively leveraged for multiple uses in nanomedicine, with a specific focus on optical and electrical phenomena.

3.2.4.1 Optical Properties of QDs

Recently, scientists have developed QDs with excellent, unique properties for use in effective biomaterials and biodevices. QDs display features that seem to be somewhere between single atoms and bulk semiconductor materials. From a practical point of view, investigation into semiconductor QDs begins with the understanding that the electronic and optical features of semiconductor particles are effectively dependent on the control of their particle size.[68] In a QD, the absorption of a photon with energy higher than the QD's bandgap energy generates an electron–hole pair. The electron and the hole together make an exciton. The average gap distance between the hole and the electron is called the exciton Bohr radius. If the particle size of a QD becomes comparable to or smaller than the exciton Bohr radius, the hole and the electron are confined by the boundaries of the materials, leading to the so-called quantum confinement.[69] The quantum confinement effect is observed in QDs due to their small size (typically smaller than 10 nm). Band-gap engineering by using chemical modification is able to produce new QDs by emitting light in a wide range of UV to infrared spectra.[70] In general, QDs' photoluminescence occurs when excited electrons relax to their original state and re-emit electromagnetic energy with the same frequency.[71] QDs' photoluminescence emissions can be maximized by charge carriers in particle size and by their utilization as fluorescent dyes for biomolecules. Current goals in QD research include: (i) the development of hydrophilic QDs that are ideally adapted to diverse medical applications; (ii) the design of robust and flexible techniques for modifying target biomolecules *in vitro* and *in vivo*; and (iii) the reduction of observed side effects and the production of more biocompatible coatings.[72–75] Owing to their improved biocompatibility and colloidal stability, nucleic acid-capped QDs are suitable candidates for cell imaging studies.[76]

Most fluorescent materials have a number of limitations: they are pH-dependent, they show self-quenching at high concentrations, they are more prone to photobleaching, their biocompatibility is low, and they are unstable in aqueous solutions over long periods of time. QDs, in contrast to other label materials, have unique features that make them preferable for use in fluorescent bio-probes. QD materials are fabricated as fluorescent bio-probes in molecular/cellular labeling for *in vivo*/*in vitro* imaging. Biosensing applications include immunoassays, clinical laboratory diagnostics, noninvasive detection of small tumors, and single-cell surface receptor imaging.[77]

Fluorescent labeling is one of the most common approaches to the visualization of biomolecules, such as peptides, bacterial cells, and pathogens.[78] *In vivo*/*in vitro* fluorescent labeling has enabled the direct monitoring of aggregation and stability in biomolecules.[79] Fluorescent labeling costs vary depending on the technique used and the type of fluorescent dye employed (such as fluorescein isothiocyanate). Organic fluorescent dye has many limitations and challenges that restrict its usefulness in biological fields. One such challenge is that it is difficult to label independent organic fluorophores simultaneously. This challenge requires different excitation wavelengths and has typically exhibited a slight Stokes shift between emission and excitation wavelengths. Thus, simultaneous excitation lines are required for simultaneous fluorescent labeling, and extra excitation lines limit the spectra available for emission collection. On the contrary, they emit over a broad domain of visible spectra. Thus, the emissions of varied fluorescent dyes overlap together.[80] In contrast, QDs can be excited by multi- and single-photon excitation light and emit fluorescence light. It is excellently separated from its emission spectra. QDs are brighter than other fluorescent dyes and are more sensitive for *in vivo* and *in vitro* applications. QDs' fluorescence labels have longer fluorescence lifetimes; this feature can decrease the general background signal emitted from bio-samples *via* time-gated detection. The fluorescence emission wavelength of the QDs is independent of the excitation wavelength, which is their best property. The biomolecule can be labeled with QDs emitting the same emission peak using the same light source.

Fluorescence resonance energy transfer (FRET) is another sensitive detection method with medical applications.[81] FRET has many benefits compared with other detection techniques: there is no need for a light source, it is simple to perform, there is a wide linear range for quantification, and it enables detection with high sensitivity. This approach is a compelling method for monitoring minor changes in the system; it is the best candidate for use in the high performance detection of biomolecules. QDs are considered as both acceptors and donors. Some of the unique features of QDs compared to conventional FRET materials include their photostability and chemical stability, their symmetrical and sharp emission spectra, and the tunability of their emission wavelengths.

Chemiluminescence resonance energy transfer (CRET) is another detection method. The term refers to the nonradiative (dipole–dipole) transfer of energy between QDs as acceptors and suitable chemiluminescent donors. CRET occurs when luminescent materials are oxidized without an excitation source.[82,83]

Nowadays, QDs as acceptors are powerful tools used in bioluminescence resonance energy transfer (BRET).[84] Furthermore, charge transfer interactions between luminescent QDs and metal–redox complexes enable the controlled quenching of QDs' emissions in a multiplexed format.

3.2.4.2 Electrical Properties of QDs

QDs are mostly called artificial atoms because their electron–hole pairs (charge carriers) occupy discrete energy states just like the electron of a single atom. As mentioned above, if a QD absorbs a photon with energy higher than its upper band gap, it forms a hole–electron pair. The generated hole and electron recombine or their charges transfer when the charge separation occurs.[84] The electrical properties of QDs depend on several influencing factors, such as the complicated interplay between their representative electronic energy states, the availability of heteroatom dopants, the existence of electronic couplings between neighboring NPs, and the presence of surface functional groups. The edge states can be a simple way to improve the electrical properties of QDs.[85] The edge states should assist in conduction, similar to the doping behavior.

Among other kinds of QDs, graphene-based QDs (GQDs) and C-dots have gained popularity for their electrical properties in bioprocess development and cell-based assays. The GQDs significantly improve their electron mobility/transport, electro-catalytic activity, mass diffusion of the analyte, and stable chemical composition in the biosensing field.[86] These materials can accelerate the transportation of electrons between the electrode and the analyte, thereby producing a sensitive electrochemical response.[87] GQD-based biosensors have been used extensively in the detection of target biomolecules in various biomedical applications.[88]

Photoelectrochemical (PEC) biosensors also make use of QDs in bioapplications. The analytical signal in the PEC biosensor is formed by photoexcitation energy, which includes oxidant molecules (QDs as electron acceptors) and reducing molecules (QDs as electron donors). PEC devices are portable and very much easier to use. Also, their responses are calculated by using external electronic device. The use of QDs in these sensors significantly improves their photo-to-current conversion efficiency in two ways: QDs facilitate charge separation, and QDs enhance the utilization of light energy.[89–91]

3.3 Application of QDs in Medical Detection/Diagnosis

3.3.1 QDs for POCT

Point-of-care testing (POCT) refers to diagnostic experiments that are carried out at the location where a patient is receiving medical care.[10,92] Diagnostic experiments are most commonly performed in medical laboratories, and a long period of time may elapse before results are ready; in POCT, in contrast, experiments can be performed by clinical staff or by patients themselves, and the results can be made available sooner and in a more straightforward manner.[92,93] Blood pressure measurement instruments, pregnancy test kits, and glucose monitoring kits are examples of everyday forms of POCT.[94] The development of nanomaterials has enabled the development of micro-scale devices for medical purposes, such as bioimaging, biosensing, drug delivery, gene delivery, and genetic disorder detection.[10,95] The introduction of microfluidic devices has boosted the development of POCT technology thanks to numerous advantages: these devices reduce sample and reagent consumption, are rapid, and allow for high-throughput analysis. On the contrary, QDs are semiconductor crystals at the nanoscale with unparalleled electronic and optical features that find various usages in biomedical sciences and bioanalytical chemistry.[96] The main characteristics that differentiate QDs from other nanomaterials are their special photoluminescence and excellent biocompatibility (that can be achieved easily by interlacing QDs with various biomolecules).[97,98] QDs play an important role in POCT systems due to their great quantum yield and higher fluorescence properties in comparison with conventional fluorophores, such as organic dyes, which facilitate their utilization in colorimetric sensing systems that rely on the naked eye.[97,99] In 2010, Yang *et al.* reported a one-step lateral flow QD-based POCT method for syphilis screening.[100] They synthesized 3.5 nm diameter thioglycolic acid (TGA)-capped CdTe QDs that had an emission at a wavelength of 610 nm and labeled them with staphylococcal protein A (SPA).[100] Afterward, a test strip was prepared by dispensing syphilis antigens on a nitrocellulose membrane. The strip was subsequently immersed in a test tube containing the sample and the QDs and incubated for 10 min. The color variations that were detected by the naked eye were used as the analytical signal. The sensitivity and selectivity of the developed test strip were 100% and 91%, respectively.[100] In another study in 2017, Panesar *et al.* evaluated the applicability of two QDs with emission wavelengths of 655 and 705 nm as a highly sensitive and straightforward POCT device for the diagnosis of breast cancer at early stages by the detection of breast-cancer-specific miRNAs at the pico-molar level in serum samples by using a complementary sequence binding method. Many studies have been conducted on immunoassay microfluidic POCT systems for the detection of different biomarkers; however, the potential utility of QDs as markers for the recognition of biomarkers and significant biomolecules is widely ignored. This issue must be considered in future research.[101]

3.3.2 Bioimaging Applications of QDs

Bioimaging has become an important field in the recent decades due to growing worldwide concerns about cancer and malignancies,[102,103] which are relatively widespread health issues in which early diagnosis plays a crucial role in patients' chances of survival.[103,104] QDs are heterogeneous NPs with a number of features that make them good candidates for bioimaging applications: they have a wide absorption wavelength area; their emission spectra are symmetrical, finite, and size-adjustable; they are highly resistant to photobleaching; and they have higher brightness and greater sensitivity (10- to 50-fold greater intensity) in comparison with organic dyes.[2,105] In addition, heterologous QDs can have simultaneous fluorescence emissions in the same conditions, which enables scientists to monitor different biomarkers concurrently.[106] QDs' fluorescence emission is inversely proportional to their size. Indeed, QDs that are smaller in size emit light in the bluer region, while larger QDs emit light in the redder region.[106,107] In one study, Wang *et al.* used QD605 to detect carbohydrate antigen 125 (CA 125) (an applicable biomarker for the diagnosis of ovarian cancer) in different types of samples, including tumor xenografts, tissue sections, and fixed cells. They compared the performance of QD605 with that of fluorescein isothiocyanate; their findings indicated that QD605 was brighter, more specific, and more stable than fluorescein isothiocyanate.[108] In another study, Yong *et al.* synthesized $CuInS_2/ZnS$ QDs in a non-aqueous medium and conjugated them with folic acid. They evaluated these QDs' effectiveness at targeted and multiplexed optical tumor imaging in mice. Their results indicated that the luminescence signals of QDs of multiple colors could be completely discriminated from the background-related autofluorescence originating from body tissues in entire animal optical imaging.[109] Caires *et al.* synthesized a ZnS QD conjugated with carboxymethylcellulose (CMC). This conjugate exhibited fluorescence emissions that were specific to the pH of the medium. Because tumors' microenvironments are more acidic than other parts of the body, the synthesized QDs could be a promising active fluorescent nanoprobe for use in cancer cell bioimaging.[110] In another study, Mansur *et al.* synthesized CMC-cored $AgInS_2$ and $AgInS_2$-ZnS QDs with average sizes of 2.2 and 4.3 nm, respectively, in an aqueous medium using a one-pot system. They evaluated these QDs' suitability for multiplexed bioimaging of malignant glioma cells. The QDs showed red and green light emissions, respectively, and were successfully used in the bioimaging and bio-labeling of contained glioma cells using confocal microscopy.[111] Although QDs have been used extensively in *in vitro* studies, their applications in *in vivo* conditions are limited due to the exudation of heavy metal ions from the nucleus as a result of photolysis and oxidation. This is especially the case for II–IV, IV–VI, and III–V QDs; the high toxicity of cadmium, lead, and arsenic can be problematic in biological systems if these QDs' surfaces are not entirely coated with shells or ligands. An emerging trend in research in this field is the search for new alternatives with lower toxicity, such as luminescent π-conjugated polymer dots and silicon NPs with sizes smaller than 10 nm. Prior research has shown that Si QDs degrade to silicic acid, which is eliminated *via* urine. *In vitro* studies have also shown that Si QDs are 10 times safer than QDs containing Cd. However, there are two main drawbacks that limit the use of Si QDs: their low solubility in aqueous systems and their impermanent luminescence. In order to overcome these challenges, Tilley and Yamamoto recommended covering the surfaces of Si QDs with allylamine and polyacrylic acid.[112] Semiconductor-conjugated polymers can exhibit an extremely vivid and constant fluorescence under one- to two-photon excitations. This fluorescence occurs due to the delocalization of π electrons along the polymeric chain and consequently the movement of mobile charge between the π and π* orbitals. A new horizon has therefore emerged in the field of bioimaging that is based on the development of techniques for fluorescent polymer encapsulation and functionalization to obtain conjugated polymer dots. In addition to better biocompatibility, higher brightness, and greater quantum yield, conjugated polymer dots have several other important advantages over other QDs with the same size under same defined conditions. Koralli *et al.* synthesized the conjugated polymers that comprised thiophene (T) as the electron-donating unit and quinoxaline (Q) as the electron deficient segment. The TQ-based conjugated polymers are denoted as TQs. These polymers were synthesized with different molar ratios and various localizations of fluorine atoms in each repeating unit.[113] These QDs showed strong absorption in the UV-visible range and fluorescence emission in the far-red and NIR range. The highest quantum yield was achieved by QDs that were doped with 3 and 4 fluorine atoms on the repeat unit of their polymers. Overall, the synthesized conjugated polymer QDs demonstrated outstanding promise as novel fluorescent bioimaging probes.

GQDs are another more biocompatible alternative to inorganic semiconductor QD-based fluorophores.[114,115] Feng *et al.* synthesized a GQD with two-photon fluorescence features and an NIR emission at 660 nm upon exciting with 810 nm femtosecond pulses. They employed this QD as a nanoprobe for bioimaging interior ascorbic acid in living cells and for deep tissue imaging.[116] The nanoprobe demonstrated a very low detection limit (270 nM), negligible background fluorescence, excellent selectivity, and high resolution. Strong luminescence, high photostability, and efficient renal elimination are other advantages of GQDs.

Despite this focus on *in vivo* imaging, it should also be noted that many important diagnostic experiments are conducted on *in vitro* or *ex vivo* clinical patient specimens, as in the analysis of cells, tissues, and solution-based biomarkers. In these analyses, toxicity is not a concern.[102]

3.3.3 QDs in Sensors and Biosensors

QDs have been used widely in the development of effective optical and electrochemical biosensors. Their distinct photoluminescence properties and their ability to conjugate with various biomolecules make them good candidates for use in biocompatible detection systems.[117] They have therefore demonstrated great potential and have attracted intense investigation in biomedical and bioanalytical research. They have been put to good use in the development of molecular and immunological assays for a variety of biomarkers and pathogens. Furthermore, the use of QDs in conjunction with their unique optical properties and microfluidics could aid in the development of sensitive bioanalysis systems for use in POCT. The use of QD-conjugated bioelements increases detection sensitivity and reliability. An example of this is the QD-based affinity for showing increased biosensor sensitivity. Recently, QDs have received much attention for their usefulness as novel fluorescent probes for the detection of microbes and viral infections. Several QD-based optical and electrochemical biosensors are used for the detection of viral pathogens (Table 3.2 and Table 3.3, respectively). For example, Zhao *et al.* used a lead sulfide (PbS) CQD-modified electrode to develop an all-solid-state SARS-CoV-2 protein biosensor. The electronic labeling of SARS-CoV-2 proteins on an Au-working electrode was made possible by the surface ligand exchangeability of CQDs, which converts the specific binding of SARS-CoV-2 antigens and antibodies into significant current signals. The all-solid-state SARS-CoV-2 protein biosensor distinguishes between patient and normal samples with roughly 90% accuracy.[134]

Because of their accessible surfaces and their small size, QDs may be functionalized with a variety of biocompatible compounds. QD-based fluorescence resonance energy transfer (FRET) biosensing devices, for example, can be designed to interact with multiple energy transfer partners. Amphiphilic polymers containing chemically reactive groups, such as amines and carboxylic acids, are used to functionalize QDs with biomolecules. Peptides, antibodies, DNA, and siRNA are examples of biomolecules that can interact with these functional groups to create covalent connections mediated by different coupling reagents.[135] Using a FRET experiment, a fluorescent CdTe QDs-DNA nanosensor was created for the effective detection of a specific target complementary DNA or RNA from the SARS CoV-2 virus using a conventional "sandwich" configuration. The complementary DNA (target DNA) sequence is based on a large portion of the SARS CoV-2 viral genome, and the QDs-DNA nanoprobe oligonucleotides are intended to complement it. Water-soluble CdTe QDs-DNA compounds were synthesized by exchanging the TGA on the surfaces of the QDs with capturing DNA (thiolated DNA) using a ligand exchange technique. Sandwiched hybrids were then created by mixing complementary DNA (target DNA) and quencher DNA (BHQ2-labeled DNA) into the QDs-DNA conjugates. The resulting assembly placed BHQ2-labeled DNA (as the acceptor) and QDs (as the donor) into close contact, causing the donor QDs' fluorescence emission to be quenched through the FRET process. This nanosensor was used to detect SARS-CoV-2 RNA in actual samples with results equivalent to those obtained with a reverse transcription-polymer-

Table 3.2 Studies on QDs for optical detection of viruses.

Carbon-based fluorescent material	Precursors	Synthesis method	Quantum yield	Target virus	Target analyte	Recognition matrix	Linear range	LOD	Application	Ref.
C-dots	Citric acid and ethylenediamine	Hydrothermal method	95.7%	HIV	p24 antigen	Biotinylated anti-p24 antibody	30–1000 pg mL^{-1}	30 pg mL^{-1}	Human plasma samples	118
Carbon NPs oxide	Candle soot	Oxidation	—	HIV	HIV-DNA sequences	ssDNA-AgNCs	1–50 nM	0.40 nM	PBS solution pH 7.5	119
C-dots	Citric acid and thiourea	Solvothermal method	43.02%	H7N9	Hemagglutinin7 (H7) gene and neuraminidase 9 (N9)	DNA	10×10^{-15} – 10×10^{-12} M	4.6×10^{-15} M 3.4×10^{-15} M	Diluted serum samples	120
C-dots	*o*-Phenylenediamine	Hydrothermal method	—	HTLV-I	DNA	CDs-DNA probe	10–320 nM	< 10 nM	—	121
Graphitic carbon nitride QDs	Citric acid and melamine		44%	H1N1 H3N2	H1N1 H3N2	Specific antibody against influenza A virus	0–10^5 pg mL^{-1} for H1N1 45–25,000 PFU mL^{-1} for H3N2	0.25 and 0.9 pg mL^{-1} for H1N1	Deionized water and human serum	122

(*continued*)

Table 3.2 (*continued*)

Carbon-based fluorescent material	Precursors	Synthesis method	Quantum yield	Target virus	Target analyte	Recognition matrix	Linear range	LOD	Application	Ref.
								45 PFU mL^{-1} for H3N2		
Sulfur-doped graphitic carbon nitride QDs	Citric acid and 4-amino-3-hydrazino-5-mercapto-1,2,4-triazole	Solvothermal method	~33%	H1N1 H3N2	H1N1 H3N2	Specific antihuman influenza A monoclonal antibody conjugated S-gCNQDs	10 fg mL^{-1} –1.0 ng mL^{-1} for H1N1 10^2–5×10^4 PFU mL^{-1} for H3N2	5.5 fg mL^{-1} for H1N1 100 PFU mL^{-1} for H3N2	Clinical samples and human serum	123
C-dots	Citric acid and *N*-β-(aminoethyl)-γ-aminopropyltrimethoxysilane	Solvothermal method	—	Zika virus	Zika NS1 protein	ZK01 antibody-conjugated fluorescent CD-based silica (FCS) spheres	0 pg mL^{-1} – 1 μg mL^{-1}	10 pg mL^{-1}	Clinical samples and human serum	124

(*continued*)

Table 3.2 (*continued*)

Carbon-based fluorescent material	Precursors	Synthesis method	Quantum yield	Target virus	Target analyte	Recognition matrix	Linear range	LOD	Application	Ref.
Boron and nitrogen co-doped single-layered GQDs (BN-SGQDs)	*p*-Phenylenediamine and 3-formylphenylboronic acid	Solvothermal method	36.5%	HIV	HIV-DNA	BN-SGQDs-DNA probe	0–20 nM	0.5 nM	Tris-HCl buffer (pH: 7.4) and living Hela cells	125
C-dots	Citric acid and *N*-β-(aminoethyl)-γ-aminopropyltrimethoxysilane	Solvothermal method	56.3%	SFTSV	SFTSV nucleoprotein	C-dots/SiO_2 nanosphere–antibody conjugates	10 pg mL^{-1}–1 μg mL^{-1}	10 pg mL^{-1}	SFTSV standard solution and human serum	126

DNA: deoxyribonucleic acid; HIV: human immunodeficiency virus; H1N1: Influenza virus A/New Caledonia (20/99/IVR/116); H3N2: Clinically isolated influenza virus A/Yokohama (110/2009); H7N9: (hemagglutinin7 gene and neuraminidase 9 gene) Avian influenza A virus; HTLV-I: human T-lymphotropic virus type 1; SFTSV: thrombocytopenia syndrome virus.

Table 3.3 Studies on QDs for electrochemical detection of viruses.

Carbon-based fluorescent material	Precursors	Synthesis method	Quantum yield	Target virus	Target analyte	Recognition matrix	Linear range	LOD	Application	Ref.
GQDs	Graphene oxide	Microwave irradiation	—	ALVs-J	ALVs-J	ALVs-J antibodies	$10^{2.08}$–$10^{4.50}$ $TCID_{50}$ mL^{-1}	115 TCID$_{50}$ mL^{-1}	Electrolyte solution	127
Thiol graphene QDs (GQD-SH)	MWCNTs	Oxidative cutting	—	HCV	HCV core antigen	Anti-HCV core antibody	0.05 pg mL^{-1}–60 ng mL^{-1}	3 fg mL^{-1}	PBS solution pH 7.0 and	128
Nitrogen-doped graphene QDs	Citric acid and dicyandiamide	Hydrothermal method	—	HBV	HBV surface antigen	Anti-HBV surface antibody	10 fg mL^{-1}–50 ng mL^{-1}	3.3 fg mL^{-1}	Human serum	129
GQDs	Citric acid	Pyrolysis	—	HCV	HCV core antigen	Core-specific aptamer	Two linear concentration ranges: 10–70 pg mL^{-1} 70–400 pg mL^{-1}	3.3 pg mL^{-1}	Human serum	130
GQDs	Citric acid	Pyrolysis	—	HBV	HBV-DNA	DNA probe	10–500 nM	1 nM	Electrolyte solution	131

(*continued*)

Table 3.3 (*continued*)

Carbon-based fluorescent material	Precursors	Synthesis method	Quantum yield	Target virus	Target analyte	Recognition matrix	Linear range	LOD	Application	Ref.
GQDs	*L*(+)-ascorbic acid and graphene nanopowder	Hydrothermal method	11.26%	FAdVs	FAdV-9	FAdVs-specific antibodies	10–10000 PFU mL^{-1}	8.75 PFU mL^{-1}	Chicken blood	132
Nitrogen- and sulfur-co-doped GQDs	Thiourea to citric acid	Hydrothermal method	—	HEV	HEV-like particles and HEV	Specific anti-HEV antibody-conjugated	10^2–10^7 RNA copies L^{-1}	96.7 RNA copies L^{-1}	Fecal specimen samples collected from G7 HEV-infected monkey	133

ALVs-J: avian leukosis virus subgroup J (ALVs-J); HCV: hepatitis C virus core antigen; MWCNTs: multiwalled carbon nanotubes; HBV: hepatitis B; DNA: deoxyribonucleic acid; FAdVs: fowl adenoviruses; FAdV-9: Fowl adenovirus-9 virus; PFU: Plaque forming unit; HEV: Hepatitis E virus.

ase chain reaction; good specificity and sensitivity (LOD was around 2.52×10^{-9} mol L^{-1}) were reported.[136]

3.3.4 QDs in Theranostics

QDs are fluorescent NCs that serve as a multifunctional imaging platform after the regulated administration of medicinal medicines, proteins, peptides, oligonucleotides, and genes. These fluorescent probes combine imaging and carrier functions; they have been shown to be useful tools for molecular diagnostics and theranostics, with a wide range of therapeutic applications. QDs' versatile surface chemistry, unique optical features, high sensitivity, and multiplexing capabilities make them an extremely promising tool for customized treatment. These nanomaterials' biocompatibility and their interactions with target tissues can be modulated by variations in size and surface chemistry. QDs may play an important part in theranostic research in the near future as the desire for enhanced disease monitoring and detection and improved chemotherapies drives research into therapeutically translatable novel nanomaterials. Among QDs, CQDs (or C-dots) are a promising candidate for use in cancer theranostics because of their low toxicity, simplicity of modification, water solubility, and excellent biocompatibility, which enables drug carrier approaches. To enhance chemodynamic therapy (CDT), Li *et al.* used an "off–on" nanomedical detection technique for fluorescence imaging-guided depletion of glutathione (GSH). Their nanomedical technique relied on tumor microenvironment (TME) stimuli-responsive theranostics. Fe (III) quenched the fluorescence of QDs *via* redox reactions, enabling the fluorescence imaging of GSH depletion.

Furthermore, the reduced Fe (II) displayed a significant Fenton catalytic capacity to create more reactive oxygen species (ROS), indicating that CDT cytotoxicity in cancer cells might be a feasible alternative. CQDs are easy to make, and they also enable CDT reagents to perform fluorescence imaging.[137]

GQDs have the potential to improve cell internalization and cargo monitoring. Structural variations between the surfaces of malignant and normal cells enable researchers to design and engineer cancer therapies with precision. Integrin, one of the cell surface receptors that are overexpressed in cancer cells, might be a good choice for cancer cell targeting. In one study, innovative nano-carriers based on developed MiRGD peptides and GQDs were employed to carry doxorubicin (Dox) and curcumin (Cur) as hydrophilic and hydrophobic drug models, respectively. Dox and Cur could be efficiently loaded into a proposed prospective drug delivery system based on the MiRGD peptide and GQDs. Drug packaging, cell internalization, endosomal escape, and targeted administration of therapeutic pharmaceuticals to αv-integrin-overexpressed cells are all possible with multifunctional MiRGD peptides (see Figure 3.1). Overall, the cellular

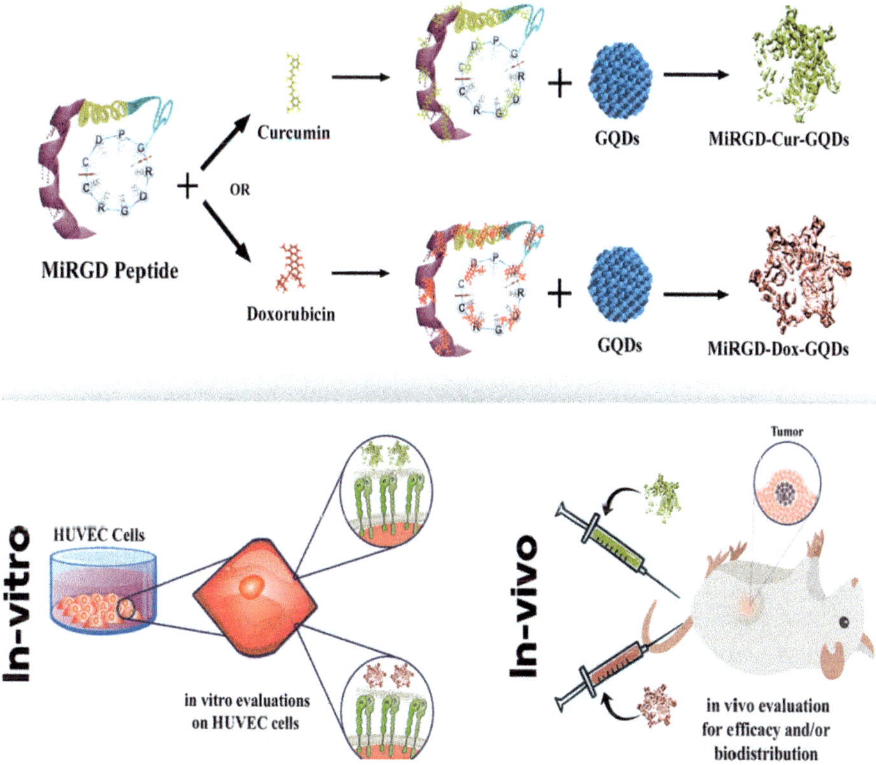

Figure 3.1 Scheme of assembling Dox/Cur-MiRGD-GQDs peptideticles through noncovalent interactions, transfection into αv-integrin-positive HUVEC cells, and intravenous injection into 4 T1-induced breast cancer BALB/c mice. Reproduced from ref. 138 with permission from Elsevier, Copyright 2022.

uptake and fluorimetric assays were investigated in HUVEC and HFF cells as models of αv-integrin-overexpressed cancer and normal cells, respectively. Here, the theranostic technique was used to study targeted cancer therapy. The theranostic is the new therapeutic method, which provides a diagnosis, treatment, and continuous follow-up of a disorder. The results of an *in vivo* study on breast cancer-bearing BALB/c mice demonstrated that the prepared nano-composites are suitable multifunctional theranostic peptideticles for targeted drug delivery and tracking.[138]

QDs have received special attention in the field of cancer imaging. Their distinct size and fluorescence properties make them promising candidates for biomedical applications, such as *in situ* multiplex imaging. To discover the special features of QDs, *in situ* QD-based multiplex imaging systems for quantitative analysis of tumor tissue specimens have been developed. The importance of early diagnosis and marker identification in targeted therapy cannot be overstated.[139] Noninvasive imaging approaches for cancer diagnosis have sparked a lot of attention in the scientific community. Optical coherence tomography (OCT) is a noninvasive, high-speed, high-resolution three-dimensional imaging technology based on the interference of low-coherent light. This technology allows the microstructures of biological tissues to be examined with micron-level resolution at the cellular and sub-cellular levels. Photothermal OCT (PT-OCT) is a newer type of OCT that uses temperature changes to cause changes in optical phase transitions in tissues under additional laser irradiation. However, several biosafety problems exist with existing PT-OCT contrast agents, such as gold-based contrast agents. As a result, innovative PT-OCT probes that are both biocompatible and biodegradable are in high demand. Chen *et al.* developed a nanoagent based on black phosphorus QDs (BPQDs) that may be used as a new PT-OCT probe and drug carrier for cancer theranostics. In brief, BPQDs are combined with cysteine-based poly(disulfide amide) (Cys-PDSA) to produce a stable and biodegradable nanoagent. BPQD/Cys-PDSA NPs can be used as a new contrast agent for high-resolution PT-OCT bioimaging due to their superior photothermal properties. *In vitro* and *in vivo*, the Cys-PDSA may rapidly respond to glutathione and release BPQDs and medicines (see Figure 3.2). Furthermore, the generated NPs have a high NIR photothermal transduction efficiency and drug transport capacity. These qualities make them a promising therapeutic platform, because they enable the use of lower doses of chemotherapy medication and may therefore reduce side effects. Both the BPQD and the polymer are biodegradable, indicating that this platform is a rare biodegradable PT-OCT agent. This biodegradable and biocompatible black phosphorus-based nanoagent may therefore be used to diagnose and treat cancer.[140]

In another study, Kumawat *et al.* used electrostatic layer-by-layer assembly with a polyethyleneimine (PEI) bridge to produce a stable complex of graphene oxide (GO) and GQDs (GO-PEI-GQDs). When the GO-PEI-GQDs were exposed to an 808 nm laser (0.5 W cm^2) for 5 min at a concentration of up to 50 g mL^{-1}, they showed an excellent photothermal response (44–49 °C). GO-PEI-GQDs have synergistic properties, such as improved photothermal and cytotoxic activities on cancer cells and stable fluorescence imaging. Composite materials composed of GO and GQDs with various properties can improve the specific therapeutic systems in theranostics.[141]

3.4 Practical Challenges in the Clinical Applications of QDs

New advancements in nanomaterial synthesis and nanotechnology research have had a dramatic impact on science, engineering, and healthcare. QDs are one of the most common types of NPs used in a variety

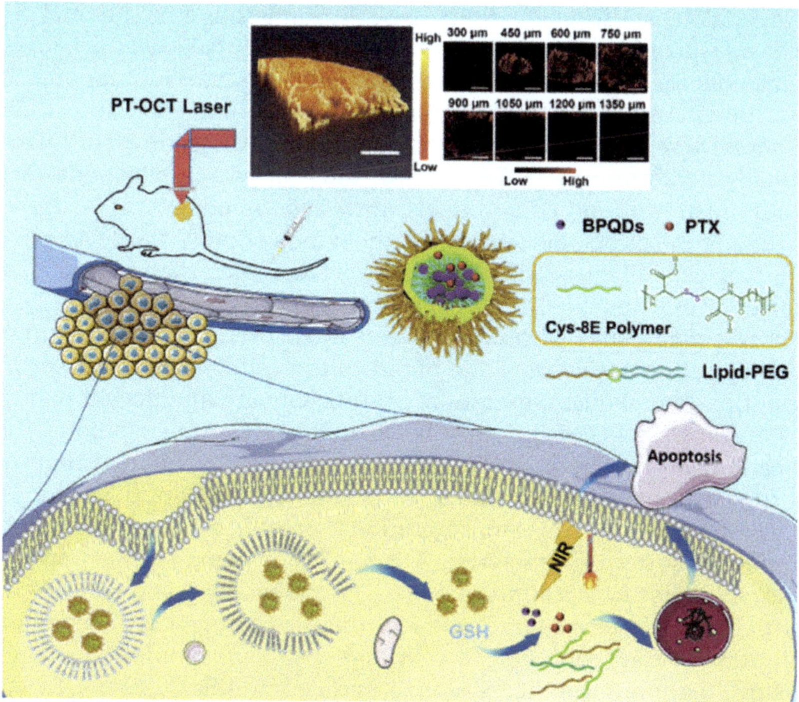

Figure 3.2 Schematic illustration of the black phosphorus-based theranostic platform and its therapeutic application. Reproduced from ref. 140 with permission from Elsevier, Copyright 2021.

of medicinal applications.[42] The unique optical features of QDs, such as broad-range excitation, size-tunable narrow emission spectra, and great photostability, enable a wide range of medical uses.[142] However, before QDs can be used in clinical settings, their cytotoxicity must be investigated and minimized. QDs are made up of potentially hazardous metal atoms and can cause unanticipated cytotoxicity due to colloidal effects and photon-induced free radical production. QDs have been proposed as a possible optical probe to replace organic fluorophores in biological applications because of their beneficial features. However, because of QDs' limitations, many researchers have concluded that these materials should be used to complement existing organic fluorophores rather than replace them in many applications. There are several main challenges associated with using QDs: (i) there is a lack of commercially available QDs in different colors; (ii) variation in the surface chemistry of QDs hinders their surfaces from being effectively coated with bio-recognition molecules; (iii) there is currently a lack of fundamental research on the optical characteristics of QDs in biological environments; (iv) nonspecific binding may lower QDs' detection sensitivity; and (v) there is a limitation of instrumentation for

QDs detection. Additionally, there are still numerous fundamental limits to the widespread usage of QDs as bioimaging agents.[143] QDs act similarly to nano-colloids when compared to traditional organic fluorophores. This makes long-term use of QDs in biological environments more difficult and poses new safety issues. The physicochemical features of QDs will eventually determine their fate *in vivo*. Most *in vivo* applications involve the administration of QDs directly into the bloodstream by systemic intravenous injection. Because QDs are nano-colloids, their behavior in this biological milieu is substantially more complex than that of typical chemical dyes. When QDs are administered to the circulatory system, they face several lines of resistance and must overcome several substantial biological obstacles from the organ to the cellular level.[18] Absorption, distribution, metabolism, excretion, and toxicity are all affected by a wide range of features of QDs and must be thoroughly researched before clinical application. Moreover, it is challenging to limit the bio-distribution of QDs to target organs and reduce long-term accumulation in the body. Also, QDs produced using more traditional organic techniques often have hydrophobic surface characteristics, resulting in poor aqueous solubility and durability for *in vivo* applications. QDs' composition and features, such as particle size, surface charge, and hydrophilic coating, can be engineered to improve contact with target cells and facilitate elimination through the body's natural clearance systems. Surface modification of QDs using biological materials, such as aptamers, peptides, antibodies, and antigens that target specific receptors, can enhance their water solubility. Another important concern in biomedical applications is QDs' long-term toxicity.[144] There are four primary factors behind QDs' toxicity: (i) heavy metals are frequently used in the particles' cores; (ii) their surface chemistries and coatings may release hazardous components, such as Cd, Se, and Hg; (iii) free radicals or ROS may be produced by stimulation; and (iv) nano-colloids may interact with tissues in biological environments due to their small particle size, positive surface charge, and chemical composition.[18] Another important application of QDs is targeted tumor imaging because the enhanced permeability and retention (EPR) characteristic of QDs increases tumor accumulation through leaky vasculature near tumor locations.[145,146] If QDs are to be used for tumor imaging, diagnostics, or therapy, there are additional hurdles and biological constraints to overcome. A complete understanding and characterization of the tumor microenvironment is required to comprehend particle penetration and accumulation in the tissue.[147]

3.5 Conclusions and Future Perspectives

This chapter has provided an overview of QDs' characteristics and applications, with a focus on their use in medical detection and diagnosis. A brief summary of types of QDs, the general method of their preparation, their functionalization, and their applications has been provided.

QDs are semiconductor NCs whose electrons and holes are quantum-confined in all three spatial dimensions. The electron–hole pair, known as an exciton, is the most fascinating phenomenon related to QDs; it is formed through the input of external energy (*e.g.*, light) in materials such as semiconductors and insulators. QDs can be classified according to chemical composition into one of three types: core-type, shell-core, or alloy. The properties of a QD depend not only on its size but also on its composition and structure. Generally speaking, however, QDs share several characteristics that make them excellent candidates for use in bioimaging: they have a wide absorption wavelength area; their emission spectrum is symmetrical, finite, and size-adjustable; they are highly resistant to photobleaching; and they are brighter and more highly sensitive compared to organic dyes. In addition, heterologous QDs can have simultaneous fluorescence emissions in the same conditions. Hence, the monitoring of different biomarkers concurrently is able by using heterologous QDs.

QDs' unique optical features make them suitable as optical probes or optically trackable biomolecule carriers for *in vitro* and *in vivo* biological research. They can be used to target specific areas *in vitro* and *in vivo* through the conjugation of relevant functional biomolecules onto their surfaces. Various functionalizations and surface changes enable QDs' use in specific pharmaceutical applications; they have been employed as nanoplatforms for delivering pharmacological, therapeutic, and diagnostic nanoprobes. Additionally, QDs' unique optical and microfluidic capabilities make them well suited for use in POCT systems.

Depending on their intended application, QDs can be synthesized as dispersions in either organic or aqueous solutions. QDs are employed in biological applications in solution or colloidal form. QDs must be water soluble to be used in biological applications; QDs that are manufactured as colloidal dispersions in nonpolar chemical media must undergo surface modification in order to be soluble in aqueous biological media. Ligand exchange and encapsulation of QDs in hydrophilic/amphiphilic compounds or polymers are typical methods for converting hydrophobic QDs to water-dispersible QDs.

To date, Cd-based QDs have been used in the majority of QD-related studies in biology and medicine, sometimes with additional protective shells. The main problem with the use of metal-based QDs is their toxicity due to the presence of heavy metal ions such as cadmium. Scientists have therefore turned to heavy Cd-free QDs (*e.g.*, silicon QDs, C-dots, and GQDs) as a safer alternative for use in biological fields.

There are several key issues in QD research on which future efforts should focus. First, various techniques to reduce QDs' toxicity are available, but further research is required to produce QDs with lower toxicity and enhance pharmacological responses. Second, the creation of hydrophilic

QDs is a recent development; more research is needed to produce a variety of water-soluble QDs for biological applications and cell labeling. Finally, little research has been conducted thus far on QDs' bio-distribution, metabolism, and excretion in biological systems; more research on their intracellular delivery is needed.

References

1. K. Morigakiand C. Ogihara, *Springer Handbook of Electronic and Photonic Materials*, Springer, 2017.
2. T. R. Pisanic, Y. Zhang and T. H. Wang, *Analyst*, 2014, **139**, 2968–2981.
3. A. J. Nozik, M. C. Beard, J. M. Luther, M. Law, R. J. Ellingson and J. C. Johnson, *Chem. Rev.*, 2010, **110**(11), 6873–6890.
4. A. M. Iga, J. H. P. Robertson, M. C. Winslet and A. M. Seifalian, *J. Biomed. Biotechnol.*, 2007, **2007**, 76087.
5. R. M. Penner, *Acc. Chem. Res.*, 2000, **33**(2), 78–86.
6. G. Xu, S. Zeng, B. Zhang, M. T. Swihart, K.-T. Yong and P. N. Prasad, *Chem. Rev.*, 2016, **116**(19), 12234–12327.
7. Y. Wang, R. Hu, G. Lin, I. Roy and K.-T. Yong, *ACS Appl. Mater. Interfaces*, 2013, **5**(8), 2786–2799.
8. A. Bruneau, M. Fortier, F. Gagne, C. Gagnon, P. Turcotte, A. Tayabali, T. A. Davis, M. Auffret and M. Fournier, *Environ. Toxicol.*, 2015, **30**(1), 9–25.
9. R. E. Bailey, A. M. Smith and S. Nie, *Phys. E*, 2004, **25**, 1–12.
10. S. Singh, A. Dhawan, S. Karhana, M. Bhat and A. K. Dinda, *Micromachines*, 2020, **11**, 1058.
11. B. Zhang, R. Hu, Y. Wang, C. Yang, X. Liu and K.-T. Yong, *RSC Adv.*, 2014, **4**, 13805–13816.
12. I. L. Medintz, *Nat. Mater.*, 2005, **4**, 435.
13. E. Cho, H. Jang, J. Lee and E. Jang, *Nanotechnology*, 2013, **24**, 215201.
14. M. Koneswaran and R. Narayanaswamy, *Sens. Actuators B Chem.*, 2015, **210**, 811–816.
15. D. Bera, L. Qian, T.-K. Tseng and P. H. Holloway, *Materials*, 2010, **3**, 2260–2345.
16. D. Gammon, *Nature*, 2000, **405**, 899–900.
17. S. Kargozar, S. J. Hoseini, P. B. Milan, S. Hooshmand, H. W. Kim and M. Mozafari, *Biotechnol. J.*, 2020, **15**, 2000117.
18. A. M. Wagner, J. M. Knipe, G. Orive and N. A. Peppas, *Acta Biomater.*, 2019, **94**, 44–63.

19. B. O. Dabbousi, J. Rodriguez-Viejo, F. V. Mikulec, J. R. Heine, H. Mattoussi, R. Ober, K. F. Jensen and M. G. Bawendi, *J. Phys. Chem. B*, 1997, **101**, 9463–9475.
20. E. S. Speranskaya, N. V. Beloglazova, P. Lenain, S. De Saeger, Z. Wang, S. Zhang, Z. Hens, D. Knopp, R. Niessner and D. V. Potapkin, *Biosens. Bioelectron.*, 2014, **53**, 225–231.
21. T. Liu, R. Xing, Y.-F. Zhou, J. Zhang, Y.-Y. Su, K.-Q. Zhang, Y. He, Y.-H. Sima and S.-Q. Xu, *Biomaterials*, 2014, **35**, 2942–2951.
22. D. Wang, A. L. Rogach and F. Caruso, *Nano Lett.*, 2002, **2**, 857–861.
23. A. R. AbouElhamd, K. A. Al-Sallal and A. Hassan, *Energies*, 2019, **12**, 1058.
24. M. Safari, S. Najafi, E. Arkan, S. Amani and M. Shahlaei, *Microchem. J.*, 2019, **146**, 293–299.
25. E. Soheyli, B. Ghaemi, R. Sahraei, Z. Sabzevari, S. Kharrazi and A. Amani, *Mater. Sci. Eng. C*, 2020, **111**, 110807.
26. S. B. Brichkin and V. F. Razumov, *Russ. Chem. Rev.*, 2016, **85**, 1297.
27. M. Mozafari, F. Moztarzadeh, A. M. Seifalian and L. Tayebi, *J. Lumin.*, 2013, **133**, 188–193.
28. C. M. Tyrakowski and P. T. Snee, *Phys. Chem. Chem. Phys.*, 2014, **16**, 837–855.
29. D. Wang, T. Xie and Y. Li, *Nano Res.*, 2009, **2**, 30–46.
30. W. Chen, G. Lv, W. Hu, D. Li, S. Chen and Z. Dai, *Nanotechnol. Rev.*, 2018, **7**, 157–185.
31. X. Xu, R. Ray, Y. Gu, H. J. Ploehn, L. Gearheart, K. Raker and W. A. Scrivens, *J. Am. Chem. Soc.*, 2004, **126**, 12736–12737.
32. M. A. Farzin and H. Abdoos, *Talanta*, 2021, **224**, 121828.
33. S. Y. Lim, W. Shen and Z. Gao, *Chem. Soc. Rev.*, 2015, **44**, 362–381.
34. P. Namdari, B. Negahdari and A. Eatemadi, *Biomed. Pharmacother.*, 2017, **87**, 209–222.
35. S. Bak, D. Kim and H. Lee, *Curr. Appl. Phys.*, 2016, **16**, 1192–1201.
36. X. Li, M. Rui, J. Song, Z. Shen and H. Zeng, *Adv. Funct. Mater.*, 2015, **25**, 4929–4947.
37. P. Zheng and N. Wu, *Chem. Asian J.*, 2017, **12**, 2343–2353.
38. D. Vasudevan, R. R. Gaddam, A. Trinchi and I. Cole, *J. Alloys Compd.*, 2015, **636**, 395–404.
39. R. E. Bailey and S. Nie, *J. Am. Chem. Soc.*, 2003, **125**, 7100–7106.
40. J. He, L. Sun, S. Chen, Y. Chen, P. Yang and J. Chu, *J. Alloys Compd.*, 2012, **511**, 129–132.
41. R. Hardman, *Environ. Health Perspect.*, 2006, **114**, 165–172.
42. V. G. Reshma and P. V. Mohanan, *J. Lumin.*, 2019, **205**, 287–298.
43. J. Li and J.-J. Zhu, *Analyst*, 2013, **138**, 2506–2515.

44. P. Zrazhevskiy, M. Sena and X. Gao, *Chem. Soc. Rev.*, 2010, **39**, 4326–4354.
45. N. Gaponik and A. L. Rogach, *Semiconductor Nanocrystal Quantum Dots*, Springer, 2008, pp. 73–99.
46. C. Frigerio, D. S. M. Ribeiro, S. S. M. Rodrigues, V. L. R. G. Abreu, J. A. C. Barbosa, J. A. V. Prior, K. L. Marques and J. L. M. Santos, *Anal. Chim. Acta*, 2012, **735**, 9–22.
47. H. R. Rajabi, F. Shahrezaei and M. Farsi, *J. Mater. Sci.: Mater. Electron.*, 2016, **27**, 9297–9305.
48. A. A. Mansur, H. S. Mansur, F. P. Ramanery, L. C. Oliveira and P. P. Souza, *Appl. Catal. B*, 2014, **158**, 269–279.
49. K. Ozga, J. Michel, B. D. Nechyporuk, J. Ebothé, I. V. Kityk, A. A. Albassam, A. M. El-Naggar and A. O. Fedorchuk, *Phys. E*, 2016, **81**, 281–289.
50. M. Mozafari and F. Moztarzadeh, *Micro Nano Lett.*, 2011, **6**, 161–164.
51. P. Wu, T. Zhao, S. Wang and X. Hou, *Nanoscale*, 2014, **6**, 43–64.
52. J. A. Thomas, *Chem. Soc. Rev.*, 2015, **44**, 4494–4500.
53. E. Petryayeva, W. R. Algar and I. L. Medintz, *Appl. Spectrosc.*, 2013, **67**, 215–252.
54. N. Xia, B. Zhou, N. Huang, M. Jiang, J. Zhang and L. Liu, *Biosens. Bioelectron.*, 2016, **85**, 625–632.
55. M. Roushani, M. Shamsipur and H. R. Rajabi, *J. Electroanal. Chem.*, 2014, **712**, 19–24.
56. W. N. Wenger, F. S. Bates and E. S. Aydil, *Langmuir*, 2017, **33**(33), 8239–8245.
57. A. S. Karakoti, R. Shukla, R. Shanker and S. Singh, *Adv. Colloid Interface Sci.*, 2015, **215**, 28–45.
58. W. Peng, S. M. Rupich, N. Shafiq, Y. N. Gartstein, A. V. Malko and Y. J. Chabal, *Chem. Rev.*, 2015, **115**, 12764–12796.
59. X. Cheng, S. B. Lowe, P. J. Reece and J. J. Gooding, *Chem. Soc. Rev.*, 2014, **43**, 2680–2700.
60. T. K. Purkait, M. Iqbal, M. A. Islam, M. H. Mobarok, C. M. Gonzalez, L. Hadidi and J. G. C. Veinot, *J. Am. Chem. Soc.*, 2016, **138**, 7114–7120.
61. M. Massey, M. Wu, E. M. Conroy and W. R. Algar, *Curr. Opin. Biotechnol.*, 2015, **34**, 30–40.
62. T. L. Rocha, N. C. Mestre, S. M. T. Sabóia-Morais and M. J. Bebianno, *Environ. Int.*, 2017, **98**, 1–17.
63. J. Park, J. Lee, J. Kwag, Y. Baek, B. Kim, C. J. Yoon, S. Bok, S.-H. Cho, K. H. Kim and G.-O. Ahn, *ACS Nano*, 2015, **9**, 6511–6521.
64. K. K. Jain, *Clin. Chem.*, 2007, **53**, 2002–2009.
65. A. SalmanOgli, *Cancer Nanotechnol.*, 2011, **2**, 1–19.

66. Y. Zhang, N. Kaji, M. Tokeshi and Y. Baba, *Expert Rev. Proteomics*, 2007, **4**, 565–572.
67. C. Knoblauch, M. Griep and C. Friedrich, *Sensors*, 2014, **14**, 19731–19766.
68. J. B. Hoffman, H. Choi and P. V. Kamat, *J. Phys. Chem. C*, 2014, **118**, 18453–18461.
69. B. Zorman, M. V. Ramakrishna and R. A. Friesner, *J. Phys. Chem.*, 1995, **99**, 7649–7653.
70. B. Rajbanshi, S. Sarkar and P. Sarkar, *J. Mater. Chem. C*, 2014, **2**, 8967–8975.
71. F. Findeis, M. Baier, E. Beham, A. Zrenner and G. Abstreiter, *Appl. Phys. Lett.*, 2001, **78**, 2958–2960.
72. Y. Guo, D. Shi, J. Lian, Z. Dong, W. Wang, H. Cho, L. Wang, G. Liu and R. C. Ewing, *Nanotechnology*, 2008, **19**, 175102.
73. P. Maruvada, W. Wang, P. D. Wagner and S. Srivastava, *Biotechniques*, 2005, **38**, S9–S15.
74. G. Lin, F. Yin and K.-T. Yong, *Journal.*, 2014, **9**, 991–994.
75. S. Jha, P. Mathur, S. Ramteke and N. K. Jain, *Artif. Cells Nanomed., Biotechnol.*, 2018, **46**, 57–65.
76. J. Farlow, D. Seo, K. E. Broaders, M. J. Taylor, Z. J. Gartner and Y.-W. Jun, *Nat. Methods*, 2013, **10**, 1203–1205.
77. F. Pinaud, X. Michalet, L. A. Bentolila, J. M. Tsay, S. Doose, J. J. Li, G. Iyer and S. Weiss, *Biomaterials*, 2006, **27**, 1679–1687.
78. L.-J. Zhang, L. Xia, H.-Y. Xie, Z.-L. Zhang and D.-W. Pang, *Anal. Chem.*, 2018, **91**, 532–547.
79. Y. Xing and J. Rao, *Cancer Biomark.*, 2008, **4**, 307–319.
80. S. Jung and X. Chen, *Adv. Healthcare Mater.*, 2018, **7**, 1800252.
81. S.-Y. Liu, A.-M. Yan, W. Y.-Z. Guo, Y.-Y. Fang, Q.-J. Dong, R.-R. Li, S.-N. Ni, Y. Sun, W.-C. Yang and G.-F. Yang, *ACS Nano*, 2020, **14**(4), 4244–4254.
82. Z.-M. Zhou, Y. Yu and Y.-D. Zhao, *Analyst*, 2012, **137**, 4262–4266.
83. X. Huang, L. Li, H. Qian, C. Dong and J. Ren, *Angew. Chem., Int. Ed.*, 2006, **45**, 5140–5143.
84. J. Song and J. Zhang, *Sci. Rep.*, 2019, **9**, 1–7.
85. R. Dutta, S. Kakkar, P. Mondal, N. Chauhan and J. K. Basu, *J. Phys. Chem. C*, 2021, **125**, 8314–8322.
86. J. Zhao, G. Chen, L. Zhu and G. Li, *Electrochem. Commun.*, 2011, **13**, 31–33.
87. F. Faridbod and A. L. Sanati, *Curr. Anal. Chem.*, 2019, **15**, 103–123.
88. S. Campuzano, P. Yáñez-Sedeño and J. M. Pingarrón, *Nanomaterials*, 2019, **9**, 634.

89. B. Ertek, C. Akgül and Y. Dilgin, *RSC Adv.*, 2016, **6**, 20058–20066.
90. F. Divsar, *Anal. Bioanal. Chem.*, 2019, **411**, 6867–6875.
91. W. Tu, Z. Wang and Z. Dai, *TrAC, Trends Anal. Chem.*, 2018, **105**, 470–483.
92. X. Wang, F. Li and Y. Guo, *Front. Chem.*, 2020, **8**, 924.
93. P. B. Luppa, C. Müller, A. Schlichtiger and H. Schlebusch, *TrAC, Trends Anal. Chem.*, 2011, **30**, 887–898.
94. N. M. Sumita, C. E. S. Ferreira, M. D. V. Martino, C. N. Franca, A. C. L. Faulhaber, M. Scartezini, J. R. R. Pinho, C. M. Dias, K. R. Cesar and V. M. Pariz, *Clin. Lab.*, 2018, **64**, 1105–1112.
95. I. Prattis, E. Hui, P. Gubeljak, G. S. K. Schierle, A. Lombardo and L. G. Occhipinti, *Trends Biotechnol.*, 2021, **39**, 1065–1077.
96. I. V. Martynenko, A. P. Litvin, F. Purcell-Milton, A. V. Baranov, A. V. Fedorov and Y. K. Gun'ko, *J. Mater. Chem. B*, 2017, **5**, 6701–6727.
97. X. Wang, F. Li and Y. Gou, *Front. Chem.*, 2020, 924.
98. A. M. Bagher, *Sens. Transducers J.*, 2016, **198**, 37.
99. H. M. E. Azzazy, M. M. H. Mansour and S. C. Kazmierczak, *Clin. Biochem.*, 2007, **40**, 917–927.
100. H. Yang, D. Li, R. He, Q. Guo, K. Wang, X. Zhang, P. Huang and D. Cui, *Nanoscale Res. Lett.*, 2010, **5**, 875–881.
101. S. Panesar, X. Weng and S. Neethirajan, *IEEE Sens. Lett.*, **1**, 1–4.
102. B. A. Kairdolf, A. M. Smith, T. H. Stokes, M. D. Wang, A. N. Young and S. Nie, *Annu. Rev. Anal. Chem.*, 2013, **6**, 143–162.
103. A. Gomase, S. Sangale, A. Mundhe, P. Gadakh and V. Nikam, *J. Drug Delivery Ther.*, 2019, **9**, 670–672.
104. A. M. Smith, S. Dave, S. Nie, L. True and X. Gao, *Expert Rev. Mol. Diagn.*, 2006, **6**, 231–244.
105. H. Arya, Z. Kaul, R. Wadhwa, K. Taira, T. Hirano and S. C. Kaul, *Biochem. Biophys. Res. Commun.*, 2005, **329**, 1173–1177.
106. O. Mashinchian, M. Johari-Ahar, B. Ghaemi, M. Rashidi, J. Barar and Y. Omidi, *Bioimpacts*, 2014, **4**, 149.
107. S. Jin, Y. Hu, Z. Gu, L. Liu and H.-C. Wu, *J. Nanomater*, 2011, **2011**, 1–13.
108. H.-Z. Wang, H.-Y. Wang, R.-Q. Liang and K.-C. Ruan, *Acta Biochim. Biophys. Sin.*, 2004, **36**, 681–686.
109. K.-T. Yong, I. Roy, R. Hu, H. Ding, H. Cai, J. Zhu, X. Zhang, E. J. Bergey and P. N. Prasad, *Integr. Biol.*, 2010, **2**, 121–129.
110. A. J. Caires, A. A. Mansur, I. C. Carvalho, S. M. Carvalho and H. S. Mansur, *Mater. Chem. Phys.*, 2020, **244**, 122716.
111. A. A. P. Mansur, H. S. Mansur, S. M. Carvalho and A. J. Caires, *Contrast Media Mol. Imaging*, 2017, **2017**, 3896107.

112. J. H. Warner, A. Hoshino, K. Yamamoto and R. D. Tilley, *Angew. Chem. Int. Ed.*, 2005, **44**, 4550–4554.
113. P. Koralli, A. D. Nega, L. E. Vagiaki, A. Pavlou, M. G. Siskos, A. Dimitrakopoulou-Strauss, V. G. Gregoriou and C. L. Chochos, *Mater. Chem. Front.*, 2020, **4**, 2357–2369.
114. M. R. Younis, G. He, J. Lin and P. Huang, *Front. Chem.*, 2020, **8**, 424.
115. L. Zhang, Y. Xing, N. He, Y. Zhang, Z. Lu, J. Zhang and Z. Zhang, *J. Nanosci. Nanotechnol.*, 2012, **12**, 2924–2928.
116. L.-L. Feng, Y.-X. Wu, D.-L. Zhang, X.-X. Hu, J. Zhang, P. Wang, Z.-L. Song, X.-B. Zhang and W. Tan, *Anal. Chem.*, 2017, **89**, 4077–4084.
117. J. Barar and Y. Omidi, *BioImpacts*, 2014, **4**, 3–14.
118. L. A. A. Chunduri, M. K. Haleyurgirisetty, S. Patnaik, P. E. Bulagonda, A. Kurdekar, J. Liu, I. K. Hewlett and V. Kamisetti, *Microfluid. Nanofluidics.*, 2016, **20**, 1–10.
119. Y.-D. Ye, L. Xia, D.-D. Xu, X.-J. Xing, D.-W. Pang and H.-W. Tang, *Biosens. Bioelectron.*, 2016, **85**, 837–843.
120. D. Yang, Z. Guo, J. Wang, Y. Jin, Q. Mei and P. Miao, *Part. Part. Syst. Charact.*, 2019, **36**, 1900273.
121. M. Zarei-Ghobadi, S.-H. Mozhgani, F. Dashtestani, A. Yadegari, F. Hakimian, M. Norouzi and H. Ghourchian, *Sci. Rep.*, 2018, **8**, 1–8.
122. O. J. Achadu, K. Takemura, I. M. Khoris and E. Y. Park, *Sens. Actuators, B*, 2020, **321**, 128494.
123. O. J. Achadu, D. X. Lioe, K. Kagawa, S. Kawahito and E. Y. Park, *Microchim. Acta*, 2020, **187**, 1–11.
124. L.-D. Xu, F.-L. Du, J. Zhu and S.-N. Ding, *Analyst*, 2021, **146**, 706–713.
125. R. S. Li, B. Yuan, J. H. Liu, M. L. Liu, P. F. Gao, Y. F. Li, M. Li and C. Z. Huang, *J. Mater. Chem. B*, 2017, **5**, 8719–8724.
126. L.-D. Xu, Q. Zhang, S.-N. Ding, J.-J. Xu and H.-Y. Chen, *ACS Omega*, 2019, **4**, 21431–21438.
127. X. Wang, L. Chen, X. Su and S. Ai, *Biosens. Bioelectron.*, 2013, **47**, 171–177.
128. A. Valipour and M. Roushani, *Biosens. Bioelectron.*, 2017, **89**, 946–951.
129. Q. Yan, Y. Yang, Z. Tan, Q. Liu, H. Liu, P. Wang, L. Chen, D. Zhang, Y. Li and Y. Dong, *Biosens. Bioelectron.*, 2018, **103**, 151–157.
130. K. Ghanbari, M. Roushani and A. Azadbakht, *Anal. Biochem.*, 2017, **534**, 64–69.
131. Q. Xiang, J. Huang, H. Huang, W. Mao and Z. Ye, *RSC Adv.*, 2018, **8**, 1820–1825.
132. S. R. Ahmed, J. Mogus, R. Chand, E. Nagy and S. Neethirajan, *Biosens. Bioelectron.*, 2018, **103**, 45–53.
133. A. D. Chowdhury, K. Takemura, T.-C. Li, T. Suzuki and E. Y. Park, *Nat. Commun.*, 2019, **10**, 1–12.

134. Y. Zhao, J. Chen, Z. Hu, Y. Chen, Y. Tao, L. Wang, L. Li, P. Wang, H.-Y. Li and J. Zhang, *Biosens. Bioelectron.*, 2022, 113974.

135. H. Zhang, D. Yee and C. Wang, *Nanomed*, 2008, **3**, 83–91.

136. G. R. Bardajee, M. Zamani and M. Sharifi, *Langmuir*, 2021, **37**, 10223–10232.

137. J. Li, Z.-E. Hu, Y.-J. We, Y.-H. Liu, N. Wang and X.-Q. Yu, *J. Colloid Interface Sci.*, 2022, **606**, 1219–1228.

138. S. M. Ghafary, E. Rahimjazi, H. Hamzehil, S. M. M. Mousavi, M. Nikkhah and S. Hosseinkhani, *Nanomedicine*, 2022, 102544.

139. R. R. Malla, *A Theranostic and Precision Medicine Approach for Female-Specific Cancers*, Elsevier, 2021, pp. 91–103.

140. H. Chen, Z. Liu, B. Wei, J. Huang, X. You, J. Zhang, Z. Yuan, Z. Tang, Z. Guo and J. Wu, *Bioact. Mater.*, 2021, **6**, 655–665.

141. M. K. Kumawat, M. Thakur, R. Bahadur, T. Kaku, R. Prabhuraj, A. Ninawe and R. Srivastava, *Mater. Sci. Eng. C*, 2019, **103**, 109774.

142. R. S. Pawar, P. G. Upadhaya and V. B. Patravale, *Handbook of Nanomaterials for Industrial Applications*, Elsevier, 2018, pp. 621–637.

143. J. M. Klostranec and W. C. W. Chan, *Adv. Mater.*, 2006, **18**, 1953–1964.

144. L. O. Cinteza, *J. Nanophoton*, 2010, **4**, 042503.

145. A. M. Wagner, D. S. Spencer and N. A. Peppas, *J. Appl. Polym. Sci.*, 2018, **135**, 46154.

146. N. Bertrand, J. Wu, X. Xu, N. Kamaly and O. C. Farokhzad, *Adv. Drug Delivery Rev.*, 2014, **66**, 2–25.

147. I. Brigger, C. Dubernet and P. Couvreur, *Adv. Drug Delivery Rev.*, 2012, **64**, 24–36.

CHAPTER 4

Quantum Dots in Imaging, Diagnosis, and Targeted Drug Delivery to Cancer Cells

ZAHRA RAMEZANI*[a], MICHAEL THOMPSON[b] AND ELAHEH MOHAMMADI[a]

[a]Toxicology Research Center, Medical Basic Sciences Research Institute, Ahvaz Jundishapur University of Medical Sciences, Ahvaz, Iran; [b]Chemistry Department, University of Toronto, Toronto, Canada
*E-mail: zramezani@ajums.ac.ir; zahramezani@gmail.com

4.1 Introduction

Cancer is a significant cause of mortality worldwide and one whose diagnosis and treatment are still difficult from a biomedical engineering perspective.[1] Early detection, achieved through early diagnosis or regular screening, greatly increases the effectiveness of treatment and diminishes the mortality rate. Simple screening tests are used to identify asymptomatic individuals, while early diagnostic tests are used to determine the progression of the disease in those who already have symptoms. Tools for detecting tumors include X-rays, computerised tomography (CT) scans, bone scans, positron emission tomography (PET) scans, ultrasound, radionuclide imaging, and magnetic resonance imaging (MRI),[2] and biopsy and cancer biomarker detection. In a biopsy, physicians collect samples of cells for laboratory testing. Many of these approaches have drawbacks.

Detection Science Series No. 22
Quantum Dots in Bioanalytical Chemistry and Medicine
Edited by Michael Thompson and Zahra Ramezani

Published by the Royal Society of Chemistry, www.rsc.org

They are unable to detect primary or metastatic sites containing a small number of cancerous cells. Additionally, imaging techniques cannot detect cancer surface biomarkers at early stages before the disease becomes life-threatening.[3]

After diagnosis, chemotherapy is one of the most widely used cancer treatment methods. However, this method kills normal cells in addition to the vigorously growing tumor cells. It also produces serious side effects. Side effects differ from patient to patient and are also dependent upon the type of medicine and the kind of treatment. They include bone marrow, hair loss, and gastrointestinal reactions. Targeted drug delivery to tumor cells minimizes damage to normal cells and reduces the required drug dose; it, thus, reduces the patient's pain and improves the treatment's efficiency. Currently, scientists are developing novel approaches to diagnosis and treatment that use nanotechnology to improve sensitivity and specificity.

The use of nanoparticles overcomes the disadvantages associated with molecular chemotherapy and photothermal chemotherapy drugs by improving drug targeting and reducing drug resistance and side effects. Nanoparticles are involved in both active and passive targeting. In active targeting, a nanoparticle's surface is modified with a specific target molecule to selectively bind to the corresponding tumor cell receptor, which improves the accumulation of the drug in the cell. The passive role of nanoparticles is observed through size-dependent increases in permeability and retention that result in higher accumulation in cancerous tissues. In addition, nanoparticles can be used as nanocarriers for gene delivery in gene therapy. In cases where the nanoparticles are illuminated, they may be used for simultaneous drug delivery and targeted site imaging so that the effectiveness of the treatment can be monitored. Alternatively, they may be used simply as an imaging tool for screening and diagnostic purposes. Nanoparticle-based nanomedicine is thus a promising new tool for screening, diagnosis, and therapy; nanoparticles are now used as versatile clinical tools to investigate and treat the primary tumor, determine the available treatment options for cancer, and evaluate the likelihood of a cure or recurrence.[4–6] The shape, size, and functional groups of a given nanoparticle affect its targeting ability and, consequently, its therapeutic efficacy. Figure 4.1 provides a schematic representation of different nanoparticles used in cancer drug delivery and treatment.

QDs are preferred over other nanocarriers for chemotherapeutics due to their unique properties. They were originally made up of metallic semiconductors from periodic element groups II–VI or III–V with physical dimensions ranging from 1 to 10 nm.[8] Subsequent developments in QD research have enabled their preparation with more environmentally friendly and biocompatible materials. Carbon-based materials such as graphene quantum dots (GQDs) and carbon quantum dots (CQDs) have recently gained prominence among the various types of nanoparticles.[9,10]

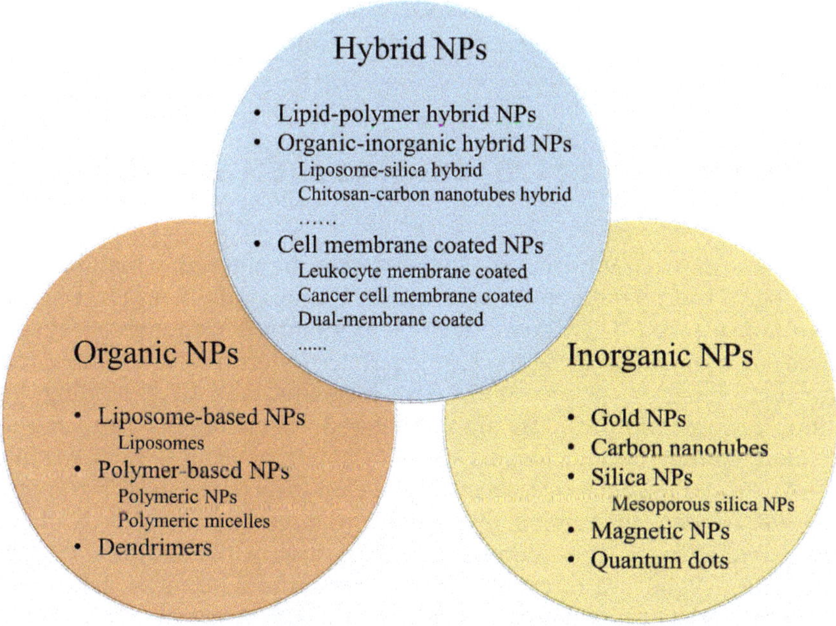

Figure 4.1 Different nanoparticles used as nanocarrier in cancer detection and treatment. Reproduced from ref. 7, https://doi.org/10.3389/fmolb.2020.00193, under the terms of the CC BY 4.0 license, https://creativecommons.org/licenses/by/4.0/.

Researchers are still examining QDs in depth for their potential applications in a wide range of cutting-edge fields related to energy and the environment, such as biosensing,[11] chemical sensing,[12] light-emitting devices,[13] photocatalysis,[14] and energy storage.[15] QDs play an important role in cancer therapy, particularly in bio-imaging,[16] drug delivery,[17] and gene delivery.[18] They have specific physiochemical properties, such as bright fluorescence, high photostability, and multiwavelength light emission; these properties make them especially useful as sensors and tools for imaging, diagnosis, and screening. However, only a few types of QDs are appropriate for biomedical imaging due to most of their issues with toxicity and clearance.[19] Moreover, they possess different surface functional groups depending on the constituent materials and the synthetic procedure; these groups allow for the conjugation of drug moieties, which in turn enables QDs' use as vehicles for controlled drug delivery and release. Although some QDs have anti-proliferative effects on cancer cells and cause cytotoxicity *via* targeting some cell apoptotic and necroptotic pathways,[20] many studies have been conducted on the use of QDs in the detection and treatment of cancer.

In this chapter, we will discuss QDs' applications in targeted drug delivery to cancer cells and their applications in imaging as diagnostic and treatment aids.

4.2 QDs in Cancer Cell Imaging and Targeted Drug Delivery

QDs are less than 10 nm in size. Their size can be altered, which enables the tuning of their band gap for a variety of applications. They emit highly stable fluorescence (FL), which makes them a better alternative to organic FL dyes for biomedical imaging. Conjugated QDs are required for *in vivo* diagnostics, imaging, passive and active targeting, and drug loading. To facilitate cell penetration, QDs are conjugated with biomolecules such as antibodies, aptamers, and biotins, or with small molecules such as folic acid (FA). Several issues arise when QD-labeled targets are used for *in vivo* and *in vitro* cancer cell imaging, diagnosis, and targeted drug delivery; these issues include tissues' autofluorescence, QDs' toxicity, penetration depth of the excitation light and intensity of the corresponding emitted light, and pharmacokinetics. Considering *in vivo* and *in vitro* imaging, QDs should have higher photostability and stronger emission and be monodispersed to provide uniform emission for *in vivo* and *in vitro* applications. QDs' toxicity is dose-dependent, and it also strongly depends on the kind and extent of surface states and coating ligands. Semiconductor QDs should be coated or functionalized so leakage of heavy metals into the bioenvironment is minimized. Metal semiconductor QDs mostly have superior emission intensities, but heavy metal leakage is highly toxic to the environment, which limits their applications. Recently, heavy metal-free QDs have been developed as an alternative to metal semiconductor QDs for imaging and targeted drug delivery purposes.

QDs were first used for bioimaging in 1998[21]; they have been the subject of active research ever since. Among metal-based QDs, cadmium-based QDs have been particularly well documented.[7,19,22] They have stable FL with high quantum yields and have been used in infrared (IR) detection imaging. Tissue autofluorescence reduces in the wavelength region of 650 to 950 nm, and the near-IR (NIR) emission of these QDs eliminates tissue FL interferences in tissue imaging and fluorescence-guided surgery (FGS). However, the QDs' high toxicity limits their biomedical applications. Indium-based QDs have been used as an alternative[23] with broader biomedical applications, especially in bioimaging, but they are still moderately toxic. Recently, carbon-based QDs have been investigated as a reliable, minimally toxic replacement for Cd-based QDs.[11,15,24–28] These QDs are environmentally friendly and can be synthesized from inexpensive carbon sources, such as waste, through simple procedures. As carbon-based QDs are relatively new, their applications in biomedical fields such as drug delivery, gene delivery, and *in vivo* and *in vitro* imaging have not yet been studied comprehensively.

Another important issue with carbon-based QDs is the minor toxicity at higher doses that should be controlled by QD surface engineering. For all QD types, improving their adsorption and their capacity to penetrate cancer cells is an important avenue for future research for imaging and targeting applications.

QDs' accumulation and penetration in the intended tissues or cells are investigated for imaging and targeted delivery purposes. As discussed above, aspects of QDs such as their charge, size, surface states, conjugated molecules, encapsulation in a larger nanoparticle, *etc.* affect their adsorption, penetration, and targeted delivery to the intended biological sites. Therefore, it is of utmost importance that researchers select appropriate synthetic methods, surface coatings, and functionalizations for specific applications.

Modified QDs have been used for various *in vitro* and *in vivo* applications. Examples of *in vitro* applications include diagnostics and prognostics using specific biomarker detection, *in vitro* imaging of cancer cells,[29,30] and simultaneous targeted drug delivery and imaging studies.[7,31] QDs' *in vivo* applications include simultaneous targeted drug delivery and imaging, follow-up treatment and diagnostics through biomarker and image evaluations,[32] and FGS.[33] The ability of QDs as imaging material and targeted drug delivery to the tumor cells are initially examined *in vitro* using different cancerous cell lines.

Two approaches are used for the *in vitro* evaluation of prepared QDs.[33] The first approach is the use of 2D cell models, in which prepared QDs are placed in contact with a monolayer of a cell line (see Figure 4.2A). 2D cell models constitute a simple, easy-to-use means of determining whether a particular QD is a good candidate for use. However, they do not provide detailed information on what may happen in a complex tumor environment; these models only evaluate the behavior of one cell line (normal or cancerous) at a time. Most studies to date have used this approach for the initial evaluation of prepared QDs for cancer cell imaging and targeted drug and gene delivery (Figure 4.2A shows 8900 studies have been performed using this model as of June 2022). The second approach involves the use of a spheroid, a 3D *in vitro* model (see Figure 4.2A and B). Spheroids mimic the structure of small avascular tumors by forming a sphere of tumors. In a spheroid, one part of the cell is in contact with the environment, while other cells are hidden inside the sphere. Cells inside the spheroids are separated into three layers: a proliferative periphery, a quiescent center, and a necrotic core. Oxygen and nutrient concentrations decrease, and the pH becomes more acidic while approaching from the outside to the core of the sphere. Figure 4.2B shows QDs' penetration and targeted delivery to the cancer cells, while stromal cells are not affected by functionalized QDs. Spheroids are highly useful for studying the influence of QDs' charge, size, and surface chemistry on adsorption and penetration. Surface modifications are an important determining factor[34]; for example, polyethylene

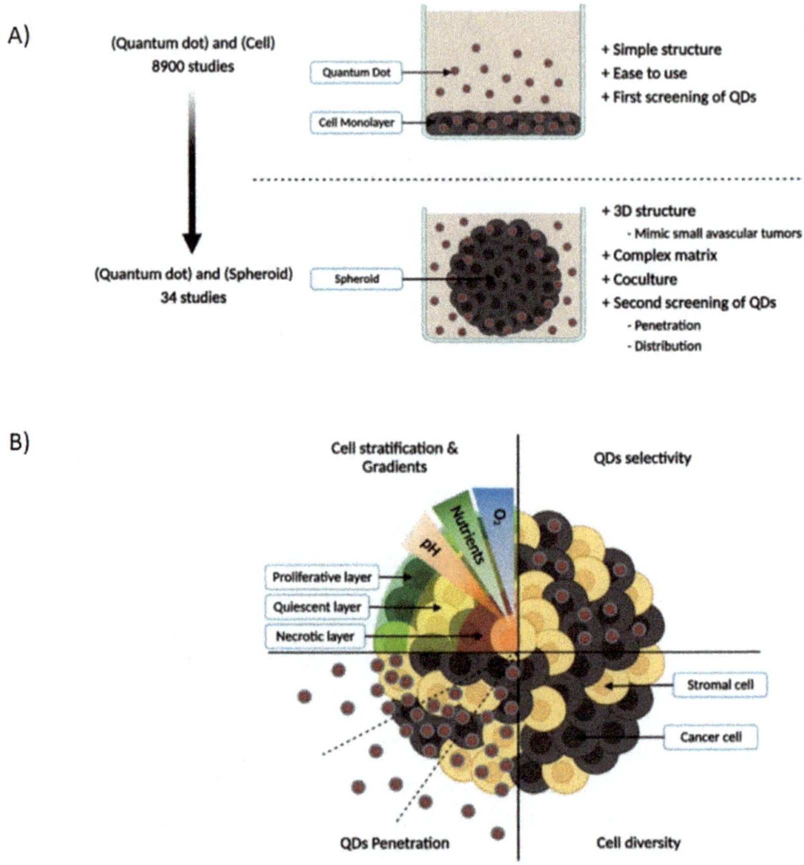

Figure 4.2 (A) *In vitro* evaluation models for QDs interaction with the cancerous cell and (B) detailed interaction between QDs and spheroid. Reproduced from ref. 33, https://doi.org/10.3390/pharmaceutics14102136, under the terms of the CC BY 4.0 license, https://creativecommons.org/licenses/by/4.0/.

glycol (PEG) coating reduces non-specific adsorption and aggregation of QDs and increases circulation lifetime.[35] However, specific targeting and drug delivery are developed by proteins, peptides, aptamers, antibodies, and other molecules that are conjugated QDs/magnetic QDs.

In conclusion, 3D spheroids are more useful than 2D monolayers for modeling interactions between QDs and normal and cancer cells; however, fewer studies have been conducted using spheroids due to the relatively high complexity of these systems (see Figure 4.2(A); 34 studies up to June 2022).

4.2.1 QDs in Cancer Cell Imaging, Prognostics, and Diagnostics

QDs' remarkable FL properties have inspired researchers to evaluate them for use in cancer prognostics and diagnostics.[36] As mentioned before, QDs' emission wavelengths can be continuously tuned by precise control over size, shape, and functional group *via* accurate selection of synthetic conditions.[37–40] The color of their emitted light varies depending on the energy radiated by the crystal. Blue light is related to higher energy, while red light is related to lower energy; larger QDs emit red light, whereas smaller QDs emit blue light.[41] QDs are ideal for *in vivo* cancer diagnosis and bio-imaging because of their wide excitation wavelength, narrow emission wavelength, long-term stability, high brightness, and deep penetration.[42]

Most diagnostic and prognostic processes involving QDs are performed through biomarker detection and quantification. The term "biomarker" (short for biological marker) refers broadly to any medical sign that is an objective indication of a medical status and can be observed by someone other than the patient. Medical signs stand in contrast with medical symptoms, which are limited to those indications of health or illness that are perceived by the patient themselves. The International Program on Chemical Safety, a joint venture on chemical safety led by the World Health Organization (WHO) and in coordination with the United Nations and the International Labor Organization, has defined a biomarker as "any substance, structure, or process that can be measured in the body or its products and influence or predict the incidence of an outcome or disease".[43] Under this definition, biomolecules released from tumor cells or produced by other body cells in response to tumors can be considered as biomarkers. Various *in vivo* and *in vitro* methods have been developed for biomarker detection using QDs as a sensing element; they are mostly based on FL spectroscopy. QDs may be engineered in different ways for the preparation of sensing elements; examples include QDs imbedded in polymers, dendrimers, liposomes, microbeads, magnetic QDs, and molecularly imprinted polymers (MIPs).

To identify cancer at an earlier stage, it is essential to develop biosensors that are highly sensitive and selective. These sensors must also be label-free, rapid, portable, and multiplexed in their detection of tumor biomarkers and cancer cells. In particular, biosensors have shown significant potential for *in vivo* sensing of disease-specific biomarkers.[44,45] Some of these biomarkers are found in body fluids, such as blood, saliva, and urine. Biomarkers include proteins, nucleic acids, transferrin receptors, vascular endothelial growth factor (VEGF), albumin, messenger RNA (mRNA), non-coding RNAs, iron, and circulating tumor cells (CTCs). Various metabolites of glycolysis, the tricarboxylic acid cycle, and oxidative phosphorylation are also used most frequently as biomarkers in the early detection of cancer.

Most QD-based biosensors are developed for *in vitro* evaluation and take advantage of different modified QDs. The conjugation of molecularly imprinted polymers (MIPs), mimicked antibodies to QDs, increases the optical detection system's sensitivity to ranges within clinical significance[46]; it also improves QDs' selectivity. Carbohydrate antigen 19–9 (CA 19–9) is released by pancreatic tumor cells. It is a glycosylated mucin-type protein biomarker with an average molecular weight of 210 kDa. Piloto *et al.*[46] successfully assembled CA 19–9 MIP around red-emitting cadmium telluride QDs (MIP@QDs). MIP@QDs were dispersed in hydrogel and used for fluorometric detection and quantification of CA19-9 to evaluate pancreatic cancer. The MIP@QDs retained their structure inside the binding site of the hydrogel, with an excitation wavelength of 477 nm and an emission range between 530 and 780 nm. The FL signal was measured at the wavelength with maximum intensity (about 630 nm). The imprinted hydrogel showed a linear response below the cut-off values for pancreatic cancer diagnosis, with a limit of detection of $1.58 \times 10–3$ U mL^{-1}. The toxicity of Cd inspired the researchers to synthesize indium phosphide QDs (InP-QDs), an environmentally friendly alternative to Cd-based QDs, by using a low-toxic phosphorus source for *in vitro* diagnostics and bioimaging applications. This InP-QD-based fluorescent sensor was able to detect alpha-fetoprotein in the widest range of 1–1000 ng mL^{-1} with a limit of detection of 0.58 ng mL^{-1}. Furthermore, as-prepared InP-QDs displayed excellent performance in labeling liver cancer cells and *in vivo* tumor imaging in mice.[47] Carbon-based QDs were also considered as a more reliable, environmentally friendly, and biocompatible in cancer diagnosis and imaging. GQDs were introduced as an effective ratiometric fluorescence resonance energy transfer (FRET) for the specific and sensitive detection of miRNA-21 with a detection limit of 0.5 pM in human serum samples.[48] Similarly, Kurniawan *et al.* developed photoluminescence (PL) microplasma-tunable nitrogen-doped GQDs (NGQDs) for ultrasensitive detection of FA as a cancer biomarker with a single material. In this work, they synthesized NGQDs from chitosan, a biocompatible material, and showed that the FL of NGQDs was quenched in the presence of FA. The NGQDs' emission quenching was higher when FA concentration increased. Having considered this relationship, they achieved a selective FA detection and quantification method through the precise regulation of solution pH in a broad concentration range from 0.8 to 80 μM of FA.[49] Researchers also used GQD in a Forster resonance energy transfer (FRET) probe for the detection of human epididymis protein 4 (HE4) as an ovarian cancer (OVC) biomarker in a wide variety of protein mixtures. The upregulated protein HE4 was used as a biomarker to diagnose OVC malignancies. Two different GQDs were synthesized: one had a strong red emission at 615 nm (GQDs615) and the other had a strong green emission at 515 nm (GQDs515). Both QDs were conjugated with HE4 antibodies. The emission wavelength of GQDs515 overlaps with the absorption wavelength of GQDs615; therefore, GQDs515 acts as the donor while GQDs615 acts as the receptor in the FRET probe. When HE4 antigen is introduced to the HE4 antibody-conjugated GQD solution, the antigen binds with two antibodies, thereby bringing the GQDs515 and GQDs615 in proximity to produce FRET-induced red emissions. This procedure was able to detect 4.8 pM HE4 with a wide dynamic range (up to 300 nM).[50]

Additionally, exosomes carry molecular contents and hold great potential as a source of biomarkers for non-invasive cancer detection and monitoring. They can be obtained from bodily fluids such as blood, saliva, and urine. Vinduska *et al.* reported a method for exosome surface protein detection using QDs coupled with immunomagnetic capture and enrichment. In this method, exosomes were captured by magnetic beads based on CD81 protein expression. Surface protein markers of interest were recognized by the primary antibody and then detected by secondary antibody-conjugated QDs with fluorescent spectroscopy. This QD-based exosome method is rapid and only requires microliters of diluted plasma without pre-purification.[51] In another study conducted by Kim *et al.*, a multi-QD-embedded silica-encapsulated nanoparticle (M-QD-SN) lateral flow immunoassay was designed for exosome detection. CdSeZnS/ZnS QDs with an alloyed structure were used because they were more stable after surface modification (*i.e.*, after being covered with a layer of silicon) and exhibited a higher FL quantum yield. Anti-CD63 antibodies were immobilized on M-QD-SNs (M–QD-SNs-CD63 Ab) and applied to the test strip to identify exosomes released from human foreskin fibroblasts (HFFs).[52] A nitrocellulose (NC) membrane was assembled on a backing card with the test line and control line sprayed with CD3 Ab and monoclonal rabbit IgG Ab, respectively. The strip was placed in the well of a 96-well plate containing a mixture of HFF exosomes and M–QD-SNs-CD63 Ab. After 10 min, the strip was analyzed. This highly selective and sensitive method quantified exosomes in a wide concentration range from 100 to 1000 exosomes per microliter with a detection limit of 117.94 exosomes per microliters, which was 11 times lower than the lowest previously reported limit.

Several cancers and molecular targets have been examined *in vivo* with QD-based probes.[53,54] In 2021, Ryu *et al.*[32] showed that attenuation-based plain QDs X-ray imaging could overcome the penetration limits of FL-based imaging techniques. In this study, X-rays successfully detected 5-mm-sized Panc-1 tumor cells grown in mice. Detection was achieved using cesium lead bromide ($CsPbBr_3$) QDs (CPB QDs) dispersed in water and injected from the tail (see Figure 4.3). CPB QDs were encapsulated with SiO_2 to prevent decomposition and aggregation. They were then conjugated with antibodies (CPB-SiO_2@SiO_2-Ab) for targeting and consequently recognizing the pancreatic cells, as indicated in Figure 4.3. *In vivo* mouse experiments revealed that a bright spot at the location of the tumor was observed when a tiny amount of CPB-SiO_2@SiO_2-Ab was injected. The particles were biocompatible and non-toxic. The X-ray signal was clearly observed from CPB QDs in tissue with a thickness of 2 cm, while their FL signal was not detected at all.[32]

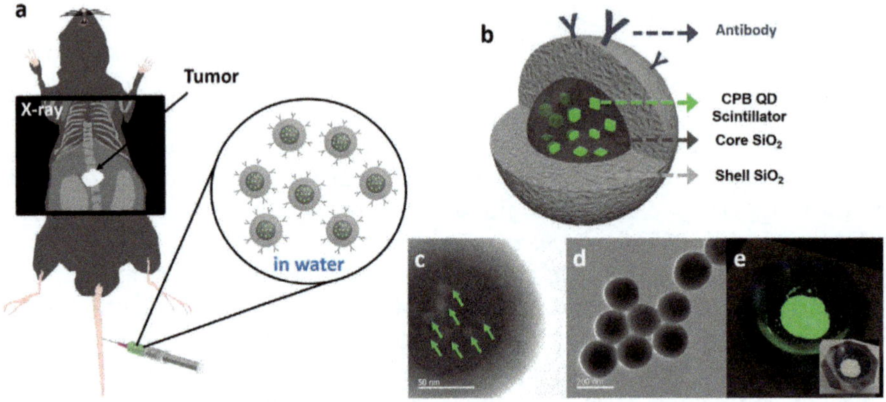

Figure 4.3 Plain X-ray imaging of cancer using CPB QD scintillators. (a, b) Schematic illustration of cancer detection and CPB–SiO_2@SiO_2–Ab NP structure. (c, d) TEM images of CPB–SiO_2@SiO_2 NPs at different magnifications. The CPB QDs are marked with green arrows. (e) Photographs of CPB–SiO_2@SiO_2 NP powder under UV irradiation and daylight (inset). Reproduced from ref. 32 with permission from John Wiley and Sons, Copyright 2021.

The gold standard detection methods for bladder cancer (BC), the tenth most common cancer worldwide, are cystoscopy and urine cytology. Cystoscopy is invasive and expensive, while urine cytology is less sensitive and not suitable for different stages of BC diagnosis. Therefore, the invention of a reliable, sensitive, and economic method is of utmost importance. Abnormal microRNA (miRNA) production closely correlates with tumor development, including tumors associated with BC. Circulating miRNA has, therefore, been used as a biomarker for monitoring the progression of cancer through the analysis of blood and urine.[55] Xu *et al.*[56] synthesized an miRNA detection method based on core-shell-shell bifunctional magnetic fluorescence microbeads (MFBs) with the integration of catalytic hairpin assembly (CHA) and terminal deoxynucleotidyl transferase (TdT)-mediated DNA polymerization. CdSSe/ZnS QDs capable of emitting 620 nm (QDs620) and 450 nm (QDs450) were used. MFBs included both QDs and magnetic iron oxide nanoparticles and were used in FL encoding and magnetic separation. They evaluated their corresponding microscopic images for the multiplexed detection of bladder cancer. This strategy can detect a single miRNA species at femtomolar levels with a wide linear dynamic range.

4.2.1.1 Multimodal Detection

According to recent findings, CD44-expressing phenotypes and stem cell antigens are associated with many metastatic tumors and CTCs.[57] Identifying CTCs and performing liquid biopsy of metastatic cells are two approved methods of early-stage detection, isolation, and genomic identification of various carcinomas; these methods can also help to guide approaches to treatment.[58] They are also painless and non-invasive tools for obtaining carcinoma cell samples for molecular profiling. Sobhanan *et al.* reported efficient detection and collection of cancer cells in blood samples by combining stem cell antigen (CD44)-specific immune-silica particles and immunofluorescent QDs with spectrally and temporally resolved single-photon counting. They collected MCF7 and Hella cells from blood samples using immuno-silica microparticles targeting the mesenchymal antigen CD44. The large silica nanoparticles separated the cancer cells easily, with no need for centrifugation or magnetic separation. It also prevented endocytosis, a common phenomenon observed for magnetic or silica particles. The collected cancer cells were characterized by triple FL staining followed by multicolor imaging, FL spectra deconvolution, and decay profile evaluations (see Figure 4.4). The three staining reagents were Syto13/25 nucleus-staining dye, a 585 nm CdSe/ZnS QDs (QD585) conjugated anti-EpCAM Ab, and a QD655 conjugated anti-CD44 Ab. They detected 1–10 cells among 100 cancer cells of the breast, lungs, or cervix in 1 mL of blood samples. In addition, the bright and narrow band emission of CdSe/ZnS QDs enabled temporally and spectrally resolved photon counting for multiplexed cancer cell detection. The cancer cell-specific and large immune-silica particles helped the group to collect the specific cells.[29]

Single biomarker detection has limited diagnostic sensitivity and specificity for early detection purposes; the development of multiplex biomarker sensing devices is key to improving cancer screening and early diagnosis. Recently, an early liver cancer detection method was established[59] using multiplex biomarkers: alpha-fetoprotein (AFP), glypican-3 (GPC3), and dickkopf-1 (DKK1). In this method, three different QD sizes with emissions at 565, 605, and 655 nm were conjugated with GPC3, DKK1, and AFP antibodies, respectively. The antibody-conjugated QDs were used as a triple FL nanoprobe for simultaneously assaying these three liver cancer biomarkers. The specific antibodies encode their corresponding antigen, and through evaluation of changes in the emitted light of the corresponding QDs, decision is made on the stage of the liver cancer. The probe showed a linear range of 0.625–2.5 ng mL^{-1} for the entire biomarker panel. This method of simultaneous determination merits further clinical validation *via* biomarker panels in human serum samples.

Suspension microarray technology (SMT) is a high-throughput, large-scale, and multiplex screening platform used in molecular biology. SMT has been applied to genomic and proteomic research, such as genetic disease screening, gene expression profiling, and clinical diagnosis. SMT uses microbeads to prepare arrays of various antibodies to detect corresponding antigens. Each microbead has a unique identification based on its optical properties, such as its FL wavelength. Recently, a novel

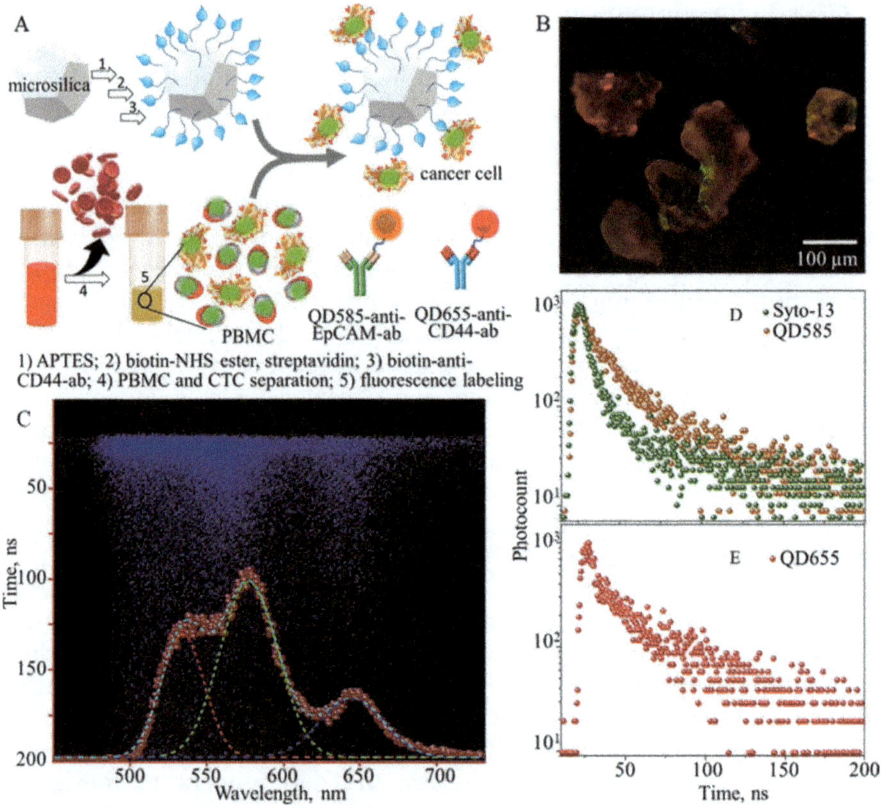

Figure 4.4 Cancer cell isolation and detection. (A) A scheme showing silica particle functionalization and cancer cell separation from a blood sample. (B) Fluorescence image of silica particles with MCF7 cells. (C) A streak camera image and the corresponding fluorescence spectra of cells attached to silica particles. The broken lines represent the Lorentzian fits for the Syto-13 dye (515 nm), QD-CD44-ab conjugate (585 nm), and QD-EpCAM-ab conjugate (655 nm). (D, E) The fluorescence decays based on the deconvoluted data shown in C for the three labels. Reproduced from ref. 29, https://doi.org/10.1038/s41427-021-00353-5, under the terms of the CC BY 4.0 license, http://creativecommons.org/licenses/by/4.0/.

suspension microarray system using dual-color QD nanobeads as FL labels was developed for the simultaneous detection of prostate-specific antigen (PSA) biomarkers.[60] CdSe@ZnS QDs with red and green emissions were prepared to make the dual color array; free PSA (f-PSA) antibodies were conjugated with green QDs, and complexed PSA (c-PSA) antibodies were conjugated with red QDs. Total PSA (t-PSA) antibody-conjugated magnetic beads were prepared and dispersed in serum samples to separate both antigens, f-PSA, and c-PSA (see Figure 4.5). After magnetic separation, the particles were redispersed into a solution containing both conjugated QDs, as is demonstrated in Figure 4.5. The magnetic FL hybrid was purified from the mixture by an external magnet prior to flow cytometry analysis.[60] This SMT is a novel QD-based method for simultaneously detecting f-PSA and c-PSA antigens, which are overexpressed in the sera of prostate cancer patients. Quantification ranges were 0.2 ng mL^{-1} to 50.0 ng mL^{-1} and 0.64 ng mL^{-1} to 80.0 ng mL^{-1} for c-PSA and f-PSA antigens, respectively; the detection sensitivity of c-PSA was 45 pg mL^{-1}, whereas that of f-PSA was 22 pg mL^{-1}.

4.2.2 Targeted Drug Delivery

Drug delivery research focuses on the development of clinically effective formulations and controlled drug delivery methods for treating cancerous cells or tumors with minimal side effects for patients. Thereafter, significant improvements were observed. Targeted drug delivery (TDD) has attracted significant attention because of high cancer mortality, impossible tumor removal by surgery in metastatic cancer cases, and several shortcomings of conventional cancer therapy strategies.[61–63] Common chemotherapy drugs have non-specific targeting and release behaviors, and their poor physiochemical and pharmacokinetic properties increase side effects and reduce therapeutic efficacy. Nanotechnology addressed these limitations by providing highly traceable nanocarriers that release drugs in response to the pH of the environment, called pH-sensitive nanocarriers.[7] QDs have outstanding physiochemical and optical properties.

Recently, QDs have been used in the design of nanocarriers for TDD, cell imaging, targeted chemotherapy, phototherapy, gene therapy, and many other related applications.[40,48,63–66] An effective TTD system must have sufficient blood residence time to reach the intended sites of the body and release the drug within the timeframe required for its efficient function. The drug delivery route determines the delivery systems for drug targeting; these systems are categorized as passive and active. In passive TDD, the drug accumulates in the intended sites of the body by improving its penetration and retention. Active targeting describes a specific interaction between ligands on the drug's (or the drug carrier's) surface and the cell's receptors. FA is taken up by most tumor cells and usually used for modifications of the drug carrier's surface. Aptamers, antibodies, and peptides are applied to modify the drug or carriers' surface and allow us to distinguish one cell from another.[67,68] TDD minimizes the effective doses of the drug and reduces cytotoxicity by changing the surface functional groups and the precise selection of the drug carriers.[64]

Cd-based QDs have shown notable results in TDD.[69,70] However, they are not biodegradable and have high cell and environmental toxicity due to heavy metal leakage. Carbon-based QDs, including CQDs and GQDs,

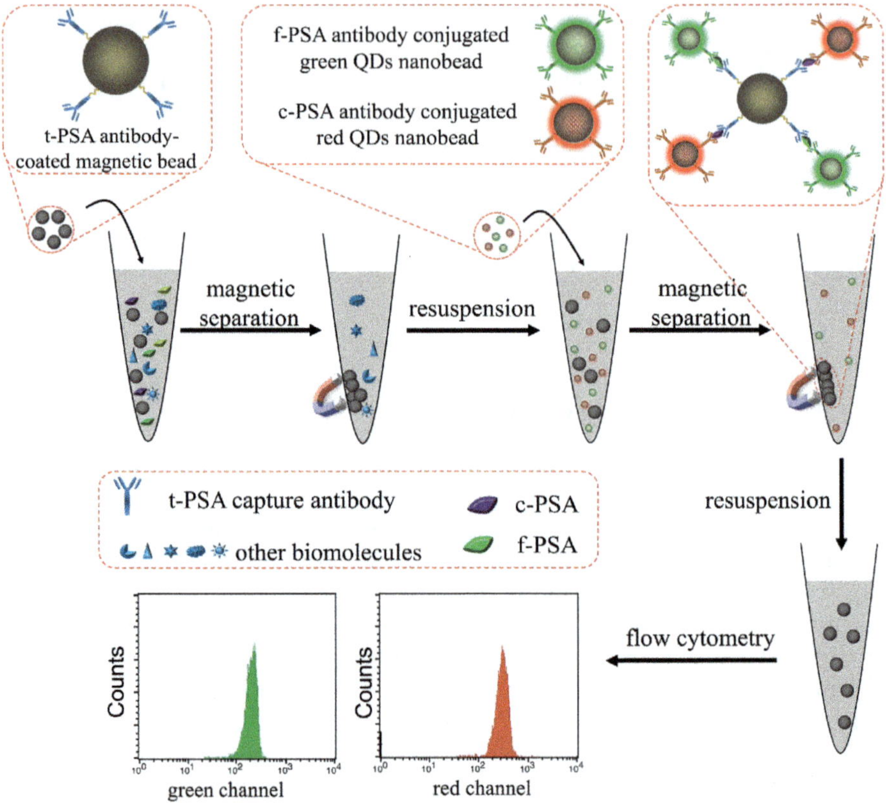

Figure 4.5 Outline of dual-plex detection process of f- and c-PSA antigen proteins by dual-color signal reporters. Magnetic beads was functionalized with t-PSA antibody to prepare the capture probe prior to incubation with sera containing both antigens. After magnetic separation with a magnet and washing with PBS, those beads were resuspended in PBS/BSA solution, into which a mixture of f-PSA antibody-conjugated green QDs nanobeads and c-PSA antibody-conjugated red QDs nanobeads were added. Those magnetic beads were magnetic separated and washed again before flow cytometry analysis. The fluorescence intensities of those magnetic beads in both fluorescence channels were decoded and collected by this method. Reproduced from ref. 60 with permission from Elsevier, Copyright 2022.

are bio- and environmentally friendly, easily prepared, and functionalized with inexpensive precursors and simple procedures.[71] They are, therefore, a promising alternative to Cd-based QDs in biomedical sciences and especially in TDD-based cancer therapy. See Figure 4.6.

Cell death occurs through two mechanisms: programmed cell death and necrosis. Programmed cell death happens in response to a variety of

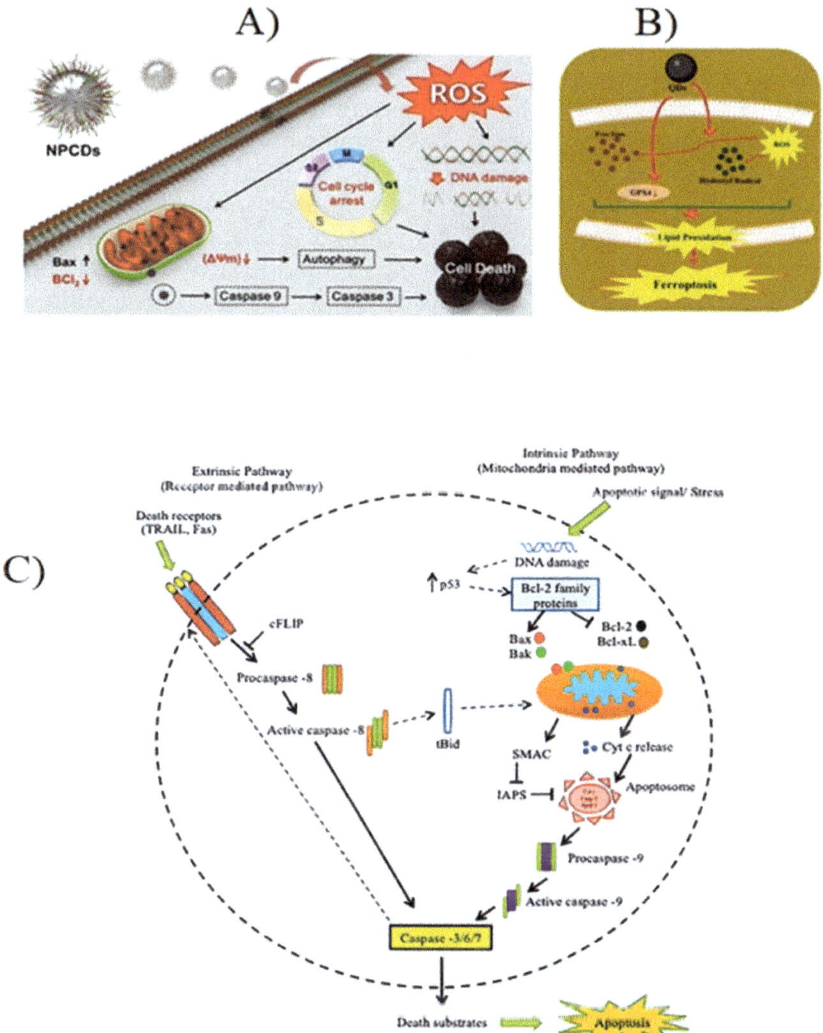

Figure 4.6 (A) Autophagic anti-cancer mechanism of nitrogen-/phosphorous-doped carbon dots (NPCDs) against B16F10 melanoma cells. Reproduced from ref. 72, https://doi.org/10.7150/thno.42291, under the terms of the CC BY 4.0 license, https://creativecommons.org/licenses/by/4.0/. (B) Proposed schematic diagram for QDs-induced ferroptosis. GPX 4: Glutathione peroxidase 4, ROS: Reactive Oxygen Species. (C) Apoptosis pathway. Reproduced from ref. 73, https://doi.org/10.15171/apb.2019.024, under the terms of the CC BY 4.0 license, http://creativecommons.org/licenses/by/4.0/.

extracellular and intracellular signals. It may be caused by physical agents (such as radiation, physical trauma, or chemotherapy drugs) or by infectious agents (such as viruses or bacterial toxins). Its mechanisms can be classified as apoptosis (a morphologically distinct pattern of cell death) and non-apoptosis. Apoptosis is the mechanism of regulated cell death. It is characterized by different mechanisms, such as DNA fragmentation in the nucleus, dynamic membrane blebbing, cell shrinkage, condensation of chromatin material, loss of adhesion of extracellular matrices, and caspases (*i.e.*, the activation of cysteine aspartyl proteases). Depending on their design, QD-based TDD systems can target cancer cells *via* apoptotic or non-apoptotic pathways. Figure 4.6 illustrates the apoptotic and non-apoptotic cell death mechanisms associated with currently available QD-based TDDs.

4.2.2.1 *Apoptotic QD-Based TDD and Therapy*

Various types of QDs have been used for TDD. They are capable of inducing cancer cell apoptosis while exhibiting low cytotoxicity toward normal cells because their surface functional groups and/or conjugated molecules enable selective cell uptake. For example, FA-conjugated QDs accumulate and internalize into the cells with FA receptors cells. However, QDs are decorated with molecules such as transferrin, FA, proteins, peptides, aptamers, and antibodies and loaded with a drug for TDD. QDs have also been used in bimodal therapies such as simultaneous chemotherapy and photothermal treatment of cancer cells.

QDs may also be used in drug-free targeted therapies, in which QDs target cells and prevent their growth. Davenport *et al.*[74] showed that drug-free InP/ZnS QDs-induced apoptosis in HeLa cervical cancer cells through oxidative stress *via* the production of reactive oxygen species (ROS). They found that CQD-treated samples induced more cell death than those of untreated cells. Table 4.1 lists several recently developed label-free CQDs and describes their mechanisms of targeted cell apoptosis. Depending on the preparation methods and carbon sources, CQDs have different surface functional groups that facilitate cellular uptake. They are mostly non-toxic to normal cells but show some degree of toxicity to cancer cells. Recently, 75 CQDs were synthesized from oleander extracts. They were shown to have an anti-cancer effect on MCF7 cancer cells *via* DNA damage and cell cycle arrest.

Researchers can modify QDs to target cancer cells more specifically by binding them with biomolecules to enable recognition or with drugs for the purpose of targeted chemotherapy. Manan *et al.*[79] prepared a mitomycin C (MMC)-encapsulated Mn: ZnS QDs–chitosan (MMC@CS-Mn:ZnS) nanocarrier system for TDD. The system actively targeted bladder cancer cells with MMC, a water-soluble anti-cancer agent whose cell penetration is normally limited due to the hydrophobic nature of the cell membrane. The zeta potential of as-prepared MMC@CS-Mn:ZnS was positive; it bonded readily

Table 4.1 Some CQDs used for targeted cancer cell therapy. Reproduced from ref. 75, https://doi.org/10.1038/s41598-020-70796-3, under the terms of the CC BY 4.0 license, http://creativecommons.org/licenses/by/4.0/.

Carbon source	Size (nm)	Application	Cytotoxicity	Ref.
Citric acid and L-cysteine	3–6.5	*In vitro* and *in vivo* luminescence lifetime thermometry	*Cell Type*: NIH/3T3, standard mouse, fibroblasts, HeLa, human cervical cancer cells low cytotoxicity, no significant effect on the cell cycle of HeLa cells, dose-dependent G0/G1 arrest slightly on NIH/3T3 cells	76
Glucosamine-HCl and *m*-phenylenediamine	2.42 ± 0.55	LED-activated nucleus targeting and photothermal therapy	*Cell Type*: HDF and HeLa cells less cytotoxicity on HDF than HeLa cells, nuclear localization in HeLa cell line, nucleus targeting, and imaging ability have been shown *in vitro*. CDs-based or LED-induced cell death of cancer cells were not found to be associated with ROS production, no genotoxicity evaluation	77
Nerium oleander	2.05 ± 0.22	Anti-cancer therapy	*Cell Type*: MCF-7, human breast cancer cells, HDFa, human primer dermal fibroblast cells dose-dependent cytotoxicity on MCF-7 cells, no cytotoxic effects on HDFa cells genotoxicity, clastogenicity and G0/G1 arrest on MCF-7 cells	75
Ginger juice	8.2 ± 0.6	Treatment of liver cancer	*Cell Type*: HepG2, human hepatocellular carcinoma cell line; MCF-10A, normal mammary epithelial cell line; FL83B, normal liver cell line; A549, human lung cancer cell line; MDA-MB-231, human breast cancer cell line dose-dependent cytotoxicity, higher selectivity inhibition toward HepG2 cells, effective on HepG2 cell cycle (increase in SubG1 phase), did not cause significant differences in cell cycles for the other four non-cancerous cell lines	78

with negatively charged mucosal membranes, which facilitated cellular uptake, enabling the drug's effective delivery to the tumor sites of invasive bladder diseases. The drug was released at a pH of 6.8 (*i.e.*, within the acidic tumor environment). Another example of QD-based TDD is the use of doxorubicin (DOX)-conjugated QDs. Although many reports are available on the successful application of DOX in different cancer treatments, its administration causes various side effects that make targeted delivery to tumors desirable. To this end, scientists synthesized DOX-loaded NGQDs[71] and used them for TDD in HeLa cells (a cervical cancer cell line), MCF-7 (a breast cancer cell line), and non-cancerous control HEK293 cells (human embryonic kidney cells). The NGQDs-DOX TDD was label-free; DOX was non-covalently bonded to NGQDs through the π-stacking of an aromatic DOX platform onto the NGQD's graphitic backbone. The NGQDs were biocompatible and penetrated the cells due to their small size as well as the oxygen and nitrogen functional groups available on their surface. Cell viability tests showed that these QDs were not cytotoxic to HEK293 cells even at 1 mg mL^{-1} concentration. The TDD system showed about a 10% reduction in the DOX dose required for chemotherapy; IC50% is 0.97 μg mL^{-1} while that of DOX alone is 1.77 μg mL^{-1}.

QDs have also been developed for use as nanocarriers for tumor-targeting moieties capable of more specifically targeting tumor cells. A multifunctional Ag–In–Zn–S–QDs was prepared with the addition of L-cysteine, 11-mercaptoundecanoic acid (MUA), and lipoic acid and was decorated with FA.[80] Nine different compounds of FA and DOX-conjugated Ag–In–Zn–S QDs were prepared: QD–MUA, QD–Cys, and QD–LA nanocrystals; QD–MUA–FA, QD–Cys–FA, and QD–LA–FA nanocarriers; and QD–MUA–FA–DOX, QD–Cys–FA–DOX, and QD–LA–FA–DOX nanoconjugates. The authors evaluated the effect of all these particles on the apoptosis of adenocarcinomic human alveolar basal epithelial cells (A549). Biological activity screening showed that the QD–FA–DOX nanoconjugates exhibited higher cytotoxicity against A549 cells but were ineffective against NIH/3T3 cells (folate receptor-negative), indicating that DOX was delivered to human A549 cells *via* folate receptors (which are overexpressed in cancer cells). In an *in vivo* and *in vitro* study conducted by Weng *et al.*,[81] QDs conjugated with immunoliposome systems (QD-ILs) were developed for DOX delivery to breast cancer cells *via* HER2-receptor targeting. DOX was encapsulated in liposomes, and the QDs were conjugated with anti-HER2 single-chain Fv fragments and attached to the liposomes' surfaces for targeted DOX delivery to cancer cells with overexpressed HER2 receptors. *In vivo* results showed that QD-ILs localized at tumor sites in MCF-7/HER2 xenograft mouse models. HER2 is overexpressed in breast cancer cells, and it is the focus of targeted drug delivery and targeted therapy. For HER2 receptors, Herceptin (trastuzumab) is the most specific antibody and is conjugated to QDs by different types of coupling mechanisms (QD-mAb). QD-mAb was injected into the blood vessel and entered the intracellular space by recognizing HER2 on cancer cells. They have been used for both imaging and cell apoptosis. They trigger apoptosis in cancer cells through the release of free radicals and antibodies.[82]

Antibodies are expensive and have limited stability; researchers have therefore evaluated the effectiveness of other small molecules as alternatives for breast cancer cell targeting. For this purpose, GQDs) were functionalized with glucosamine (GlcN) as a targeting agent, loaded with curcumin as an anti-cancer agent (GQDs-GlcN-Cur), and used to kill MCF-7 breast cancer cells.[83] A pH-responsive drug release behavior was observed for GQDs-GlcN-Cur with much higher releases in an acidic condition. The curcumin was released at 37% at pH 5.5% and 17% at pH 7.4. GQDs-GlcN-Cur exhibited much more cytotoxicity against MCF-7 cells than Cur-GQDs.

Phototheranostics refers to the use of light for multimodal diagnostic imaging and effective cancer therapy. Some assemblies for light-induced tumor diagnosis and treatment make use of QDs or modified QDs. As an example, a multifunctional nanoplatform of gold nanorod silica-carbon quantum dots (GNR SiO_2-CQDs) was synthesized as a photo-theragnostic agent in which GNRs were used for photoacoustics (PA) and photothermal therapy (PTT), while CQDs were used as FL imaging and photodynamic therapy (PDT) agents. GNR@SiO_2-CDs were administered *via* I.V. and gradually accumulated in B16-F0 tumor-bearing nude mice with a tumor size of approximately 100 mm^3 (the animal model). They showed high sensitivity and produced high spatial resolution images.[84] The high sensitivity and good spatial resolution of FL/PA imaging enabled its use as a guide for PDT/PTT treatments. Furthermore, Murali and co-workers produced hematoporphyrin (HP) photosensitizer-encapsulated CQDs (HP-CQDs) to target human breast cancer cells (MCF-7). The HP-CQDs were shown to have potent ROS generation abilities, and they exhibited significantly increased phototoxicity to MCF-7 cells.[85] In another study, polydopamine (PDA) particles loaded with CdTe QDs capped with mercaptopropionic acid (CdTe@MPA QDs) were evaluated for use in oxidative stress inducement and photothermal therapy. The particles interacted with the human breast adenocarcinoma cell line MDA-MB-231; significant particles co-localized into the lysosomes. Additionally, the internalized rodlike CdTe@MPA-PDA remarkably enhanced ROS generation and apoptosis in MDA-MB-231 cells.[86] In another study, glutathione (GSH)-functionalized Ag_2S-QDs (2.0–5.8 nm) were prepared using a simple aqueous precipitation method with different heating times. These Ag_2S QDs had broad absorption spectra and narrow FL emission spectra, and they exhibited high photothermal conversion efficiency with stable, effective, size-dependent photothermal effects. Cell viability assays showed that these QDs were non-cytotoxic to Hela- and 293 T-cells, which made them ideal theragnostic probes for photothermal therapy.[87]

In a photothermal immunotherapy study, Zhao *et al.*[88] targeted anti-PD-L1 with cancer cell membrane-coated biomimetic black phosphorus quantum dots (BBPQDs) to treat metastasized triple-negative breast cancer. *In vitro* and *in vivo* results showed that the BBPQDs were effective at targeting and killing tumor cells through photothermal effects under NIR irradiation. They not only killed tumors directly through photothermal effects but also induced the maturation of dendritic cells. In addition, the anti-PD-L1(αPD-L1) treatment blocked PD-1/PD-L1 pathways and enhanced T-cells' immune responses, which enabled the T-cells to recognize and kill tumor cells. Both PTT and αPD-L1 immunotherapy inhibit tumor metastasis and can be applied in many cancer treatments.

ROS are cell signaling molecules produced by normal cells for normal biological processes. When produced in excess, however, ROS can induce apoptosis, necroptosis, and autophagic cell death. Cell death occurs due to oxidative stress, which damages lipids, proteins, and DNA. ROS-induced cell death is an attractive strategy for cancer therapy, though its implementation requires the development of cancer cell-specific ROS generation. Recent studies on ROS-based cancer treatment have focused mostly on the effects of QDs on cell organelles and intracellular metabolic signaling through the generation of ROS. For example, Wang *et al.*[89] showed that defect-rich GQDs could efficiently catalyze the generation of hydroxyl radicals from hydrogen peroxide, especially in the H_2O_2-rich tumor microenvironment. These GQDs induced an anti-cancer effect even without the addition of any chemotherapeutic drugs. Similarly, Yadav *et al.* used graphitic carbon nitride QDs, with their intrinsic capacity to generate electron–hole pairs under visible light illumination, to generate ROS in glioma cells. These nanoparticles were capable of inducing apoptosis in cancer cells with no effect on normal cells.[90] Another treatment strategy takes advantage of the hypoxic conditions within tumors, which make solid tumors susceptible to oxygen-dependent PDT. Ding *et al.*[91] engineered *Escherichia coli* (*E. coli*) such that they expressed catalase and then linked the bacteria to black phosphorus QDs (BPQDs) to form an *E. coli*/BPQDs system. They were able to target hypoxic tumor tissues after intravenous injection of *E. coli*/BPQDs into mice. BPQDs generated ROS on 660 nm laser irradiation; this destroyed the bacteria's membranes, released catalase, and degraded H_2O_2 to produce O_2. The increased O_2 levels improved hypoxia and enhanced photodynamic therapy, thus killing tumor cells *in vivo*.

As the development of multifunctional nano-theranostic is important for achieving accurate diagnosis and treatment of tumors, Wu *et al.*[92] synthesized black phosphorus QDs encapsulated within biodegradable hollow mesoporous MnO_2 nanoprobes (H-MnO_2/DOX/BPQDs) for cancer imaging and synergistic chemo-phototherapy. A cationic polymer [poly allylamine hydrochloride (PAH)] and an anionic polymer [poly acrylic acid (PAA)] were layered onto the surface of the H-MnO_2 nanoparticles in a specific order

(H-MnO_2-PAH-PAA). H-MnO_2-PAH-PAA was covalently grafted with BPQDs-PEG-NH_2 *via* a carbodiimide cross-linking reaction and then loaded with the anti-cancer drug DOX to form H-MnO_2/DOX/BPQDs. The H-MnO_2/DOX/BPQDs degraded in the tumor microenvironment, releasing DOX, BPQDs, and HMnO_2. DOX acted as a chemotherapy and FL imaging agent, and BPQDs gave the nanoprobe PDT and PTT capabilities when irradiated by a laser at 630 and 808 nm. H-MnO_2 was used as a contrast agent in MRI and as a catalyst to produce tumor hypoxemia and thereby increase PDT efficiency. *In vivo* and *in vitro* results indicated that H-MnO_2/DOX/BPQDs acted as dual-modality MRI/FL imaging and synergistic chemotherapy/PDT/PTT, which ultimately enhanced the accuracy of cancer diagnosis and therapeutic performance.[92]

Recent studies in the field of targeted drug delivery have incorporated both molecular and magnetic targeting systems. Given the limited knowledge of suitable biomarkers for efficient molecular targeting, cooperative targeting by a combination of molecular biomarkers with superparamagnetic magnetic nanoparticles (MNPs) carriers is a particularly promising strategy for cancer therapy. Magnetic nanocarriers need to be stable in water at neutral pH and normal saline for biological, medical diagnostic, and therapeutic applications. Iron oxide MNPs are widely used as the core of magnetic nanocarriers; however, their surfaces must be coated to prevent the formation of large aggregates and to provide functional groups (amines or carboxylic acids) for bioconjugation to anti-cancer drugs and/or targeted ligands. Various surface coatings have been used, including lipids, liposomes, proteins, polymers, and dextran. MNPs are also commonly used as an MRI T2 contrast agent for contrast enhancement and signal amplification as they are much more efficient promoters of relaxation than the MRI T1 contrast agent gadolinium-diethylenetriaminepentaacetic acid, and their magnetic properties can be managed by controlling their size and surface coatings. The most important consideration for the clinical application of MNPs is their non-toxicity. MNPs based on magnetite (Fe_3O_4) and maghemite with satisfactory biocompatibility have therefore been under intense investigation to meet criteria such as a long blood retention time, biodegradability, and low toxicity. In a recent study, GQD-modified magnetic chitosan-loaded DOX was designed for simultaneous photothermal therapy and chemotherapy.[26] DOX was bonded to the magnetic chitosan, and GQDs were attached to the chitosan *via* amide groups. A TLS11a aptamer with a strong affinity for H22 mouse liver cancer cells29 was employed to functionalize the nanoparticles, resulting in DOX@Fe_3O_4@CS@GQD-Apt. The DOX@Fe_3O_4@CS@GQD-Apt was used for targeted simultaneous photothermal therapy and chemotherapy in hepatocellular carcinoma (HCC). HCC aptamers enabled the targeting of HCC; the GQDs acted as photosensitizers, moderating the photothermal therapy. Both *in vivo* and *in vitro* experiments confirmed the multimodal ability of DOX@Fe_3O_4@CS@GQD-Apt.

4.2.2.2 QDs and Non-apoptotic Cell Death

4.2.2.2.1 Autophagy. Recent findings suggest that autophagy is a double-edged sword: it may induce either cell death or cell protection. Autophagy clears damaged organelles and molecules and maintains healthy cell function. However, when a malignancy develops, autophagy allows the reprogramming of the tumor microenvironment and gives the tumor the ability to become resistant or sensitive to chemotherapy-induced toxicities (see Figure 4.6). Inhibiting autophagy may therefore enhance the efficacy of chemotherapy. Current studies suggest that autophagy may be a viable alternative to apoptosis in the treatment of cancers that are resistant to apoptosis.[93] Therefore, autophagy is a non-apoptotic form of cell death that occurs as a survival mechanism in response to cellular stress.[94] QDs can significantly affect cellular signaling states[24,27]; in this section, we focus on QDs' autophagy-inducing effects in cancer therapy.

Duman *et al.* synthesized Ag_2S-QDs functionalized with Cetuximab antibodies and loaded with 5-fluorouracil, an anti-cancer drug, for targeted NIR imaging and treatment of epidermal growth factor receptor (EGFR)-overexpressing cell lines. The cetuximab antibody QDs selectively delivered 5-fluorouracil to A549 cells and induced apoptosis. While free 5-fluorouracil confers resistance to cell death by activating autophagy, these nanoparticles significantly overcome drug resistance by suppressing autophagy.[95] Similarly, Xie and colleagues examined the autophagy and cytotoxicity induction capabilities in lung carcinoma cells (A549) of three types of GQDs: COOH-GQDs, OH-GQDs, and NH_2-GQDs. Results showed that OH-GQDs were the most toxic and induced significant cell death. Only COOH-GQDs and NH_2-GQDs induced autophagy, indicating that autophagy protected cells from the toxicity of the GQDs. In conclusion, cGQDs showed excellent biocompatibility, and autophagy induction may reflect the stress status leading to diseases.[96] Cd-based QDs also induced autophagy in mouse renal adenocarcinoma cells. QDs can penetrate the endoplasmic reticulum, endosomes, mitochondria, and lysosomes. An antioxidant agent (*N*-acetylcysteine) decreased ROS, reduced autophagy in the presence of QDs, and enhanced cell death.[97] Bajpai *et al.* showed that nitrogen–phosphorous-doped CDs induced autophagy by downregulating p62 and upregulating ATG-5 and LC3-II in melanoma cancer cells (B16F10).[72] Moreover, Yang *et al.* showed that MoS_2QDs in microglia triggered ROS generation and NLRP3 inflammasome activation and also induced the release of pro-inflammatory cytokines and subsequent caspase-1-dependent pyroptosis. Additionally, MoS_2QDs activated autophagy, and autophagy inhibitors enhanced the pyroptotic death response induced by MoS_2QDs. On the other hand, ROS scavengers can reduce MoS_2QDs-induced NLRP3 inflammasome activation and microglial pyroptosis.[98] Likewise, Peynshaert and co-workers evaluated the effect of two different QD surface coatings, 3-mercaptopropionic (MPA) acid and polyethylene glycol (PEG), on the autophagy pathway. The results showed that MPA-coated QDs were more biocompatible, and PEGylated-QDs, due to ROS production, lysosomal damage, and subsequently autophagy dysfunction, were very toxic to HeLa cells.[99] Figure 4.6 shows the effects of QD on autophagy mechanisms and related protein in cancerous cells.

4.2.2.2.2 Ferroptosis. Ferroptosis is a type of programmed cell death that is dependent on iron. It is characterized by the accumulation of lipid peroxides and is genetically and biochemically distinct from other forms of regulated cell death.[100] Ferroptosis is initiated by the **failure of the glutathione-dependent antioxidant defenses**, which result in unchecked lipid peroxidation, oxidative stress, and eventual cell death (see Figure 4.6B). Lipophilic antioxidants and iron chelators can prevent ferroptosis-based cell death. Currently, ferroptosis has attracted great attention.[101] Ferroptosis is mostly caused by the depletion of glutathione (GSH); this leads to the inactivation of glutathione peroxidase 4 (GPX4) and the subsequent accumulation of ROS and lipid peroxidation.[102] Ferroptosis has recently attracted scientific attention as an effective approach for cancer treatment. There are a few published studies on QD-induced ferroptosis in cancer cells and the stimulation of anti-cancer immune responses. Wu *et al.*[103] showed that nitrogen-doped GQDs could cause ferroptosis by increasing cytosolic iron and decreasing glutathione, leading to lipid peroxidation and ROS production in microglia. Furthermore, ferroptosis inhibitors, deferoxamine mesylate (DFO), and ferrostatin-1 (Fer-1) alleviated ferroptosis biomarkers in these cells. This study was performed to evaluate how the chemical functionalization of GQDs affects their toxicity and cell death characteristics.

Nanozymes, nanomaterials that display intrinsic enzymelike properties, have attracted interest for their potential applications in targeted therapy, diagnosis, and biosensing. CQD-based nanozymes have been shown to induce ferroptosis in tumors, which in turn activates an anti-tumor immune response. Yao *et al.*[25] evaluated the anti-tumor activity of CQD-based biocompatible nanozymes prepared from chlorogenic acid (ChA-CQDs) in a 3D tumor model of HepG2 cells. The ChA-CQDs exerted significant GSH oxidase-like activity by catalyzing reduced GSH to oxidized glutathione. They then induced HepG2 cell ferroptosis by promoting an imbalanced redox reaction due to GSH depletion and GPX4 inactivation. Results showed that intracellular ROS production was elevated and GSH was redacted; subsequently, ferroptosis markers GPX4, NRF2, and SLC7A11 levels were downregulated, and COX-2 expression increased. The ChA-CQDs did not alter the expression of apoptotic and necrotic proteins (BCL2, BAX, p-RIP1, and p-RIP3), indicating that they induced cell death selectively *via* ferroptosis. As the authors indicated that mitochondria play an important role in ferroptosis, they provided a proof of concept for the use of CQDs in tumor growth inhibition by noticeably enhancing ferroptosis-dependent immune cells' infiltration.

4.3 Conclusion

QDs have become a subject of recent scientific interest as nanocarriers for targeted drug delivery. They have the potential to enhance drug targeting and bioavailability, facilitate early detection and monitoring, and improve the localized treatment of specific diseases. In addition, the use of QDs as nanocarriers can improve the stability of drugs, lengthen their circulation time *in vivo*, enhance targeted absorption, and improve the distribution and metabolism processes of drugs in organisms. Their broad absorption spectra, narrow tunable emissions, and increased photostability compared with organic dyes make them highly attractive materials for bioimaging. Their large surface-to-volume ratio and physiochemical properties make them excellent payload carriers for therapeutic cargoes and improve the bioavailability and therapeutics at the target site. The modification and careful control of QDs' surface physiochemistry allow for the recognition of specific surface receptor proteins on targeted cells. Many recent developments in tumor-targeting drug delivery have leveraged the unique properties of QDs. CQDs, a recently discovered type of QD, are especially useful as scaffolds for accommodating anti-cancer drugs through physical immobilization or chemical linkage, leading to the formation of nanostructures with integrated imaging and therapy functions. Some of these emerging technologies are currently being improved and integrated into oncology clinical practice; they may soon have significant implications for cancer therapy. Application of QDs in TDD is still active research.

Abbreviations

Ab	Antibody
AFP	Alpha-fetoprotein
BBPQDs	Biomimetic black phosphorus quantum dots
BC	Bladder cancer
BPQDs	Black phosphorus quantum dots
CA 19–9	Carbohydrate antigen 19–9
CdTe@MPA QDs	CdTe QDs capped with mercaptopropionic acid
Ce6-RCDs	Ce6 loaded onto amino-rich red emissive CDs
CHA	Catalytic hairpin assembly
ChA-CQDs	CQDs-based biocompatible nanozyme prepared from chlorogenic acid
CPB QDs	Cesium lead bromide ($CsPbBr_3$) QDs
c-PSA	Complexed PSA
CQDs	Carbon quantum dots
CsPbBr3	Cesium lead bromide
CT	Computerized tomography
CTCs	Circulating tumor cells
DFO	Deferoxamine mesylate

DKK1	Dickkopf-1
DOX	Doxorubicin
FA	Folic acid
Fer-1	Ferrostatin-1
FGS	Fluorescence-guided surgery
f-PSA	Free PSA
FRET	Forster resonance energy transfer
GNR SiO2-CDs	Gold nanorod silica-CDs
GPC3	Glypican-3
GPX4	Glutathione peroxidase 4
GQDs	Graphene quantum dots
GQDs515	GQDs emitted 515 nm (green emitted)
GQDs615	GQDs emitted 615 nm (red emitted)
GSH	Glutathione
HCC	Hepatocellular carcinoma
HE4	Human epididymis protein 4
HFFs	Human foreskin fibroblasts
H-MnO_2/DOX/BPQDs	Black phosphorus quantum dots encapsulated biodegradable hollow mesoporous MnO_2 nanoprobe
HP	Hematoporphyrin
HP-CQDs	Hematoporphyrin photosensitizer-encapsulated carbon quantum dots
InP-QDs	Indium phosphide QDs
IR	Infra-red
MFBs	Magnetic fluorescence microbeads
MIP	Molecularly imprinted polymer
MIP@QDs	MIP around a red emitted cadmium telluride QDs
miRNA	MicroRNA
MPA	3-Mercaptopropionic acid
M-QD-SNs	Multi-QD-embedded silica-encapsulated nanoparticle
M–QD-SNs-CD63 Ab	Anti-CD63 antibodies immobilized on M-QD-SNs
MRI	Magnetic resonance imaging
mRNA	Messenger RNA
MUA	11 Mercaptoundecanoic acid
NC	Nitrocellulose
NGQDs	Nitrogen-doped GQDs
NIR	Near infra red
OVC	Ovarian cancer
PAA	Poly acrylic acid
PAH	Poly allylamine hydrochloride
PDA	Polydopamine
PEG	Polyethylene glycol
PET	Positron emission tomography

PL	Photoluminescence
PSA	Prostate specific antigen
QD - ILs	QDs conjugated immunoliposome system
QD585	A 585 nm CdSe/ZnS QDs
QD655	A 655 nm CdSe/ZnS QDs
QDs	Quantum dots
QDs450	CdSSe/ZnS QDs capable of emitting 450 nm
QDs620	CdSSe/ZnS QDs capable of emitting 620 nm
ROS	Reactive oxygen species
SMT	Suspension microarray technology
TDD	Targeted drug delivery
TdT	Terminal deoxynucleotidyl transferase
t-PSA	Total PSA
VEGF	Vascular endothelial growth factor
WHO	World Health Organization

References

1. G. A. Kwong, S. Ghosh, L. Gamboa, C. Patriotis, S. Srivastava and S. N. Bhatia, Synthetic biomarkers: a twenty-first century path to early cancer detection, *Nat. Rev. Cancer*, 2021, **21**(10), 655–668.
2. S. Ali, S. Fenerty and P. Jonnalagadda, Imaging Modalities in the Detection and Diagnosis of Metastatic Disease, in *Cancer Metastasis Through the Lymphovascular System*, Springer, 2022, pp. 283–293.
3. D. Albano, M. Benenati, A. Bruno, F. Bruno, M. Calandri, D. Caruso, D. Cozzi, R. De Robertis, F. Gentili and I. Grazzini, Imaging side effects and complications of chemotherapy and radiation therapy: a pictorial review from head to toe, *Insights into Imaging*, 2021, **12**(1), 1–28.
4. X. Kong, R. Cheng, J. Wang, Y. Fang and K. C. Hwang, Nanomedicines inhibiting tumor metastasis and recurrence and their clinical applications, *Nano Today*, 2021, **36**, 101004.
5. W. Sun, Y. Du, X. Liang, C. Yu, J. Fang, W. Lu, X. Guo, J. Tian, Y. Jin and J. Zheng, Synergistic triple-combination therapy with hyaluronic acid-shelled PPy/CPT nanoparticles results in tumor regression and prevents tumor recurrence and metastasis in 4T1 breast cancer, *Biomaterials*, 2019, **217**, 119264.
6. M. Zhao, D. Mi, B. E. Ferdows, Y. Li, R. Wang, J. Li, D. Patel, N. Kong, S. Shi and W. Tao, State-of-the-art nanotechnologies for the detection, recovery, analysis and elimination of liquid biopsy components in cancer, *Nano Today*, 2022, **42**, 101361.
7. Y. Yao, Y. Zhou, L. Liu, Y. Xu, Q. Chen, Y. Wang, S. Wu, Y. Deng, J. Zhang and A. Shao, *Front. Mol. Biosci.*, 2020, **7**, 193.

8. Y. Yao, Z. Chen, T. Zhang and M. Tang, Adverse reproductive and developmental consequences of quantum dots, *Environ. Res.*, 2022, 113666.
9. S. Campuzano, P. Yáñez-Sedeño and J. M. Pingarrón, Carbon dots and graphene quantum dots in electrochemical biosensing, *Nanomaterials*, 2019, **9**(4), 634.
10. J. Xiao, R. Momen and C. Liu, Application of carbon quantum dots in supercapacitors: A mini review, *Electrochem. Commun.*, 2021, **132**, 107143.
11. G. Chellasamy, S. K. Arumugasamy, S. Govindaraju and K. Yun, Green synthesized carbon quantum dots from maple tree leaves for biosensing of Cesium and electrocatalytic oxidation of glycerol, *Chemosphere*, 2022, **287**, 131915.
12. I. A. Revesz, S. M. Hickey and M. J. Sweetman, Metal ion sensing with graphene quantum dots: Detection of harmful contaminants and biorelevant species, *J. Mater. Chem. B*, 2022, **10**(23), 4346–4362.
13. X. Mei, D. Jia, J. Chen, S. Zheng and X. Zhang, Approaching high-performance light-emitting devices upon perovskite quantum dots: Advances and prospects, *Nano Today*, 2022, **43**, 101449.
14. C. Sanjayan, M. S. Jyothi and R. G. Balakrishna, Stabilization of CsPbBr 3 quantum dots for photocatalysis, imaging and optical sensing in water and biological medium: a review, *J. Mater. Chem. C*, 2022, **10**(18), 6935–6956.
15. M.-H. Jo, K.-H. Kim and H.-J. Ahn, P-doped carbon quantum dot graft-functionalized amorphous WO3 for stable and flexible electrochromic energy-storage devices, *Chem. Eng. J.*, 2022, **445**, 136826.
16. H. Lian, Y. Li, S. Saravanakumar, H. Jiang, Z. Li, J. Wang, L. Xu, W. Zhao and G. Han, Metal halide perovskite quantum dots for amphiprotic bio-imaging, *Coord. Chem. Rev.*, 2022, **452**, 214313.
17. H. Song, J. Wang, B. Xiong, J. Hu, P. Zeng, X. Liu and H. Liang, Biologically Safe, Versatile, and Smart Bismuthene Functionalized with a Drug Delivery System Based on Red Phosphorus Quantum Dots for Cancer Theranostics, *Angew. Chem.*, 2022, **134**(22), e202117679.
18. C. Huang, Y. Sun, Y. Zhao, J. Li, L. Qu, R. Yang and Z. Li, Visual Monitoring of Nucleic Acid Dynamic Structures during Cellular Ferroptosis Using Rationally Designed Carbon Dots with Robust Anti-Interference Ability to Reactive Oxygen Species, *ACS Appl. Bio Mater.*, 2022, **5**(6), 2703–2711.
19. B. Gidwani, V. Sahu, S. S. Shukla, R. Pandey, V. Joshi, V. K. Jain and A. Vyas, Quantum dots: Prospectives, toxicity, advances and applications, *J. Drug Delivery Sci. Technol.*, 2021, **61**, 102308.
20. P. Yadav, Mimansa, K. Kailasam and A. Shanavas, Nontoxic Metal-Free Visible Light-Responsive Carbon Nitride Quantum Dots Cause

Oxidative Stress and Cancer-Specific Membrane Damage, *ACS Appl. Bio. Mater.*, 2022, **5**(3), 1169–1178.

21. L. Ren, L. Wang, M. Rehberg, T. Stoeger, J. Zhang and S. Chen, Applications and Immunological Effects of Quantum Dots on Respiratory System, *Front. Immunol.*, 2021, **12**, 795232.
22. M. X. Zhao and E. Z. Zeng, Application of functional quantum dot nanoparticles as fluorescence probes in cell labeling and tumor diagnostic imaging, *Nanoscale Res. Lett.*, 2015, **10**, 171.
23. A. M. Derfus, W. C. W. Chan and S. N. Bhatia, Probing the Cytotoxicity of Semiconductor Quantum Dots, *Nano Lett.*, 2004, **4**(1), 11–18.
24. N. Zhao, X. Gui, Q. Fang, R. Zhang, W. Zhu, H. Zhang, Q. Li, Y. Zhou, J. Zhao and X. Cui, Graphene quantum dots rescue angiogenic retinopathy via blocking STAT3/Periostin/ERK signaling, *J. Nanobiotechnol.*, 2022, **20**(1), 1–15.
25. L. Yao, M.-M. Zhao, Q.-W. Luo, Y.-C. Zhang, T.-T. Liu, Z. Yang, M. Liao, P. Tu and K.-W. Zeng, Carbon Quantum Dots-Based Nanozyme from Coffee Induces Cancer Cell Ferroptosis to Activate Antitumor Immunity, *ACS Nano*, 2022, **16**(6), 9228–9239.
26. L. Chen, W. Hong, S. Duan, Y. Li, J. Wang and J. Zhu, Graphene quantum dots mediated magnetic chitosan drug delivery nanosystems for targeting synergistic photothermal-chemotherapy of hepatocellular carcinoma, *Cancer Biol. Ther.*, 2022, **23**(1), 281–293.
27. X. Yang, Q. Zhao, J. Chen, J. Liu, J. Lin, J. Lu, W. Li, D. Yu and W. Zhao, Graphene Oxide Quantum Dots Promote Osteogenic Differentiation of Stem Cells from Human Exfoliated Deciduous Teeth via the Wnt/β-Catenin Signaling Pathway, *Stem Cells Int.*, 2021, **2021**, 8876745.
28. S. Yang, X. Wang, P. He, A. Xu, G. Wang, J. Duan, Y. Shi and G. Ding, Graphene Quantum Dots with Pyrrole N and Pyridine N: Superior Reactive Oxygen Species Generation Efficiency for Metal-Free Sonodynamic Tumor Therapy, *Small*, 2021, **17**(10), 2004867.
29. J. Sobhanan, Y. Takano, S. Sugino, E. Hirata, S. Yamamura and V. Biju, Multimodal CTC detection using stem cell antigen-specific immunosilica particles and immunofluorescent quantum dots, *NPG Asia Mater.*, 2022, **14**(1), 1–8.
30. P. Zhang, M. S. Draz, A. Xiong, W. Yan, H. Han and W. Chen, Immunoengineered magnetic-quantum dot nanobead system for the isolation and detection of circulating tumor cells, *J. Nanobiotechnology*, 2021, **19**(1), 116.
31. G. Nifontova, F. Ramos-Gomes, M. Baryshnikova, F. Alves, I. Nabiev and A. Sukhanova, Cancer Cell Targeting With Functionalized Quantum Dot-Encoded Polyelectrolyte Microcapsules, *Front Chem.*, 2019, **7**, 34.

32. I. Ryu, J. Ryu, G. Choe, H. Kwon, H. Park, Y. Cho, R. Du and S. Yim, In Vivo Plain X-Ray Imaging of Cancer Using Perovskite Quantum Dot Scintillators, *Adv. Funct. Mater.*, 2021, **31**(34), 2102334.
33. L. Dirheimer, T. Pons, F. Marchal and L. Bezdetnaya, Quantum Dots Mediated Imaging and Phototherapy in Cancer Spheroid Models: State of the Art and Perspectives, *Pharmaceutics*, 2022, **14**(10), 2136.
34. J. Bugno, H.-J. Hsu, R. M. Pearson, H. Noh and S. Hong, Size and Surface Charge of Engineered Poly(amidoamine) Dendrimers Modulate Tumor Accumulation and Penetration: A Model Study Using Multicellular Tumor Spheroids, *Mol. Pharm.*, 2016, **13**(7), 2155–2163.
35. L. Shi, J. Zhang, M. Zhao, S. Tang, X. Cheng, W. Zhang, W. Li, X. Liu, H. Peng and Q. Wang, Effects of polyethylene glycol on the surface of nanoparticles for targeted drug delivery, *Nanoscale*, 2021, **13**(24), 10748–10764.
36. R. Mohammadi, H. Naderi-Manesh, L. Farzin, Z. Vaezi, N. Ayarri, L. Samandari and M. Shamsipur, Fluorescence sensing and imaging with carbon-based quantum dots for early diagnosis of cancer: A review, *J. Pharm. Biomed. Anal.*, 2022, 114628.
37. M. Borovaya, I. Horiunova, S. Plokhovska, N. Pushkarova, Y. Blume and A. Yemets, Synthesis, properties and bioimaging applications of silver-based quantum dots, *Int. J. Mol. Sci.*, 2021, **22**(22), 12202.
38. T. Bai, X. Shi, M. Liu, H. Huang, J. Zhang and X.-H. Bu, gC 3 N 4/ZnCdS heterojunction for efficient visible light-driven photocatalytic hydrogen production, *RSC Adv.*, 2021, **11**(60), 38120–38125.
39. L. Zhu, D. Shen, C. Wu and S. Gu, State-of-the-art on the preparation, modification, and application of biomass-derived carbon quantum dots, *Ind. Eng. Chem. Res.*, 2020, **59**(51), 22017–22039.
40. K. Hagiwara, S. Horikoshi and N. Serpone, Photoluminescent Carbon Quantum Dots: Synthetic Approaches and Photophysical Properties, *Chem. Eur. J.*, 2021, **27**(37), 9466–9481.
41. D. V. Talapin and J. Steckel, Quantum dot light-emitting devices, *MRS Bull.*, 2013, **38**(9), 685–691.
42. S. Chinnathambi and N. Shirahata, Recent advances on fluorescent biomarkers of near-infrared quantum dots for in vitro and in vivo imaging, *Sci. Technol. Adv. Mater.*, 2019, **20**(1), 337–355.
43. O. World Health, *International Programme on Chemical S. Biomarkers in risk assessment: validity and validation*, World Health Organization, Geneva, 2001.
44. M. R. Hasan, P. Sharma, R. Pilloton, M. Khanuja and J. Narang, Colorimetric biosensor for the naked-eye detection of ovarian cancer biomarker PDGF using citrate modified gold nanoparticles, *Biosensors. and Bioelectronics.: X*, 2022, **11**, 100142.

45. S. I. Kaya, G. Ozcelikay, F. Mollarasouli, N. K. Bakirhan and S. A. Ozkan, Recent achievements and challenges on nanomaterial based electrochemical biosensors for the detection of colon and lung cancer biomarkers, *Sens. Actuators, B*, 2022, **351**, 130856.
46. A. M. L. Piloto, D. S. M. Ribeiro, S. S. M. Rodrigues, J. L. M. Santos, P. Sampaio and M. G. F. Sales, Cellulose-based hydrogel on quantum dots with molecularly imprinted polymers for the detection of CA19-9 protein cancer biomarker, *Microchim. Acta*, 2022, **189**(4), 1–13.
47. Y. Zhang, Y. Lv, L. Li, X. Zhao, M. Zhao and H. Shen, Aminophosphate-precursors for the synthesis of near-unity emitting InP quantum dots and theirapplication in liver cancer diagnosis, *Exploration, Wiley Online Library*, 2022, 20220082.
48. D. Bahari, B. Babamiri, A. Salimi and A. Rashidi, Graphdiyne/graphene quantum dots for development of FRET ratiometric fluorescent assay toward sensitive detection of miRNA in human serum and bioimaging of living cancer cells,*J. Lumin.*, 2021, **239**, 118371.
49. D. Kurniawan, R.-C. Jhang, K. K. Ostrikov and W.-H. Chiang, Microplasma-tunable graphene quantum dots for ultrasensitive and selective detection of cancer and neurotransmitter biomarkers, *ACS Appl. Mater. Interfaces*, 2021, **13**(29), 34572–34583.
50. G. Bharathi, F. Lin, L. Liu, T. Y. Ohulchanskyy, R. Hu and J. Qu, An all-graphene quantum dot Förster resonance energy transfer (FRET) probe for ratiometric detection of HE4 ovarian cancer biomarker, *Colloids Surf., B*, 2021, **198**, 111458.
51. V. Vinduska, C. E. Gallops, R. O'Connor, Y. Wang and X. Huang, Exosomal Surface Protein Detection with Quantum Dots and Immunomagnetic Capture for Cancer Detection, *Nanomaterials*, 2021, **11**(7), 1853.
52. H.-M. Kim, C. Oh, J. An, S. Baek, S. Bock, J. Kim, H.-S. Jung, H. Song, J.-W. Kim and A. Jo, Multi-quantum dots-embedded silica-encapsulated nanoparticle-based lateral flow assay for highly sensitive exosome detection, *Nanomaterials*, 2021, **11**(3), 768.
53. M. Sun, T. Wang, L. Li, X. Li, Y. Zhai, J. Zhang and W. Li, The application of inorganic nanoparticles in molecular targeted cancer therapy: EGFR targeting, *Front. Pharmacol.*, 2021, **12**, 702445.
54. S. Kundu, M. Ghosh and N. Sarkar, State of the Art and Perspectives on the Biofunctionalization of Fluorescent Metal Nanoclusters and Carbon Quantum Dots for Targeted Imaging and Drug Delivery, *Langmuir*, 2021, **37**(31), 9281–9301.
55. P. Khetrapal, M. W. L. Lee, W. S. Tan, L. Dong, P. de Winter, A. Feber and J. D. Kelly, The role of circulating tumour cells and nucleic acids in blood for the detection of bladder cancer: A systematic review, *Cancer Treat. Rev.*, 2018, **66**, 56–63.

56. J. Xu, X. Wei, X. Zhang, Z. Cai, Y. Wei, W. Liu, J. Yang, Y. Li, X. Cai and T. Bai, Multiplexed detection of bladder cancer microRNAs based on core-shell-shell magnetic quantum dot microbeads and cascade signal amplification, *Sens. Actuators, B*, 2021, **349**, 130824.
57. J. Yu, X. Wu, J. Yan, H. Yu, L. Xu, Z. Chi, X. Sheng, L. Si, C. Cui, J. Dai, M. Ma, T. Xu, Y. Kong and J. Guo, Anti-GD2/4-1BB chimeric antigen receptor T cell therapy for the treatment of Chinese melanoma patients, *J. Hematol. Oncol.*, 2018, **11**(1), 1.
58. L. Keller and K. Pantel, Unravelling tumour heterogeneity by single-cell profiling of circulating tumour cells, *Nat. Rev. Cancer*, 2019, **19**(10), 553–567.
59. N. Cheng and J. Fu, An Approach to the Simultaneous Detection of Multiple Biomarkers for the Early Diagnosis of Liver Cancer Using Quantum Dot Nanoprobes, *Infect. Microbes Dis.*, 2022, **4**(1), 34–40.
60. X. Min, S. Huang and C. Yuan, Dual-color quantum dots nanobeads based suspension microarray for simultaneous detection of dual prostate specific antigens, *Anal. Chim. Acta*, 2022, **1204**, 339704.
61. X. Sang, Q. Yang, G. Shi, L. Zhang, D. Wang and C. Ni, Preparation of pH/redox dual responsive polymeric micelles with enhanced stability and drug controlled release, *Mater. Sci. Eng. C*, 2018, **91**, 727–733.
62. F. Nasrollahi, Y. R. Koh, P. Chen, J. Varshosaz, A. A. Khodadadi and S. Lim, Targeting graphene quantum dots to epidermal growth factor receptor for delivery of cisplatin and cellular imaging, *Mater. Sci. Eng. C*, 2019, **94**, 247–257.
63. Y. H. Bae and K. Park, Targeted drug delivery to tumors: Myths, reality and possibility, *J. Controlled Release*, 2011, **153**(3), 198–205.
64. V. V. Veselov, A. E. Nosyrev, L. Jicsinszky, R. N. Alyautdin and G. Cravotto, Targeted Delivery Methods for Anticancer Drugs, *Cancers*, 2022, **14**(3), 622.
65. H. Y. Wang, X. W. Hua, H. R. Jia, P. Liu, N. Gu, Z. Chen and F. G. Wu, Enhanced cell membrane enrichment and subsequent cellular internalization of quantum dots via cell surface engineering: illuminating plasma membranes with quantum dots, *J. Mater. Chem. B*, 2016, **4**(5), 831–843.
66. Y. Zhang, B. Liu, Z. Liu and J. Li, Research progress in the synthesis and biological application of quantum dots, *New J. Chem.*, 2022, **46**(43), 20515–20539.
67. S. Zununi Vahed, N. Fathi, M. Samiei, S. Maleki Dizaj and S. Sharifi, Targeted cancer drug delivery with aptamer-functionalized polymeric nanoparticles, *J. Drug Target.*, 2019, **27**(3), 292–299.
68. A. Farasat, F. Rahbarizadeh, D. Ahmadvand, S. Ranjbar and S. Khoshtinat Nikkhoi, Effective suppression of tumour cells by oligoclonal

HER2-targeted delivery of liposomal doxorubicin, *J. Liposome Res.*, 2019, **29**(1), 53–65.

69. H. N. Abdelhamid, H. M. El-Bery, A. A. Metwally, M. Elshazly and R. M. Hathout, Synthesis of CdS-modified chitosan quantum dots for the drug delivery of Sesamol, *Carbohydr. Polym.*, 2019, **214**, 90–99.
70. J. Kim, B. T. Huy, K. Sakthivel, H. J. Choi, W. H. Joo, S. K. Shin, M. J. Lee and Y.-I. Lee, Highly fluorescent CdTe quantum dots with reduced cytotoxicity-A Robust biomarker, *Sens. Bio-Sens. Res.*, 2015, **3**, 46–52.
71. M. Frieler, C. Pho, B. H. Lee, H. Dobrovolny, G. R. Akkaraju and A. V. Naumov, Effects of doxorubicin delivery by nitrogen-doped graphene quantum dots on cancer cell growth: experimental study and mathematical modeling, *Nanomaterials*, 2021, **11**(1), 140.
72. V. K. Bajpai, I. Khan, S. Shukla, S.-M. Kang, F. Aziz, K. M. Tripathi, D. Saini, H.-J. Cho, N. S. Heo and S. K. Sonkar, Multifunctional N-P-doped carbon dots for regulation of apoptosis and autophagy in B16F10 melanoma cancer cells and in vitro imaging applications, *Theranostics*, 2022, **10**(17), 7841.
73. R. Jan and G.-E.-S. Chaudhry, Understanding Apoptosis and Apoptotic Pathways Targeted Cancer Therapeutics, *Adv. Pharm. Bull.*, 2019, **9**(2), 205–218.
74. V. Davenport, C. Horstmann, R. Patel, Q. Wu and K. Kim, An Assessment of InP/ZnS as Potential Anti-Cancer Therapy: Quantum Dot Treatment Increases Apoptosis in HeLa Cells, *J. Nanotheranostics*, 2021, **2**(1), 16–32.
75. S. Şimşek, A. A. Şüküroğlu, D. Yetkin, B. Özbek, D. Battal and R. Genç, DNA-damage and cell cycle arrest initiated anti-cancer potency of super tiny carbon dots on MCF7 cell line, *Sci. Rep.*, 2020, **10**(1), 13880.
76. S. Kalytchuk, K. Poláková, Y. Wang, J. P. Froning, K. Cepe, A. L. Rogach and R. Zbořil, Carbon Dot Nanothermometry: Intracellular Photoluminescence Lifetime Thermal Sensing, *ACS Nano*, 2017, **11**(2), 1432–1442.
77. S. A. Hill, S. Sheikh, Q. Zhang, L. Sueiro Ballesteros, A. Herman, S. A. Davis, D. J. Morgan, M. Berry, D. Benito-Alifonso and M. C. Galan, Selective photothermal killing of cancer cells using LED-activated nucleus targeting fluorescent carbon dots, *Nanoscale Adv.*, 2019, **1**(8), 2840–2846.
78. C.-L. Li, C.-M. Ou, C.-C. Huang, W.-C. Wu, Y.-P. Chen, T.-E. Lin, L.-C. Ho, C.-W. Wang, C.-C. Shih, H.-C. Zhou, Y.-C. Lee, W.-F. Tzeng, T.-J. Chiou, S.-T. Chu, J. Cang and H.-T. Chang, Carbon dots prepared from ginger exhibiting efficient inhibition of human hepatocellular carcinoma cells,*J. Mater. Chem. B*, 2014, **2**(28), 4564–4571.
79. F. A. A. Manan, N. A. Yusof, J. Abdullah, F. Mohammad, A. Nurdin, L. S. Yazan, S. K. Khiste and H. A. Al-Lohedan, Drug release profiles of

mitomycin c encapsulated quantum dots–chitosan nanocarrier system for the possible treatment of non-muscle invasive bladder cancer, *Pharmaceutics*, 2021, **13**(9), 1379.

80. M. Ruzycka-Ayoush, P. Kowalik, A. Kowalczyk, P. Bujak, A. M. Nowicka, M. Wojewodzka, M. Kruszewski and I. P. Grudzinski, Quantum dots as targeted doxorubicin drug delivery nanosystems in human lung cancer cells, *Cancer Nano.*, 2021, **12**(1), 1–27.
81. K. C. Weng, C. O. Noble, B. Papahadjopoulos-Sternberg, F. F. Chen, D. C. Drummond, D. B. Kirpotin, D. Wang, Y. K. Hom, B. Hann and J. W. Park, Targeted tumor cell internalization and Imaging of multifunctional quantum dot-conjugated immunoliposomes in vitro and in vivo, *Nano Lett.*, 2008, **8**(9), 2851–2857.
82. I. Fatima, A. Rahdar, S. Sargazi, M. Barani, M. Hassanisaadi and V. K. Thakur, Quantum dots: Synthesis, antibody conjugation, and HER2-receptor targeting for breast cancer therapy, *J. Funct. Biomater.*, 2021, **12**(4), 75.
83. N. Ghanbari, Z. Salehi, A. A. Khodadadi, M. A. Shokrgozar and A. A. Saboury, Glucosamine-conjugated graphene quantum dots as versatile and pH-sensitive nanocarriers for enhanced delivery of curcumin targeting to breast cancer, *Mater. Sci. Eng. C*, 2021, **121**, 111809.
84. Q. Jia, J. Ge, W. Liu, S. Liu, G. Niu, L. Guo, H. Zhang and P. Wang, Gold nanorod@silica-carbon dots as multifunctional phototheranostics for fluorescence and photoacoustic imaging-guided synergistic photodynamic/photothermal therapy, *Nanoscale*, 2016, **8**(26), 13067–13077.
85. G. Murali, B. Kwon, H. Kang, J. K. R. Modigunta, S. Park, S. Lee, H. Lee, Y. H. Park, J. Kim and S. Y. Park, Hematoporphyrin Photosensitizer-Linked Carbon Quantum Dots for Photodynamic Therapy of Cancer Cells, *ACS Appl. Nano Mater.*, 2022, **5**(3), 4376–4385.
86. G. A. Ortega, S. Del Sol-Fernández, Y. Portilla, E. Cedeño, E. Reguera, S. Srinivasan, D. F. Barber, E. Marin and A. R. Rajabzadeh, Rodlike Particles of Polydopamine-CdTe Quantum Dots: An Actuator As a Photothermal Agent and Reactive Oxygen Species-Generating Nanoplatform for Cancer Therapy, *ACS Appl Mater Interfaces*, 2021, **13**(36), 42357–42369.
87. M. Gao, H. Zhao, Z. Wang, Y. Zhao, X. Zou and L. Sun, Controllable preparation of Ag2S quantum dots with size-dependent fluorescence and cancer photothermal therapy, *Adv. Powder Technol.*, 2021, **32**(6), 1972–1982.
88. P. Zhao, Y. Xu, W. Ji, S. Zhou, L. Li, L. Qiu, Z. Qian, X. Wang and H. Zhang, Biomimetic black phosphorus quantum dots-based photothermal therapy combined with anti-PD-L1 treatment inhibits recurrence and metastasis in triple-negative breast cancer, *J. Nanobiotechnol.*, 2021, **19**(1), 1–18.

89. X. Wang, C. Hu, Z. Gu and L. Dai, Understanding of catalytic ROS generation from defect-rich graphene quantum-dots for therapeutic effects in tumor microenvironment, *J. Nanobiotechnol.*, 2021, **19**(1), 1–12.
90. P. Yadav, Mimansa, K. Kailasam and A. Shanavas, Nontoxic Metal-Free Visible Light-Responsive Carbon Nitride Quantum Dots Cause Oxidative Stress and Cancer-Specific Membrane Damage, *ACS Appl. Bio Mater.*, 2022, **5**(3), 1169–1178.
91. S. Ding, Z. Liu, C. Huang, N. Zeng, W. Jiang and Q. Li, Novel engineered bacterium/black phosphorus quantum dot hybrid system for hypoxic tumor targeting and efficient photodynamic therapy, *ACS Appl. Mater. Interfaces*, 2021, **13**(8), 10564–10573.
92. Y. Wu, Z. Chen, Z. Yao, K. Zhao, F. Shao, J. Su and S. Liu, Black Phosphorus Quantum Dots Encapsulated Biodegradable Hollow Mesoporous MnO2: Dual-Modality Cancer Imaging and Synergistic Chemo-Phototherapy, *Adv. Funct. Mater.*, 2021, **31**(41), 2104643.
93. R. F. Zaarour, B. Azakir, E. Y. Hajam, H. Nawafleh, N. A. Zeinelabdin, A. S. T. Engelsen, J. Thiery, C. Jamora and S. Chouaib, Role of hypoxia-mediated autophagy in tumor cell death and survival, *Cancers*, 2021, **13**(3), 533.
94. J. Chen, Z. Wei, K. Fu, Y. Duan, M. Zhang, K. Li, T. Guo and R. Yin, Non-apoptotic cell death in ovarian cancer: Treatment, resistance and prognosis, *Biomed. Pharmacother.*, 2022, **150**, 112929.
95. F. D. Duman, Y. Akkoc, G. Demirci, N. Bavili, A. Kiraz, D. Gozuacik and H. Y. Acar, Bypassing pro-survival and resistance mechanisms of autophagy in EGFR-positive lung cancer cells by targeted delivery of 5FU using theranostic Ag 2 S quantum dots, *J. Mater. Chem. B*, 2019, **7**(46), 7363–7376.
96. Y. Xie, B. Wan, Y. Yang, X. Cui, Y. Xin and L.-H. Guo, Cytotoxicity and autophagy induction by graphene quantum dots with different functional groups, *J. Environ. Sci.*, 2019, **77**, 198–209.
97. Y.-H. Luo, S.-B. Wu, Y.-H. Wei, Y.-C. Chen, M.-H. Tsai, C.-C. Ho, S.-Y. Lin, C.-S. Yang and P. Lin, Cadmium-based quantum dot induced autophagy formation for cell survival via oxidative stress, *Chem. Res. Toxicol.*, 2013, **26**(5), 662–673.
98. P. Yang, S. Ke, L. Tu, Y. Wang, S. Ye, S. Kou and L. Ren, Regulation of autophagy orchestrates pyroptotic cell death in molybdenum disulfide quantum Dot-induced microglial toxicity, *ACS Biomater. Sci. Eng.*, 2020, **6**(3), 1764–1775.
99. K. Peynshaert, S. J. Soenen, B. B. Manshian, S. H. Doak, K. Braeckmans, S. C. De Smedt and K. Remaut, Coating of quantum dots strongly defines their effect on lysosomal health and autophagy, *Acta Biomater.*, 2017, **48**, 195–205.

100. T. Zuo, T. Fang, J. Zhang, J. Yang, R. Xu, Z. Wang, H. Deng and Q. Shen, pH-Sensitive Molecular-Switch-Containing Polymer Nanoparticle for Breast Cancer Therapy with Ferritinophagy-Cascade Ferroptosis and Tumor Immune Activation, *Adv. Healthc. Mater.*, 2021, **10**(21), e2100683.
101. K. Totsuka, T. Ueta, T. Uchida, M. F. Roggia, S. Nakagawa, D. G. Vavvas, M. Honjo and M. Aihara, Oxidative stress induces ferroptotic cell death in retinal pigment epithelial cells, *Exp. Eye Res.*, 2019, **181**, 316–324.
102. G. C. Forcina and S. J. Dixon, GPX4 at the crossroads of lipid homeostasis and ferroptosis, *Proteomics*, 2019, **19**(18), 1800311.
103. T. Wu, X. Liang, X. Liu, Y. Li, Y. Wang, L. Kong and M. Tang, Induction of ferroptosis in response to graphene quantum dots through mitochondrial oxidative stress in microglia, *Part. Fibre Toxicol.*, 2020, **17**(1), 1–19.

CHAPTER 5

Quantum Dots in Viral and Bacterial Detection

Z. RAMEZANI[a] AND M. THOMPSON*[b]

[a]Toxicology Research Center, Medical Basic Sciences Research Institute, Ahvaz Jundishapur University of Medical Sciences, Ahvaz, Iran; [b]Chemistry Department, University of Toronto, Toronto, Canada
*E-mail: mikethom@chem.utoronto.ca; m.thompson@utoronto.ca

5.1 Introduction

Viruses and bacteria that cause infectious diseases are responsible for millions of deaths worldwide. They are transmitted *via* contaminated air, food, and water, as well as by person-to-person contact. Some of their most common infectious diseases are lower respiratory infections, hepatitis, diarrhea, influenza, tuberculosis, pneumonia, and human immunodeficiency virus. Viruses have sizes in the range of approximately 50 to 200 nm, while bacteria are approximately 0.5 to 10 µm in size; both can only be seen by microscopic techniques.[1] Recently, the severe acute respiratory syndrome coronavirus 2 (SARS-CoV-2) outbreak caused severe lower respiratory tract infections and acute respiratory distress syndromes; the virus's mortality rate was high, and it significantly affected human life, the environment, and the economy. The COVID-19 pandemic's impacts underscored the importance of developing new fast, easy, and reliable viral detection methods. The detection of viruses and bacteria in body fluids can help to prevent outbreaks, and it can facilitate the implementation of suitable treatment strategies for infectious diseases.[2–5] It is also necessary to

Detection Science Series No. 22
Quantum Dots in Bioanalytical Chemistry and Medicine
Edited by Michael Thompson and Zahra Ramezani

Published by the Royal Society of Chemistry, www.rsc.org

take viral and bacterial infections apart for accurate and on-time treatments. Various strategies have been introduced to detect and distinguish bacterial and viral infections at early stages.

Microorganisms and their biomarkers have been detected and quantified in body fluids using a variety of methods. Nucleic acid amplification tests (NAATs) are the gold standard for viral detection[6]; several multi-target assays based on this technique are commercially available. As for bacteria, the gold standard is the traditional culture and counting method. This method involves pre-concentration, selective isolation, and biochemical and serological identifications.[7,8] Although the culture and counting method is cost-effective and highly reliable, it is not suitable for fast and early bacterial detection because it takes 3 to 4 days to complete and has a detection limit of more than 10^4 CFU mL^{-1}. NAAT is sensitive and much faster than traditional culture methods for bacteria and virological tests, as the results are ready on the same day. Despite these advantages, several issues have limited NAAT's routine application: nucleic acid extraction is required, the signal must be amplified, and the reagents and equipment are expensive relative to culture methods. Scientists have thus been inspired to look for reliable alternatives. Immunoassay techniques such as immunochromatography and enzyme-linked immunosorbent assay (ELISA) are fast, highly sensitive, and specific. The only problem is that antibody preparation takes time and is expensive.[8] Polymerase chain reactions (PCRs) are highly sensitive, specific, fast, and simple. The DNA or RNA of viruses from the samples can be detected *via* the quantitative method of real-time reverse transcriptase polymerase chain reaction (RT-qPCR). RT-qPCR is sensitive but has time-consuming pre-processing and DNA/RNA extraction steps. PCR techniques require skilled operators and expensive equipment, and point-of-care (POC) applications are not possible. Figure 5.1 classifies various serological tests for detecting viral and bacterial infections. The figure indicates that nucleic acid-based tests are considered as direct bacterial or viral detection methods. It also classifies nanoparticles used for this purpose.

Nanotechnology for clinical laboratory diagnostics has the advantages of enhanced sensitivity and flexibility, reduced sample sizes (as little as a few volumes), the ability to detect and quantify based on the presence of only a few molecules or microbes, and reduced assay time and treatment costs. Various nano-based sensors have been prepared for the detection and quantification of microorganisms.[9–12] Biosensors are fast, sensitive, and simple detection methods that can easily be made in small sizes for POC uses. Recently, quantum dots (QDs), a sub-class of nanomaterials, have been used in diagnostic device design and drug and gene delivery for theragnostic purposes.[13–16] A viral infection begins when a virus enters the cell. Many studies have been conducted to evaluate methods for deactivating viruses to prevent them from entering the cell; QDs have recently been studied for their potential anti-bacterial effects.[17,18]

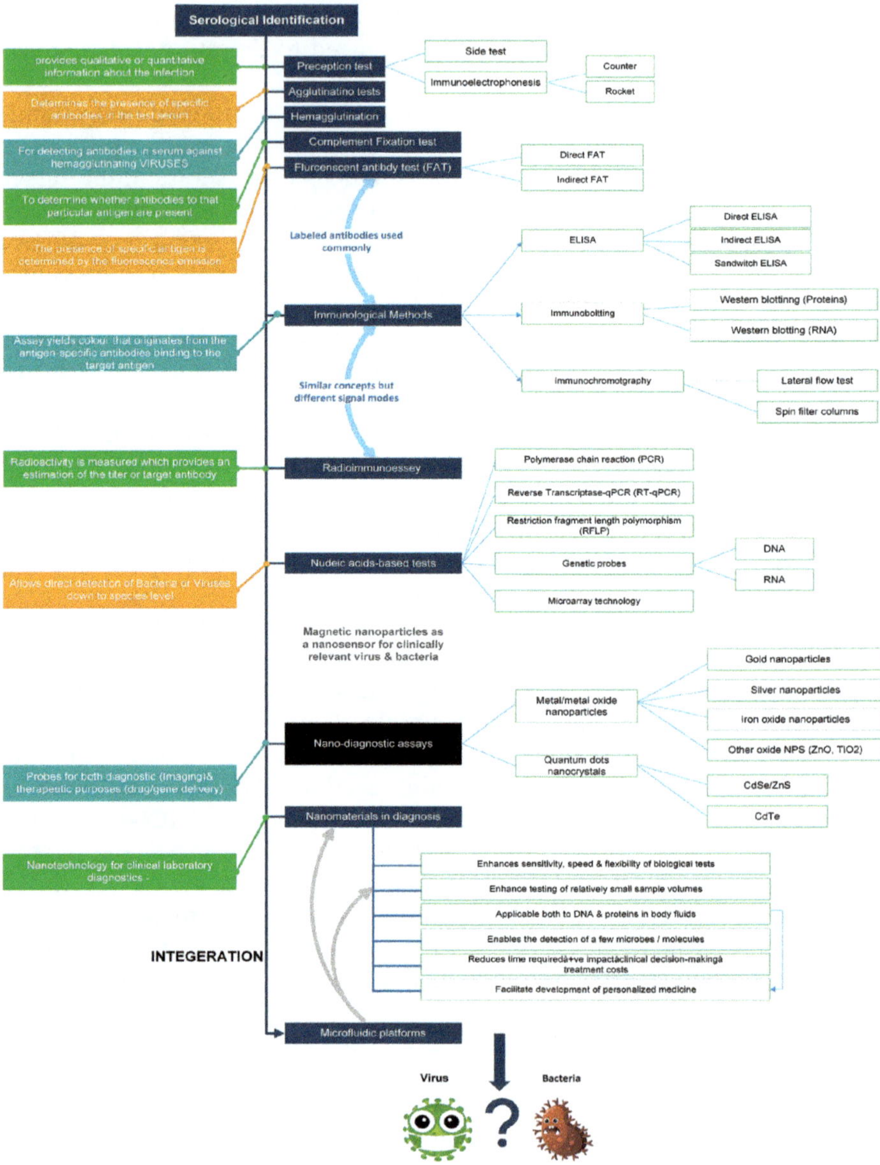

Figure 5.1 Serological methods used conventionally for the diagnosis of bacterial or viral infections and current trends involving the integration of microfluidics and nanomaterials for point-of-care for disease diagnosis. Reproduced from ref. 1 with permission from the Royal Society of Chemistry.

QDs are highly bright luminescent materials with extraordinary optical and physicochemical properties. They are an excellent alternative to organic fluorophores in various fields. QDs' biological applications have been studied since 1998.[19] Since their introduction, tremendous progress has been made in terms of QDs' sourcing, synthetic methods, and modifications for targeted biological applications with lower toxicity. QDs with high quantum yield and photostability have been used in pathogen imaging.[18,20–22] QDs may be conjugated with biomolecules such as Ab, DNA, RNA, aptamers, peptides, and small molecules to produce specific interactions with pathogens. Currently, interactions between QDs and microorganisms such as bacteria and viruses represent a subject of intense study; this interest is driven especially by certain unique properties of QDs that, in some cases, make them capable of producing bactericidal and antiviral effects.[13,14,18] Carbon-based QDs (CQDs) and metal-semiconductor QDs are widely used to produce interactions with microorganisms.

This chapter provides a historical overview of QD-based bacterial and viral detection and assay strategies. Recent work in this field has focused on the use of QDs for detecting and interacting with viruses and even tracking viruses in live cells.[2,3,23–29] We list all currently available QD-based detection and biosensing strategies and evaluate them according to speed, accuracy, sensitivity and selectivity, the possibility of state-of-care uses, cost-effectiveness, and environmental friendliness.

5.2 QDs in Microorganism Imaging

Organic dyes were originally used as fluorescence labels in cell imaging to detect pathogenic microorganisms.[30] Since the invention of nanoparticles, fluorescent nanoparticles have attracted researchers' attention as a reliable alternative to these dyes. Fluorescence QDs as a class of nanoparticles (*i.e.*, particles below 10 nm in size) are preferred as labels and amplifiers due to their outstanding features. They have extraordinary photostability, high quantum yield, size-dependent and narrow emission, and broad excitation spectra; these qualities make them promising fluorescence labels for use in multicolor *in vitro* and *in vivo* imaging of bacteria and viruses, as well as in single virus tracking (SVT) studies. Compared with organic dyes and fluorescence protein (FP), they provide more detailed information in a shorter time as the signal-to-noise ratio is high due to their high brightness and outstanding photostability. Semiconductor QDs, Si-based QDs, and CQDs are used for bacterial and viral labeling.[30] CQDs have drawn recent attention for their use as fluorescence labels to identify viruses and bacteria *via* fluorescence image evaluations (see Table 5.1). Compared to semiconductor-based QDs, CQDs are more biocompatible and environmentally friendly. Another important advantage of CQDs is that they can be prepared using simple procedures with inexpensive and easily available carbon sources.[43,52–54]

Table 5.1 Application of QDs in bacteria and viruses' fluorescence imaging.

QDs	Virus/Bacteria	Application	Ref.
QD-containing virus-like particles	Simian virus 40 (SV40)	Tracing viruses in living cell	31
CdSe/ZnS QD-labeled SA	JEV SA-14-14-2 and WT JEV AT31	SVT	32
NIR SA-QDs	Influenza H5N1p	*In vivo* dynamic tracking	33
NIR-QDs	H5N1p	*In vivo* virus tracking	34
SA-QDs	Influenza A virus (H9N2)	*In vitro* virus tracking	35
SA-QDs	Pseudorabies virus	*In vitro* virus tracking	36
QDM_S	SARS-COV-2	*In vitro* detection	37
PbS QDs	*S. aureus*, *S. epidermidis*, *E. coli*, and *S. anginosus*	*In vitro* and *in vivo* bacterial infection	38
ZnCuInSe QDs	*S. aureus*	Bacterial discrimination and imaging	39
CdSe QDs	*E. coli*	Imaging	40
CQDs	*S. aureus*	Fluorescence imaging	22
Streptavidin-coated QDs	*E. coli* mutants	*In vivo* imaging	41
Ab-coated QSH620	*E. coli* O157:H7	Optical (fluorescence microscope images)	42
CQDs from two carbon sources	*E. coli*	Optical (fluorescence microscope images)	43
CQDs	*Pseudomonas aeruginosa*	Confocal fluorescence microscope	44
FC3N4 QDs	*E. coli*	Confocal fluorescence microscope	45
W-CQDs	*E. coli*	Confocal fluorescence microscope	46
Tris-CQDs	*E. coli*	Confocal fluorescence microscope	47
CQDs	*E. coli*; *S. aureus*	Confocal fluorescence microscope	48
CQDs	*E. coli*; *S. aureus*	Optical (fluorescence microscope images)	49
CQDs-605 from L. Plantarum LLC-605 as carbon source	*E. coli*, *S. oneidensis*, *P. aeruginosa*, *S. aures*, and *T. reesei*	Biofilm confocal fluorescence microscope imaging	50
Ab-CQDs	Methicillin-resistant *S. aureus* and *Salmonella enterica*	Multicolor fluorescence imaging	51

5.2.1 Virus Labeling and Tracking

Infection patterns differ between viruses. They depend mainly on viruses' major sites of replication and damage. SVT can clarify researchers' understanding of the mechanisms of viral infection by providing images of a virus during its attachment and entry to the cell; it can even be used to observe viral replication in live cells. Originally, viruses were labeled with fluorophores such as cyanine5 or with FPs such as green fluorescent protein. Recently, QDs with distinct optical properties have become a reliable alternative to these older labels for long-term tracking of single viruses with high sensitivity (see Table 5.1). For SVT, viruses are conjugated with QDs. Various conjugation methods are available, such as direct chemical reactions, biotin-streptavidin interactions, peptide-tag mediating labeling, oligonucleotide-guided labeling, and other labeling strategies (see Figure 5.2).[55,56] Generally speaking, viruses have a complex structure and consist of different parts. These parts include DNA and RNA, which carry genetic information; a protein coat called a capsid that protects the genetic materials; and, in some viruses, a lipid envelope that surrounds the capsid. QDs can be attached to the surface of a virus or to its interior parts. QD labeling strategies for virus tracking are often categorized into three major groups depending on the labeling site of the virus: (1) internal components; (2) external components; and (3) other components. For SVT studies, the virus activity should be maintained regardless of the conjugation strategies.

In the external method, QDs bind to viruses' external components. In this process, QDs are conjugated to biomolecules such as streptavidin to facilitate attachment. Joo *et al.* successfully tagged retroviruses with QDs through the membrane incorporation of a short acceptor peptide (AP) that was susceptible to site-specific biotinylation for further attachment of streptavidin-conjugated QDs.[57] Lentiviruses can be tracked in live cells using conjugated QDs[35] with two-step biotin-streptavidin labeling. The labeled viruses possess high efficacy and could track infections in the live cell. Direct chemical labeling is another approach to external component conjugation. Amine-carboxyl crosslinking reactions[30,58] and click chemistry[59,60] provide easy and efficient conjugation strategies for viruses' external conjugation to QDs for SVT.

Proteins on viruses' surfaces are modified to conjugate with the functionalized QDs to increase their cellular intake and obtain information on their infection mechanisms. However, viral surface modifications may reduce the efficiency of viral entry or alter virus–cell interactions, which would render it impossible to observe the infection mechanism.[61] In contrast, internal virus component conjugation with QDs is considered a reliable method. QDs' encapsulation in the viral capsids is a reliable method for maintaining the viral activity in the live cell for those viruses that have a lipid envelope. Another option is the coupling of QDs with viruses' RNA and DNA.[62,63]

Some viruses do not have complex structures but may contain structures that enable other labeling strategies. For example, viruses made of histidine (His)-rich proteins have binding sites for divalent metal ions such as Ni^{2+}. Luo and co-workers[64] introduced polyethylene glycol (PEG) interspersed with nitroacetic acid (NTC)-modified Ni^{2+} QDs for the specific labeling of the His-rich protein of the virus-like prion for its tracking in the live cells.

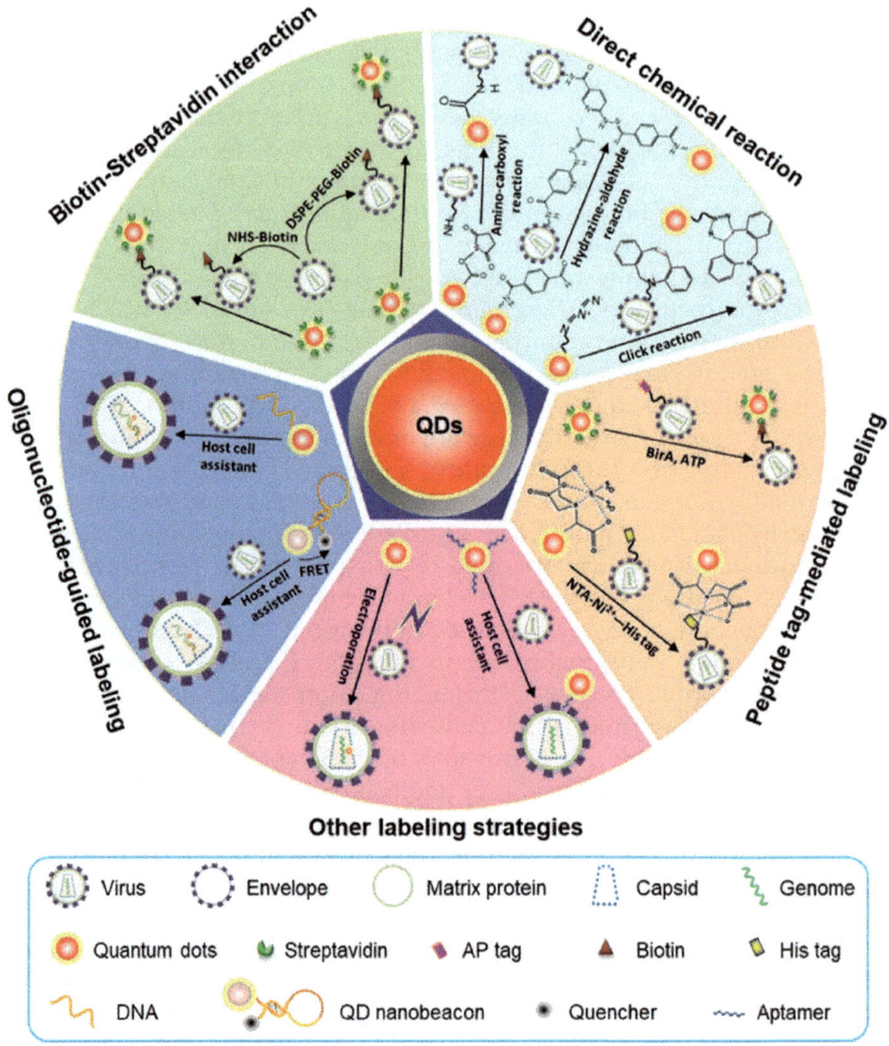

Figure 5.2 Different strategies for QDs labeling of viruses in single virus tracking. Reproduced from ref. 55 from American Chemical Society, Copyright 2021.

The conjugated QDs attached specifically to the virus and did not alter the virus's structure or function.

Table 5.1 lists QD-based fluorescence imaging methods for bacteria and viruses. Most methods use commercially available conjugated QDs. Both semiconductors and CQDs with high quantum yields and bright and stable emission have been utilized. Near-infrared (NIR) emission performs better for *in vivo* applications because it eliminates interference from complex body tissue emissions. As Table 5.1 indicates, a large number of reports are available on applications for nanomaterials, whereas relatively little research has been conducted on QDs' applications in viral imaging and SVT. More attention should be paid to QD research in this field because QDs are smaller and have a high specific surface area, which facilitates coupling to the virus with a lesser impact on viral activity.

5.2.2 Bacteria Labeling

Despite the development of many new generations of antibiotics, bacterial infection remains a threat to human health with high mortality rates. Sensitive and accurate early-stage detection of bacterial infections *in vivo* can facilitate rapid treatment and, thus, prevent disease progression and irreversible damage. *In vitro* studies such as specimen culture, PCR, and ELISA are conventional diagnostic methods for bacterial infections. They are time-consuming, have low sensitivity, and cannot produce an accurate *in vivo* picture of the infection mechanism.[65] Therefore, the development of fast, sensitive, and reliable detection protocols is of vital importance from a clinical point of view.

QDs for use in the detection of bacterial infections *in vivo* must be biocompatible with the human body. The image processing should have a high signal-to-noise ratio to avoid interference from the complex dynamic environment inside the body. Last but not least, the imaging technique should be cost-effective enough for routine clinical applications. NIR fluorescence imaging has a high signal-to-noise ratio and is, therefore, promising for bacterial detection.[66] QDs have been used as labels in NIR fluorescence imaging to show bacterial infections and real-time bacterial loads during treatment with antibiotics. Several examples of this application are listed in Table 5.1. PbS QDs produce NIR emissions and are one of several QDs that are suitable for bacteria imaging[38]; they have shown great potential for *in vivo* imaging with deep penetration and low toxicity. PbS QDs have been used to label four different types of bacteria[38] including *Escherichia coli* (*E. coli*), *Streptococcus anginosus* (*S. anginosus*), *Staphylococcus aureus* (*S. aureus*), and *Staphylococcus epidermidis* (*S. epidermidis*). The labeling efficiency was evaluated through *in vitro* experiments that tested the uptake of PbS QDs by these four bacteria. Three different concentrations of bacterial suspensions were mixed with PbS QDs and co-cultured at 37 °C under a 5% CO_2 atmosphere for 1 h. After centrifugation, bacteria agglomerated and sedimented at the bottom of the test tubes. NIR fluorescence showed bright images, indicating the uptake of the PbS QDs by the bacteria and thus demonstrating highly efficient bacterial labeling. PbS QDs were able to penetrate bacterial cells because of the electrostatic interaction between the functional groups outside the PbS QDs and the surface of the bacterial cell wall. PbS QDs were biocompatible and cleared out of the body without causing considerable harm to any major organs. These QDs are, therefore, reliable candidates for *in vivo* imaging, both for studying bacterial infection mechanisms and for evaluating the effectiveness of treatments.

Treatments for Gram-positive and Gram-negative bacterial infections are different. Rapid and accurate Gram-type identification is an urgent requirement in clinical labs, as this identification determines which antibiotics will be most effective in treating a specific bacterial infection. Gram-type identification and classification are traditionally performed by bacterial staining with different organic dyes. It is a time-consuming process and one that may produce false positives.[67] QDs can distinguish Gram-type bacteria in a fast and convenient way. Recently, CQDs were functionalized with vancomycin (CQDs@Van)[68] or lauryl betaine (CQDs@LB)[69] and used for the selective identification of Gram-positive bacteria by fluorescence microscopy and flow cytometry. CDs@Van bound to the cell wall of Gram-positive bacteria and provided CDs@Van aggregation on bacterial surfaces. CQDs@Van showed high selectivity towards Gram-positive bacteria and could detect *S. aureus* in the range of 3.18×10^5 to 1.59×10^8 CFU mL^{-1} with a detection limit of 9.40×10^4 CFU mL^{-1}. Amikacin is an antibiotic that is commonly used to treat Gram-negative bacterial infections. Amikacin-modified CQDs (CQDs@Ami) were used to detect Gram-negative bacteria,[54] and they were shown to be effective at distinguishing *E. coli* from *S. aureus*. The fluorescence intensity of CDs@amikacin was quenched on attachment to the surface of *E. coli*. *E. coli* was distinguished in the concentration range of 7.625×10^2 to 3.904×10^5 with a LOD of 552 CPU mL^{-1}. CDs@amikacin successfully detected *E. coli* in oranges, apples, and pineapples. Therefore, Gram-staining identification by QDs proved to be faster and simpler than traditional staining methods.

Biofilms, multicellular communities of microorganisms, are formed on surfaces in clinical, industrial, and natural environments. They are more resistant to antibiotics and cause serious health problems. In industrial settings, they block filter membranes, foul marine surfaces, and corrode pipes.[50] Therefore, the study of biofilm formation mechanisms and treatment is urgent from both a medical and industrial point of view. Fluorescence microscopy enables the investigation of biofilm formation in microorganisms and, therefore, provides information of practical importance for the control of biofilm formation. For fluorescence evaluations, microorganisms are labeled with fluorescent materials. QDs with high brightness and stable emissions are a good choice for this purpose. For example, CQDs prepared from *L. Plantarum* LLC-605 (CQDs-605)[50] as a carbon source have been used for *E. coli, S. oneidensis, P. aeruginosa, S. aureus,* and *T. reesei* biofilm evaluations. X-ray photoelectron spectroscopy and FTIR spectroscopy revealed that CQDs-605 contained C, H, O, N, S, and P elements and that CQDs-605 staining could discriminate between dead and live bacteria.

As Table 5.1 indicates, many studies have been conducted on bacterial detection *via* fluorescence imaging; however, research in this field is still active. To the best of our knowledge, only one article to date has reported on QDs' applications in the study of biofilm formation and evaluation. Therefore, research in biofilm evaluations using QDs for labeling is still in its infancy and requires more attention.

5.3 Biosensor Design Using QDs

Biosensors are designed for sensitive, specific, fast, and accurate detection of trace (ng or fg in mL) amounts of an analyte in biological, clinical, agricultural, and environmental samples. Conventionally, they consist of a biological recognition element (generally called the bioreceptor), a transducer component that converts the recognition process to a readable signal, and an electronic system (often combined with the transducer) that amplifies the detected signal (see Figure 5.2). Biosensors can also be miniaturized, and, thanks to their outstanding features, adapted for POC uses. They are classified according to the type of transducer or the biorecognition principle. There are four types of transducers: optical, electrochemical, thermal, and piezoelectric. Catalytic biosensors and affinity-based biosensors (ABBs) are two examples of biosensor categories according to their biorecognition principles. In catalytic biosensors, the analyte converts to the product in the presence of another compound that acts as a catalyst in a chemical reaction; the analyte is recognized by an Ab, antigen, aptamer, or DNA. ABBs are referred to as immune sensors when the recognition elements are antibodies or antigens. QDs have outstanding optical and electrical properties and can be used for the immobilization of biomolecules and amplification of the signal, as mediators, electroactive species, and the detection of nanoprobes in biosensor fabrications. QDs can be utilized in biosensors with different transducers, such as electrochemical, optical, thermoelectric, *etc.* To our knowledge, almost all current QD-based biosensors for bacterial and viral assays are based on electrochemical and optical biosensors, with QDs acting as signal amplifiers.[70,71]

Two biosensor classes have been developed for use in bacterial sensing: (1) processing-based systems, which require sample processing to achieve bacterial lysis to liberate the target bacterial component and (2) processing-free systems, which target whole bacteria. Analysis of the bacterial components, DNA,[72,73] RNA (*e.g.*, rRNA),[74,75] intracellular proteins such as enzymes[10] and secreted exotoxins,[76] indicates the presence of bacteria. The major drawback of the first system type is that it requires sample processing and extra reagents, which makes these protocols more time-consuming and costly. The direct detection of whole bacteria is much more rapid and cost-effective, which makes it a much more desirable testing method at the POC. This method is particularly desirable because the infectious dose of bacteria for many human pathogens is very low; for *E. coli* O157:H7, the dose has been reported to be as low as only 10 cells per gram of food or environmental sample.[77] See Figure 5.3.

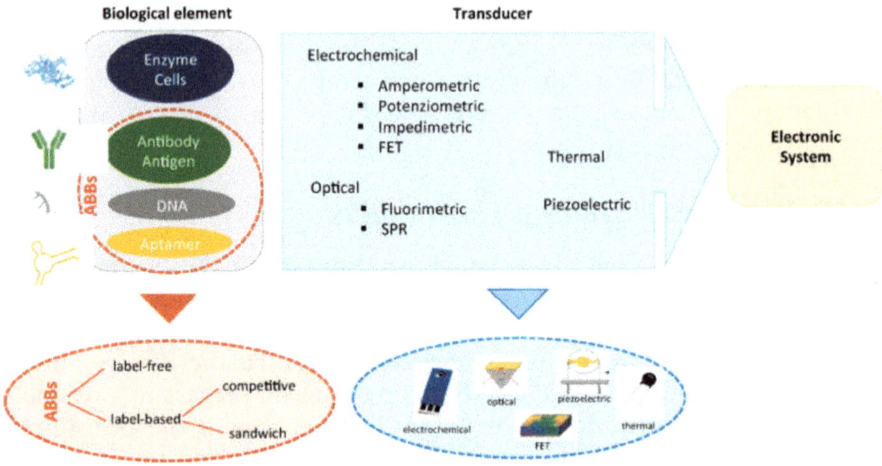

Figure 5.3 Schematic presentation of optical sensor components and classifications.[70] Reproduced from ref. 70, https://doi.org/10.1088/1742-6596/1906/1/012028, under the terms of the CC By 3.0 license, http://creativecommons.org/licenses/by/3.0.

Up to now, QD-based optical and electrochemical biosensors (EBs) have been reported for viral and bacterial detection and assays in different clinical, environmental, and food samples.

5.3.1 Optical Biosensors

Optical sensors are widely developed and used for the detection of microorganisms. Optical sensors are divided into subclasses such as fluorescence, phosphorescence, colorimetric, dispersion, reflection, IR, Raman scattering, and chemiluminescence. In an optical sensor, the sensor's optical properties change when the analyte interacts with the sensing elements. Optical sensors for biological and chemical analytes constitute a rapidly growing area of research with potential applications in many fields, such as medical diagnostics, food analysis, environmental monitoring, agriculture, security, genomics, and proteomics. Colorimetric, fluorescence, and surface-enhanced Raman scattering (SERS) sensors are commonly used as biosensors for bacterial and viral detection because they are easy to operate and low in cost. The optical sensing mechanisms are changes in solution fluorescence intensity, color intensity, or surface resonance energy of the sensing elements in the presence and absence of microorganisms. QDs' optical properties make them well-suited for use in sensor design; they have been employed in various sensors since their invention, and the introduction of new QDs with improved functionality and photochemical properties has enabled their use in a variety of contexts. Tables 5.2 and 5.3 describe QD-based optical sensors for bacteria and viruses.

Colorimetric biosensors have several advantages when used to assay microorganisms. They are the method of choice due to their rapid results, simple instrumentation, easy operation, high sensitivity, and low cost. To the best of our knowledge, while various colorimetric biosensors based on nanomaterials are available for viral and bacterial detection,[4,105,106] no reports exist yet on their QD-based counterparts. Most QD-based optical biosensors are based on fluorescence changes in the solution suspension or on a solid platform.

Fluorescence-based optical biosensors perform extraordinarily well in the detection of viruses and bacteria. Therefore, many commercial fluorescence probes are available for rapid POC. In all these probes, detection is based on variations in the fluorescence of the sensing material when it comes in contact with a virus or bacteria. However, photobleaching of the sensing element affects the biosensor's sensitivity, and for a viral or bacterial sensor, the long-term stability of the luminescent material is of utmost importance. As such, many experimental parameters, including illumination time and wavelength, must be optimized to achieve the best results. QDs have highly stable and bright luminescence, and their emission wavelength can be tuned in the UV to NIR range. These qualities make them ideal candidates for use in viral and bacterial sensor design; they have recently attracted much scientific attention (see Tables 5.2 and 5.3). Signal detection and interpretation have been performed by spectrofluorometers and fluorescence microscopes, by photos captured with mobile phone cameras, and even by the naked eye.[55,95,96,107,108]

Regardless of the sensing strategies, biorecognition elements (*e.g.*, Ab, antigens, proteins, nucleic acid sequences, RNA, DNA) are immobilized on QDs' surfaces to improve the specificity of viral and bacterial detection (Table 5.2 and Table 5.3).

Biosensor development became a subject of great interest in the wake of the recent SARS-CoV-2 pandemic, which tremendously impacted all aspects of human life. Fast, sensitive, and reliable detection was an urgent priority for healthcare providers in the fight against COVID-19. The Wu research group[37] assembled lateral flow biosensors for SARS-CoV-2 quantification. They introduced a CRISPR/Cas13a-based fluorescent nanoparticle SARS-CoV-2 (CFNS) assay that combined a Cas13a-based nucleic acid test strip with quantum dot microspheres (QDMS) (Figure 5.4A–C). They extracted nucleic acids by reverse-transcription recombinase aided amplification (RT-RAA) at 39 °C in 25 min. RNA was reversely transcribed into cDNA through an RT-RAA reaction and then amplified to obtain sufficient target nucleic acid. Next, the amplified products were specifically recognized and cleaved through the CRISPR/Cas13 reaction. In IFIC design, a streptavidin (2.5 mg mL^{-1}) solution and rabbit anti-sheep IgG polyclonal antibody (0.8 mg mL^{-1}) solution were sprayed on a nitrocellulose membrane as the test

Table 5.2 Some application of QDs as fluorescence labels, amplifiers, or electrode modifiers in biosensors for viral detection and quantification.

Sensing element	Biorecognition material	Virus	Transducer	Lr (LOD)	Ref.
Zincides/ZnS QDs	Monoclonal antibody (mAb)	Bluetongue	Fluorescence-linked immunosorbent assay (FLISA)	NA (3.91 ng mL^{-1})	25
QuantumPACK Easy	Antibody	Influenza virus A & B	Lateral flow immunoassays (LFIAs)	NA	78
NIR-QDs	Virus	H5N1p	*In vivo* fluorescence imaging	NA	34
Semiconductor QDs565	DNA	Hepatitis B virus (HBV)	On-chip virus genotyping	NA (1000 copies mL^{-1})	79
SA-QDs	Ab	Human entero virus 71 (EV71)	Optical immune fluorescence	0.42 ng mL^{-1}	80
AuNPs-CdSeSQDs	DNA	Zika	Fluorescence probe	1.7 copies mL^{-1}	81
AgNPs/thiol-CQDs	Ab	HCV	Electrochemical (DPV)	0.05 pg mL^{-1}–60 ng mL^{-1} (3 fg mL^{-1})	82
N,S-GQDs/AuNPs/PANI	Ab	HEV	Electrochemical (CV; EIS)	1 fg mL^{-1}–100 pg mL^{-1} (0.8 fg mL^{-1})	83
AuPd/n-GQDs@PS	Ab	HBsAg	Amperometry	10 fg mL^{-1}–50 ng mL^{-1} (3.3)	11
GQDs/AuNBs	Ab	FAdVs	Optoelectronic (CV)	10–50 pfu mL^{-1} (8.8)	84
Fe_3O_4/GQDs/Cu-apoferritin	Ab	ALVs-J	Electrochemical (DPV)	$10^{2.08}$–$10^{4.4}$ TCID 50 mL^{-1} (115)	85
GCE/GQDs	DNA	HBV	Electrochemical (CV)	10–500 nM (1)	86
GCE/GQDs	Aptamer	HCV	Electrochemical (EIS; CV; DPV)	10–70 pg mL^{-1}; 70–400 pg mL^{-1} (3.3)	87
SiO_2/CQDs	Ab	SFTSV	Lateral flow assay Immuno-fluorescence	10 pg mL^{-1}	88

(continued)

Table 5.2 *(continued)*

Sensing element	Biorecognition material	Virus	Transducer	Lr (LOD)	Ref.
SiO_2/CQDs	Ab	Zika	Lateral flow assay immuno-fluorescence	10 pg mL^{-1}	89
SiO_2/CQDs	Ab	SARS-CoV-2	Fluorescence in lateral flow immuno-chromatography (LFI)	10 pg mL^{-1}	90
QDMs	Sheep anti-FITC IgG antibody	SARS-CoV-2	Lateral flow assay immuno-fluorescence	10^2 to 10^{15} copies mL^{-1} (1)	37
AuNP/gC_3N_4-CdTe QDs	Ab	*J. avian* leukosis viruses (ALV-J)	Photoelectrochemical immunosensor	$10^{2.08}$–$10^{4.08}$ TCID50 mL^{-1} (85)	91
AuNPs/GO-CQDs	ssDNA	HIV-1 gene	Turn off-turn on fluorescence	50.0 fM to 1. 0 nM (15 fM)	92
GQDs	Antigen	HBV	Amperometry	10 fg mL^{-1} –50 ng mL^{-1} (3.3 fg mL^{-1})	16
CdTe QDs	DNA	HPV18	Fluorescence	1.0–50.0 nM (0.2)	93

NA stands for not assigned.

Table 5.3 Some applications of QDs as fluorescence labels, amplifiers, or electrode modifiers in biosensors for bacterial detection and quantification.

QDs	Bio conjugated molecule	Bacteria	Transducer	Lr (LOD)	Ref.
CdSe/ZnS	Streptavidin	*Salmonella*	Optical (fluorescence immunoassay)	4.9×10^4–4.9×10^7 CFU mL^{-1} (4.9×10^3)	94
Carboxylated CdSe/ZnS QDs (540, 580 and 630 nm)	Ab; peptide	*E. coli* O157:H7, *L. monocytogenes* ; *B. melitensis* 16 M	Optical (simultaneous fluorescence detection)	10^1–10^7 CFU mL^{-1}	95
Not mentioned	Aptamer	*S. aureus*	Optical (fluorescence quenching)	10–5×10^8 CFU mL^{-1} (2×10^8)	96
CQDs	cDNA	*S. aureus*	Optical (fluorescence quenching)	50–10^7 CFU mL^{-1} (8)	22
QDs-605	Avidin	*S. aureus*	Optical (increase in fluorescence)		97
PbS	RNase-A	*S. aureus*, *E. coli*, *S. epidermidis*, and *S. anginosus*	Optical (fluorescence intensity)	10^2–10^4 CFU mL^{-1}	38
CdS@ZIF-8	Ab	*E. coli* O157:H7	Electrochemistry (EIS; DPV)	10–10^8 CFU mL^{-1} (3)	98
PbS, CdS, ZnS	DNA	*Salmonella* sp. *Shigella* sp. *Vibrio cholerae*	Electrochemistry (SWASV)	Single (10–60 aM) and multiplex (50–100 aM) pathogen	99
CdTe QDs	Streptavidin	*E. coli*	Optical (Test strip) 2D visual and dual fluorescence	0.2 CFU mL^{-1}	100

(*continued*)

Table 5.3 (*continued*)

QDs	Bio conjugated molecule	Bacteria	Transducer	Lr (LOD)	Ref.
NGQDs	Poly-dopamine surface imprinted polymer and antibody	*E. coli O157:H7*	Electro-chemiluminescence	$10–10^7$ CFU mL^{-1} (8)	101
GQDs/AgNPs	Ab	*E. coli*	Electro-chemiluminescence	$10–10^7$ CFU mL^{-1} (5)	102
CQDs from plastic waste	—	*E. coli*	Optical fluorescence	10^8 CFU mL^{-1}	103
MnO_2-CdSe/ZnS QDs	Ab	*Salmonella*	Microfluidic fluorescence	$1.0 \times 10^2–1.0 \times 10^7$ CFU mL^{-1}	104

line (T-line) and control line (C-line), respectively. Each immunochromatographic strip was cut into 5 mm wide sections using a cutting machine (JinBio, Shanghai, China), followed by drying in a vacuum drying oven at room temperature overnight. QDMS were then separately labeled with sheep anti-FITC IgG Ab, mixed with the cleavage products, and added to the test strip (15 min contact). Finally, a miniature intelligent fluorescence detector read the fluorescence intensity of the T- and C-lines (Figure 5.4B, C). This biosensor successfully detected SARS-CoV-2 viruses in the range of 10^2 to 10^{15} copies mL^{-1} with a detection limit of 1 copy mL^{-1}. The severity of the COVID-19 pandemic led to the design and use of many QD-based biosensors in laboratory settings; however, further evaluations in larger populations must be conducted before these biosensors can be employed in clinical settings for early detection.

Most QD-based biosensors for bacterial detection rely on fluorescence changes (see Table 5.3). Optical QD-based biosensors have been designed for both *in vivo* and *in vitro* bacterial recognition[38] utilizing QDs with different emission wavelengths. For *in vivo* applications, QDs must be non-invasive and biocompatible with the human body; they must also have a high signal-to-noise ratio to avoid emission interference from other tissues. Thanks to their outstanding photochemical properties, NIR QDs are frequently used for *in vivo* applications. As for viral recognition, QDs have been conjugated with biorecognition molecules to specifically and selectively distinguish bacteria. The Chen group[38] introduced PbS QDs as NIR optical elements for rapid infection detection, real-time *in vivo* viral load monitoring, and follow-up after bacterial infection treatment. This fluorescence probe was highly efficient in bacterial detection and the real-time observation of bacterial infections *in vivo*; it also provided a novel approach to antibiotic treatment with respect to bacterial load. Additionally, the PbS QDs were metabolized and cleared out of the body with no significant effects on the major organs.

Li *et al.*[95] introduced a method for the simultaneous detection of three bacteria: *E. coli* O157:H7, *L. monocytogenes*, and *B. melitensis* 16 M. They prepared three different CdSe/ZnS species with different excitation wavelengths (540, 580, and 630 nm). To do this, they attached biotin to three modified peptides, and the resulting conjugated material was coupled to streptavidin-conjugated magnetic beads (MBs). The modified MBs were brought into contact with the sample, separated with a magnet, and washed several times with buffer, after which the supernatant was discarded. The three bacteria were loaded and preconcentrated on the MBs (see Figure 5.4D). Three QD probes were prepared by functionalizing their surfaces with rabbit anti-*E. coli* O157:H7 polyclonal antibodies, rabbit anti-*L. monocytogenes* polyclonal antibodies, and rabbit anti-*B. melitensis* 16 M polyclonal antibodies. The bacteria-loaded MBs were dispersed in the bio-conjugated QD solutions, as illustrated in Figure 5.4E; the three

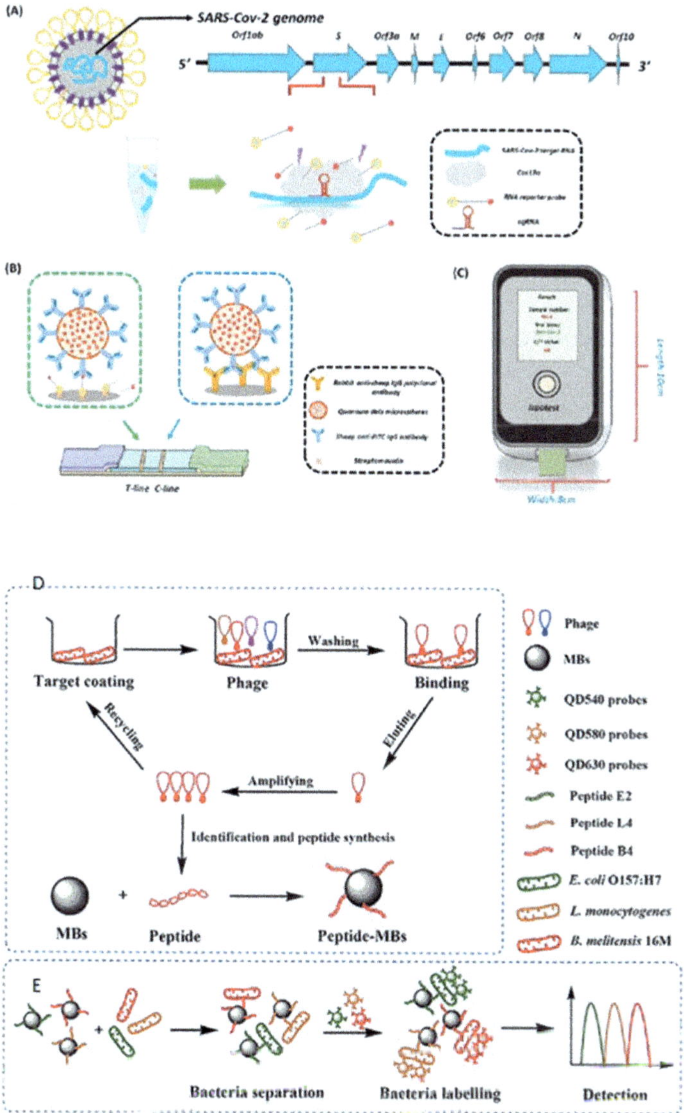

Figure 5.4 Schematic and characterization of CFNS sensor for SARS-CO19. (A) Amplification and cleavage reaction target of the CFNS detection method. (B) Ultrasensitive immunochromatographic assay of SARS-CoV-2 with QDMs. (C) Miniaturized intelligent fluorescence reader.[37] Reproduced from ref. 37 with permission from Elsevier, Copyright 2022. (D–E) Schematic diagram of the simultaneous detection for three pathogens based on phage peptides (*E. coli* O157:H7, *L. monocytogenes,* and *B. melitensis* 16 M). Reproduced from ref. 95 with permission from Elsevier, Copyright 2020.

different bacteria were detected and quantified by measuring the fluorescence intensity of the corresponding QDs in the range of 10^1–10^7 CFU mL^{-1}.

The number of publications on QD-based optical biosensors is higher for bacteria compared with viruses (compare Tables 5.2 and 5.3).

5.3.2 Electrochemical Biosensors

In EBs, the electrochemical reactions of the analyte at the surface of the sensor electrode produce an electrical signal that is linearly related to the analyte concentration. Electrochemical transducers include but are not limited to amperometry, potentiometry, conductometry, and field effect transistors (FET). EBs are sensitive, rapid, and accurate and can easily be fabricated in small sizes for POC uses. QDs are much faster than conventional immunoassays like ELISA, and their signals can be amplified. QDs have several properties that make them ideal for use in electrochemical-based biosensor design: they have a large surface area, exhibit excellent electrochemical properties, and can easily be functionalized. They can, therefore, be attached easily to the surface of an electrode and modified with selected biomolecules to provide a platform for a desired selective or specific interaction with viruses or bacteria (see Table 5.2 and Table 5.3).

There are several strategies for microorganism detection *via* EBs. These strategies include: (1) QDs are attached to the surface of an electrode and then conjugated with biomolecules; (2) the conjugated QDs remain in the solution and then interact with a virus that is attached to the surface of an electrode (a sandwich structure); and (3) QD composites are conjugated with biomolecules. The bioconjugation material is typically Ab; however, a few reports exist on DNA, RNA, and aptamer conjugation. A typical interaction between a virus and QDs in an electrochemical biosensor is illustrated in Figure 5.5.

In EC-based biosensors, QDs and/or electrode surfaces are modified with biomolecules. Modification is usually a multi-step process. Sun *et al.*[91] proposed a photoelectrochemical (PEC) immunosensor for the avian leukosis virus subgroup J (ALV-J) assay. They coated the surface of an ITO electrode with AuNPs/g-C_3N_4 as the photoactive material and dried it under a nitrogen atmosphere (see Figure 5.5). Next, the primary antibody (Ab1) was attached to the coated electrode by incubation in mercaptoacetic acid (MPA) and activation by EDC/NHS mixture (Ab1/AuNPs/g-C_3N_4/ITO), a covalent attachment. Secondary antibody (Ab2) and alkaline phosphatase (ALP) were also attached to MPA-capped CdTe QDs (ALP-CdTe-Ab2) using amino linkage by the EDC/NHS mixture. Subsequently, Ab1/AuNPs/g-C_3N_4/ITO was put in ALV-J solution for 1 h, incubated in ALP-CdTe-Ab2 solution for another 1 h, washed with washing buffer, and placed in detection water. ALV-J was attached to Ab1 (Figure 5.5) and then magnetically stirred at 37 °C for 20 min, and the photocurrents were recorded. The immune reaction between Ab2 and ALV-J determined the extent of

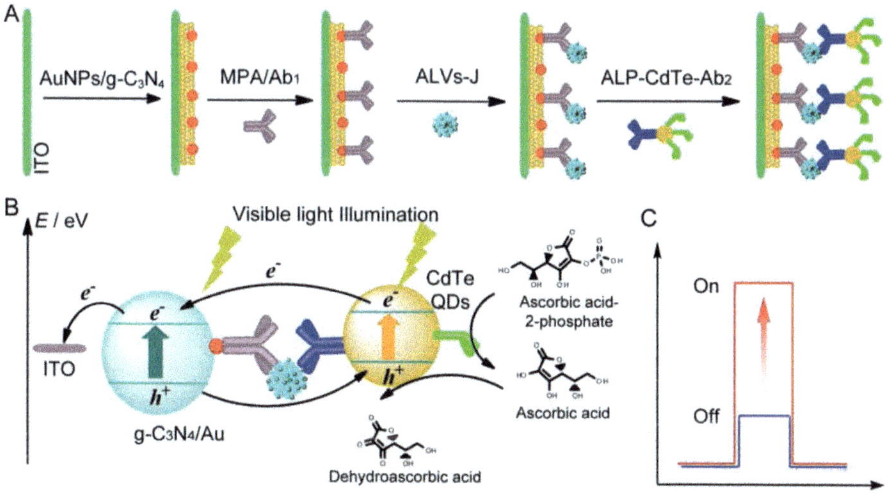

Figure 5.5 Schematic illustration for (A) the fabrication of PEC immunosensor and (B) charge carrier transfer route and enzymatic catalysis procedure in PEC detection of ALV-J. (C) Photocurrent turn-on switching in the detection of target avian virus. Reproduced from ref. 97 with permission from Elsevier, Copyright 2019.

conjugated QDs on the electrode's surface; the ALP-CdTe-Ab2 load on the electrode's surface increased with an increase in ALV-J concentration. ALP-CdTe-Ab2 as a PEC probe promoted the ultrafast transfer of photogenerated charge carriers. ALP catalyzed the conversion of ascorbic acid 2-phosphate (AAP) to ascorbic acid to enhance the photocurrent.

Metal–organic frameworks (MOFs) consisting of inorganic metal ions/clusters and organic ligands have ultrahigh porosity and a large surface area. Recently, QDs have been encapsulated in MOFs to increase their stability and modulate their optical properties. The Tan research group[98] fabricated an ultrasensitive electrochemical sensor for the detection of *E. coli* O157:H7 using CdS QDs encapsulated in zeolitic imidazolate framework-8 (ZIF-8) particles as signal-amplifying tags. They synthesized CdS@ZIF8 by *in situ* growth of ZIF-8 in the presence of CdS QDs. To fabricate the immunosensor, GCE was coated with poly(*p*-aminobenzoic acid) (PABA) by potential cycling between 0.4 to 1.2 V with a scan rate of 10 mV s^{-1} (PABA/GCE). PABA/GCE was soaked in a solution containing an equal amount of EDC and NHS for 30 min. Next, anti-*E. coli* O157:H7 Ab was dropped on the activated surface and kept at 37 °C for 1 h to immobilize the Ab. The electrode was incubated in bovine serum albumin (BSA) to block the non-specific binding sites on the electrode (BSA/Ab-PABA/GCE). The modified electrode was then immersed in a solution containing different concentrations of *E. coli* O157:H7 cells, after which the *E. coli*-loaded electrode was brought into contact with the CdS@ZIF8 solution, and electrochemical signals were recorded. The designed biosensor was successfully used for *E. coli* O157:H7 quantification in milk. Li *et al.*[102] prepared an electrogenerated chemiluminescence biosensor with graphene-based QD (GQD) nanocomposites for *E. coli* detection using a smartphone. They took advantage of the catalytic and amplification effects of GQDs on luminophores. They successfully detected *E. coli* in the concentration range of 10^1–10^7 CFU mL^{-1}. Imprinted polymer technology on the surface of the electrode was used for the selective attachment of *E. coli* to the electrode surface in an electro-chemiluminescence biosensor with NGQDs for luminescence sensing.[101]

In order to provide meaningful information for clinical treatment and preventive measures, a sensor must be able to detect multiple pathogens simultaneously. A highly sensitive and specific electrochemical biosensor assay has been reported[99] for the simultaneous detection of three gene targets using PbS, CdS, and ZnS QDs functionalized with DNA. The sensor targeted the novel non-protein coding RNA (npcRNA) sequences of *Vibrio cholerae*, *Salmonella* sp., and *Shigella* sp., which cause diarrheal diseases. Square-wave anodic stripping voltammetry (SWASV) of QD mixtures showed three resolved peaks of Pb, Cd, and Zn at 0.5, 0.75, and 1.1 V, respectively. This technique showed the attomoles' quantification range. This sensor showed great potential for use in the detection of pathogens in clinical, food, and environmental samples.

A few reports are available on bacterial electrochemical detection (Table 5.3), despite many successful electrochemistry methods for the detection of whole bacteria, bacterial by-products, metabolites, and enzymes.[12] The first QD-based electrochemical biosensor was designed in 2016, and no significant progress in this field has been made since.[12,109] This lack of progress can be attributed to the fact that these biosensors have not yet fulfilled several key requirements for use in clinical detection settings: more work must be done to increase precision, lower the detection limit, shorten the analysis time, eliminate the need for preconcentration, increase ease of operation, render the substrate disposable, and, above all, make this method more cost-effective.

5.4 Conclusion and Future Trends

The early-stage detection of viral and bacterial infections is of vital importance from a therapeutic point of view: early detection leads to early treatment, which lowers mortality rates and reduces the spread of infection within a population. As seen in the recent COVID-19 pandemic, viruses mutate quickly, which poses a significant challenge for biosensor design and necessitates frequent upgrades to sensing platforms. In the last several decades, QDs have emerged as a reliable replacement for fluorescence organic dyes in bacterial and viral detection and quantification, as well as in the study of infection mechanisms. This trend in research is due to QDs' useful optical and electrical properties; it can also be attributed to improvements in preparation and conjugation strategies, which have produced more environmentally friendly and biocompatible QDs. QDs' narrow emission profile prevents fluorescence spectra from overlapping, thus facilitating multicolor imaging. The brilliant brightness of QDs also increases the signal-to-noise ratio, which consequently improves detection sensitivity. QDs have been used in various optical and electrochemical detection systems and have shown promising results for use in follow-up after treatment for viral or bacterial infections.

SVT studies give an overview of viruses' action in living cells: they show how viruses enter cells and how they infect them. However, most QD-based SVT studies have been conducted *in vitro*, which may not accurately reflect viral infection mechanisms *in vivo*. There is, therefore, an urgent need for new labeling and imaging strategies for monitoring the infection behavior of viruses *in vivo* and in living tissues. In recent years, CQDs have attracted a great deal of attention as components in imaging and tracking systems for live cells and microbes. CQDs have also been shown to have therapeutic effects in viral infections; they may even act as antibacterial agents. CQDs thus have a variety of applications in viral and microbial research, such as detection, quantification, infection mechanism detection, and simultaneous imaging and therapy. Paper-based optical biosensors are commercially available, but there is still more work to be done to improve sensitivity and specificity. The therapeutic effects of CQDs have only recently become a subject of research; more theoretical and experimental evaluations are required.

Abbreviations

Ab	Antibody
ABBs	Affinity-based biosensors
AgNPs	Silver nanoparticles
ALVs-J	Avian leucosis virus subgroup J
AuNPs	Gold nanoparticles
BSA	Bovine serum albumin
CQDs	Carbon-based quantum dots
CQDs@Ami	Amikacin modified CQDs
CQDs@LB	Lauryl betaine conjugated CQDs
CQDs@Van	Vancomycin conjugated CQDs
CV	Cyclic voltammetry
DPV	Differential pulse voltammetry
E. coli	*Escherichia coli*
EIS	Electrochemical impedance spectroscopy
ELISA	Enzyme-linked immunosorbent assay
EV71	Enterovirus 71

FAdVs	Fowl adenovirus
FC3N4 QDs	Fluorine-doped carbon nitride quantum dots
FCS	Fluorescent carbon dot
FP	Fluorescence protein
GCE	Glassy carbon electrode
GQDs	Graphene-based quantum dots
H5N1p	Avian influenza H5N1 pseudotype virus
HBsAg	Hepatitis B surface antigen
HBV	Hepatitis B virus
HCoV-229E	Human coronavirus 229E
HCV	Hepatitis C virus
HEV	Hepatitis E virus
His	Histidine
HIV-1	Human immunodeficiency virus 1
HPV18	Human papillomavirus 18
HSV-1	Herpes simplex virus type 1
mAb	Monoclonal antibody
MBs	Magnetic beads
MOFs	Metal organic frameworks
N,S-GQDs	Nitrogen and sulfur dopped graphene quantum dots
NAAT	Nucleic acid amplification tests
N-GQDs	Nitrogen-doped graphene quantum dots
NTC	Interspersed nitro acetic acid
PABA	Poly(*p*-aminobenzoic acid)
PCR	Polymerized chain reactions
PEC	Photoelectrochemical
PEG	Polyethylene glycol
PS	Electroactive polymer nanospheres
QDM_S	Quantum dot microspheres
QDMS	Quantum dot microspheres
QDs	Quantum dots
S. anginosus	*Streptococcus anginosus*
S. aureus	*Staphylococcus aureus*
S. epidermidis	*Staphylococcus epidermidis*
SA	Streptavidin
SERS	Surface-enhanced Raman scattering
SiO_2	Silicon dioxide
ssDNA	Single stranded DNA
SVT	Single virus tracking
SWASV	Square wave anodic stripping voltammetry
Tris-CQDs	Tris (hydroxylmethyl) aminomethane CQDs
W-CQDs	Water-soluble CQDs
ZIF-8	Zeolitic imidazolate framework-8

References

1. A. Qureshi and J. H. Niazi, Biosensors for detecting viral and bacterial infections using host biomarkers: a review, *Analyst*, 2020, **145**(24), 7825–7848.
2. L. Tessaro, A. Aquino, P. de Almeida Rodrigues, N. Joshi, R. G. Ferrari and C. A. Conte-Junior, Nucleic Acid-Based Nanobiosensor (NAB) Used for Salmonella Detection in Foods: A Systematic Review, *Nanomaterials*, 2022, **12**(5), 821.
3. A. Aquino, V. M. F. Paschoalin, L. L. G. Tessaro, P. A. Raymundo-Pereira and C. A. Conte-Junior, Updating the use of nano-biosensors as promising devices for the diagnosis of coronavirus family members: A systematic review, *J. Pharm. Biomed. Anal.*, 2022, 114608.
4. J. C. Shi, Application of nanomaterials-based optical sensors for virus detections, *J. Phys.: Conf. Ser.*, 2021, 1906 012028.
5. P. J. Hernando, S. Dedola, M. J. Marín and R. A. Field, Recent Developments in the Use of Glyconanoparticles and Related Quantum Dots for the Detection of Lectins, Viruses, Bacteria and Cancer Cells, *Front. Chem.*, 2021, **9**, 668509.
6. M. G. Bloomfield, M. N. D. Balm and T. K. Blackmore, Molecular testing for viral and bacterial enteric pathogens: gold standard for viruses, but don't let culture go just yet?, *Pathol.*, 2015, **47**(3), 227–233.
7. W. Wu, Z. Jing, X. Yu, Q. Yang, J. Sun, C. Liu, W. Zhang, L. Zeng and H. He, Recent advances in screening aquatic products for Vibrio spp, *TrAC, Trends Anal. Chem.*, 2019, **111**, 239–251.
8. F. Sun, J. Zhang, Q. Yang and W. Wu, Quantum dot biosensor combined with antibody and aptamer for tracing food-borne pathogens, *Food Qual. Saf.*, 2021, **5**, fyab019.
9. J. Riu and B. Giussani, Electrochemical biosensors for the detection of pathogenic bacteria in food, *TrAC, Trends Anal. Chem.*, 2020, **126**, 115863.
10. S. Muniandy, S. J. Teh, K. L. Thong, A. Thiha, I. J. Dinshaw, C. W. Lai, F. Ibrahim and B. F. Leo, Carbon nanomaterial-based electrochemical biosensors for foodborne bacterial detection, *Crit. Rev. Anal. Chem.*, 2019, **49**(6), 510–533.
11. Q. Yan, Y. Yang, Z. Tan, Q. Liu, H. Liu, P. Wang, L. Chen, D. Zhang, Y. Li and Y. Dong, A label-free electrochemical immunosensor based on the novel signal amplification system of AuPdCu ternary nanoparticles functionalized polymer nanospheres, *Biosens. Bioelectron.*, 2018, **103**, 151–157.
12. S. Kuss, H. M. A. Amin and R. G. Compton, Electrochemical Detection of Pathogenic Bacteria-Recent Strategies, Advances and Challenges, *Chem. Asian J.*, 2018, **13**(19), 2758–2769.

13. N. Rabiee, S. Ahmadi, G. J. Soufi, A. Hekmatnia, M. Khatami, Y. Fatahi, S. Iravani and R. S. Varma, Quantum dots against SARS-CoV-2: diagnostic and therapeutic potentials, *J. Chem. Technol. Biotechnol.*, 2022, **97**(7), 1640–1654.
14. J. Belza, A. Opletalová and K. Poláková, Carbon dots for virus detection and therapy, *Microchim. Acta*, 2021, **188**(12), 430.
15. J. Belza, A. Opletalová and K. Poláková, Carbon dots for virus detection and therapy, *Microchim. Acta*, 2021, **188**, 430.
16. M. H. Nguyen, G. Wong, E. Gane, J. H. Kao and G. Dusheiko, Hepatitis B Virus: Advances in Prevention, Diagnosis, and Therapy, *Clin. Microbiol. Rev.*, 2020, **33**(2), e00046–19.
17. Z. Ramezani, M. R. Dayer, S. Noorizadeh and M. Thompson, Deactivation of SARS-CoV-2 via Shielding of Spike Glycoprotein Using Carbon Quantum Dots: Bioinformatic Perspective, *COVID*, 2021, **1**(1), 120–129.
18. S. Liu, K. Lv, Z. Chen, C. Li, T. Chen and D. Ma, Fluorescent carbon dots with a high nitric oxide payload for effective antibacterial activity and bacterial imaging, *Biomater. Sci.*, 2021, **9**(19), 6486–6500.
19. M. Bruchez, M. Moronne, P. Gin, S. Weiss and A. P. Alivisatos, Semiconductor nanocrystals as fluorescent biological labels, *Science*, 1998, **281**(5385), 2013–2016.
20. Y. Zhang, K. Liu, J. Yu, H. Chen, R. Fu, S. Zhu, Z. Chen, S. Wang and S. Lu, Single stain hyperspectral imaging for accurate fungal pathogens identification and quantification, *Nano Res.*, 2022, **15**(7), 6399–6406.
21. P. Devi, S. Saini and K.-H. Kim, The advanced role of carbon quantum dots in nanomedical applications, *Biosens. Bioelectron.*, 2019, **141**, 111158.
22. F. Cui, J. Sun, J. de Dieu Habimana, X. Yang, J. Ji, Y. Zhang, H. Lei, Z. Li, J. Zheng, M. Fan and X. Sun, Ultrasensitive Fluorometric Angling Determination of Staphylococcus aureus in Vitro and Fluorescence Imaging in Vivo Using Carbon Dots with Full-Color Emission, *Anal. Chem.*, 2019, **91**(22), 14681–14690.
23. L. Tessaro, A. Aquino, P. Panzenhagen, N. Joshi and C. A. Conte-Junior, A systematic review of the advancement on colorimetric nanobiosensors for SARS-CoV-2 detection, *J. Pharm. Biomed. Anal.*, 2022, 115087.
24. Y. Xue, C. Liu, G. Andrews, J. Wang and Y. Ge, Recent advances in carbon quantum dots for virus detection, as well as inhibition and treatment of viral infection, *Nano Convergence*, 2022, **9**(1), 1–31.
25. J. Yin, A. Wang, J. Zhou, Y. Chen, C. Liang, X. Zhu, Y. Zhang, Y. Liu, R. Jia and G. Zhang, Establishment of an Immunological Method for Detection of Bluetongue Virus by Fluorescence-Linked Immunosorbent Assay, *Microbiol. Spectr.* 2022, **10**(5), e01429-22.

26. E. Zor, F. Mollarasouli, L. Karadurmus, G. Ozcelikay and S. A. Ozkan, Carbon Dots in the Detection of Pathogenic Bacteria and Viruses, *Crit. Rev. Anal. Chem.*, 2022, 1–29.
27. A. D. Jose and A. Ghosh, Advances in nanomaterials-based biosensors for the development of virus detection, in *Advanced Biosensors for Virus Detection*, Elsevier, 2022, pp. 203–217.
28. C. Harito, M. Khalil, N. L. W. Septiani, K. K. Dewi, A. Hardiansyah, B. Yuliarto and F. C. Walsh, Trends in nanomaterial-based biosensors for viral detection, *Nano Futures*, 2022, **6**(2), 23.
29. L. Fu, Q. Chen and L. Jia, Carbon dots and gold nanoclusters assisted construction of a ratiometric fluorescent biosensor for detection of Gram-negative bacteria, *Food Chem.*, 2022, **374**, 131750.
30. G. Gao, Y.-W. Jiang, W. Sun and F.-G. Wu, Fluorescent quantum dots for microbial imaging, *Chin. Chem. Lett.*, 2018, **29**(10), 1475–1485.
31. F. Li, Z. P. Zhang, J. Peng, Z. Q. Cui, D. W. Pang, K. Li, H. P. Wei, Y. F. Zhou, J. K. Wen and X. E. Zhang, Imaging viral behavior in Mammalian cells with self-assembled capsid-quantum-dot hybrid particles, *Small*, 2009, **5**(6), 718–726.
32. L. J. Zhang, S. Wang, L. Xia, C. Lv, H. W. Tang, Z. Liang, G. Xiao and D. W. Pang, Lipid-Specific Labeling of Enveloped Viruses with Quantum Dots for Single-Virus Tracking, *mBio*, 2020, **11**(3), e00135–20.
33. J. Liu, M. Su, X. Chen, Z. Li, Z. Fang and L. Yi, Lipid-mediated biosynthetic labeling strategy for in vivo dynamic tracing of avian influenza virus infection, *J. Biomater. Appl.*, 2022, **36**(9), 1689–1699.
34. H. Pan, P. Zhang, D. Gao, Y. Zhang, P. Li, L. Liu, C. Wang, H. Wang, Y. Ma and L. Cai, Noninvasive visualization of respiratory viral infection using bioorthogonal conjugated near-infrared-emitting quantum dots, *ACS Nano*, 2014, **8**(6), 5468–5477.
35. S.-L. Liu, Z.-Q. Tian, Z.-L. Zhang, Q.-M. Wu, H.-S. Zhao, B. Ren and D.-W. Pang, High-efficiency dual labeling of influenza virus for single virus imaging, *Biomaterials*, 2012, **33**(31), 7828–7833.
36. Y. B. Yang, Y. D. Tang, Y. Hu, F. Yu, J. Y. Xiong, M. X. Sun, C. Lyu, J. M. Peng, Z. J. Tian, X. H. Cai and T. Q. An, Single Virus Tracking with Quantum Dots Packaged into Enveloped Viruses Using CRISPR, *Nano Lett.*, 2020, **20**(2), 1417–1427.
37. Q. Zhang, J. Li, Y. Li, G. Tan, M. Sun, Y. Shan, Y. Zhang, X. Wang, K. Song, R. Shi, L. Huang, F. Liu, Y. Yi and X. Wu, SARS-CoV-2 detection using quantum dot fluorescence immunochromatography combined with isothermal amplification and CRISPR/Cas13a, *Biosens. Bioelectron.*, 2022, **202**, 113978.
38. S. Feng, H. Li, C. Liu, M. Chen, H. Sheng, M. Huang, Y. Li, J. Chen, J. Zhang, Y. Hao and S. Chen, Real-Time In Vivo Detection and

Monitoring of Bacterial Infection Based on NIR-II Imaging, *Front. Chem.*, 2021, **9**, 689017.

39. H. Geng, Y. Qiao, N. Jiang, C. Li, X. Zhu, W. Li and Q. Cai, Water-soluble ZnCuInSe quantum dots for bacterial classification, detection, and imaging, *Anal. Bioanal. Chem.*, 2020, **412**(30), 8379–8389.
40. D. Mohan, A. Pathak, P. E. Resmi, P. V. Suneesh and T. G. Satheesh Babu, *IOP ConferenceSeries: Materials Science and Engineering*, 2019, Fluorescence Imaging of E. coli using CdSeQuantum Dots, vol. **577(1)**.
41. W. M. Leevy, T. N. Lambert, J. R. Johnson, J. Morris and B. D. Smith, Quantum dot probes for bacteria distinguish Escherichia coli mutants and permit in vivo imaging, *Chem. Commun.*, 2008, (20), 2331–2333.
42. P. Li, Y. Xiong, H. Wei, W. Lai, L. Wang, W. Luo and H. Xu, *Quantum dots-basedsystem for the detection of bacteria in drinking water*, in 2012 5th International Conference on Biomedical Engineering and Informatics, 16–18 Oct. 2012, 2012, pp. 772–776.
43. M. Amjad, M. Iqbal, A. Faisal, A. M. Junjua, I. Hussain, S. Z. Hussain, H. A. Ghramh, K. A. Khan and H. A. Janjua, Hydrothermal synthesis of carbon nanodots from bovine gelatin and PHM3 microalgae strain for anticancer and bioimaging applications, *Nanoscale Adv.*, 2019, **1**(8), 2924–2936.
44. B. S. B. Kasibabu, S. L. D'souza, S. Jha, R. K. Singhal, H. Basu and S. K. Kailasa, One-step synthesis of fluorescent carbon dots for imaging bacterial and fungal cells, *Anal. Methods*, 2015, **7**(6), 2373–2378.
45. N. Wang, H. Fan, J. Sun, Z. Han, J. Dong and S. Ai, Fluorine-doped carbon nitride quantum dots: Ethylene glycol-assisted synthesis, fluorescent properties, and their application for bacterial imaging, *Carbon*, 2016, **109**, 141–148.
46. N. Wang, Y. Wang, T. Guo, T. Yang, M. Chen and J. Wang, Green preparation of carbon dots with papaya as carbon source for effective fluorescent sensing of Iron (III) and Escherichia coli, *Biosens. Bioelectron.*, 2016, **85**, 68–75.
47. Y. Zhang, X. Liu, Y. Fan, X. Guo, L. Zhou, Y. Lv and J. Lin, One-step microwave synthesis of N-doped hydroxyl-functionalized carbon dots with ultra-high fluorescence quantum yields, *Nanoscale*, 2016, **8**(33), 15281–15287.
48. A. Sachdev, I. Matai, S. U. Kumar, B. Bhushan, P. Dubey and P. Gopinath, A novel one-step synthesis of PEG passivated multicolour fluorescent carbon dots for potential biolabeling application, *RSC Adv.*, 2013, **3**(38), 16958–16961.
49. M. M. F. Baig and Y.-C. Chen, Bright carbon dots as fluorescence sensing agents for bacteria and curcumin, *J. Colloid Interface Sci.*, 2017, **501**, 341–349.

50. F. Lin, C. Li, L. Dong, D. Fu and Z. Chen, Imaging biofilm-encased microorganisms using carbon dots derived from L. plantarum, *Nanoscale*, 2017, **9**(26), 9056–9064.

51. A. Pramanik, S. Jones, F. Pedraza, A. Vangara, C. Sweet, M. S. Williams, V. Ruppa-Kasani, S. E. Risher, D. Sardar and P. C. Ray, Fluorescent, Magnetic Multifunctional Carbon Dots for Selective Separation, Identification, and Eradication of Drug-Resistant Superbugs, *ACS Omega*, 2017, **2**(2), 554–562.

52. J. Yang, G. Gao, X. Zhang, Y.-H. Ma, X. Chen and F.-G. Wu, One-step synthesis of carbon dots with bacterial contact-enhanced fluorescence emission: Fast Gram-type identification and selective Gram-positive bacterial inactivation, *Carbon*, 2019, **146**, 827–839.

53. I. P.-J. Lai, S. G. Harroun, S.-Y. Chen, B. Unnikrishnan, Y.-J. Li and C.-C. Huang, Solid-state synthesis of self-functional carbon quantum dots for detection of bacteria and tumor cells, *Sensors and Actuators B: Chemical*, 2016, **228**, 465–470.

54. S. Chandra, A. R. Chowdhuri, T. K. Mahto, A. Samui and S. k. Sahu, One-step synthesis of amikacin modified fluorescent carbon dots for the detection of Gram-negative bacteria like Escherichia coli, *RSC Adv.*, 2016, **6**(76), 72471–72478.

55. Z.-G. Wang, S.-L. Liu and D.-W. Pang, Quantum Dots: A Promising Fluorescent Label for Probing Virus Trafficking, *Acc. Chem. Res.*, 2021, **54**(14), 2991–3002.

56. H.-Y. Liu, Z.-G. Wang, S.-L. Liu and D.-W. Pang, Single-virus tracking with quantum dots in live cells, *Nat. Protoc.*, 2023, **18**, 458–489.

57. K.-I. Joo, Y. Lei, C.-L. Lee, J. Lo, J. Xie, S. F. Hamm-Alvarez and P. Wang, Site-Specific Labeling of Enveloped Viruses with Quantum Dots for Single Virus Tracking, *ACS Nano*, 2008, **2**(8), 1553–1562.

58. M. Abdel-Salam, B. Omran, K. Whitehead and K. H. Baek, Superior Properties and Biomedical Applications of Microorganism Derived Fluorescent Quantum Dots, *Molecules*, 2020, **25**(19), 4486.

59. H.-S. Han, N. K. Devaraj, J. Lee, S. A. Hilderbrand, R. Weissleder and M. G. Bawendi, Development of a Bioorthogonal and Highly Efficient Conjugation Method for Quantum Dots Using Tetrazine–Norbornene Cycloaddition, *J. Am. Chem. Soc.*, 2010, **132**(23), 7838–7839.

60. I. L. Medintz, T. Pons, J. B. Delehanty, K. Susumu, F. M. Brunel, P. E. Dawson and H. Mattoussi, Intracellular Delivery of Quantum Dot–Protein Cargos Mediated by Cell Penetrating Peptides, *Bioconjug. Chem.*, 2008, **19**(9), 1785–1795.

61. T. Du, K. Cai, H. Han, L. Fang, J. Liang and S. Xiao, Probing the interactions of CdTe quantum dots with pseudorabies virus, *Sci. Rep.*, 2015, **5**(1), 16403.

62. L.-L. Huang, G.-H. Lu, J. Hao, H. Wang, D.-L. Yin and H.-Y. Xie, Enveloped Virus Labeling via Both Intrinsic Biosynthesis and Metabolic Incorporation of Phospholipids in Host Cells, *Anal. Chem.*, 2013, **85**(10), 5263–5270.
63. P. Zhou, Z. Zheng, W. Lu, F. Zhang, Z. Zhang, D. Pang, B. Hu, Z. He and H. Wang, Multicolor labeling of living-virus particles in live cells, *Angew. Chem., Int. Ed. Engl.*, 2012, **51**(3), 670–674.
64. K. Luo, S. Li, M. Xie, D. Wu, W. Wang, R. Chen, L. Huang, T. Huang, D. Pang and G. Xiao, Real-time visualization of prion transport in single live cells using quantum dots, *Biochem. Biophys. Res. Commun.*, 2010, **394**(3), 493–497.
65. A. Ahmed, J. V. Rushworth, N. A. Hirst and P. A. Millner, Biosensors for Whole-Cell Bacterial Detection, *Clin. Microbiol. Rev.*, 2014, **27**(3), 631–646.
66. H. M. Gil, T. W. Price, K. Chelani, J.-S. G. Bouillard, S. D. J. Calaminus and G. J. Stasiuk, NIR-quantum dots in biomedical imaging and their future, *iScience*, 2021, **24**(3), 102189.
67. R. P. Nugent, M. A. Krohn and S. L. Hillier, Reliability of diagnosing bacterial vaginosis is improved by a standardized method of gram stain interpretation,*J. Clin. Microbiol.*, 1991, **29**(2), 297–301.
68. D. Zhong, Y. Zhuo, Y. Feng and X. Yang, Employing carbon dots modified with vancomycin for assaying Gram-positive bacteria like Staphylococcus aureus, *Biosens. Bioelectron.*, 2015, **74**, 546–553.
69. M. J. Meziani, X. Dong, L. Zhu, L. P. Jones, G. E. LeCroy, F. Yang, S. Wang, P. Wang, Y. Zhao, L. Yang, R. A. Tripp and Y.-P. Sun, Visible-Light-Activated Bactericidal Functions of Carbon "Quantum" Dots, *ACS Appl. Mater. Interfaces*, 2016, **8**(17), 10761–10766.
70. R. Antiochia, Developments in biosensors for CoV detection and future trends, *Biosens. Bioelectron.*, 2021, **173**, 112777.
71. L. Liu, J. Hong, W. Wang, S. Xiao, H. Xie, Q. Wang and N. Gan, Fluorescent aptasensor for detection of live foodborne pathogens based on multicolor perovskite-quantum-dot-encoded DNA probes and dual-stirring-bar-assisted signal amplification, *J. Pharm. Anal.*, 2022, **12**(6), 913–922.
72. N. Paniel and J. Baudart, Colorimetric and electrochemical genosensors for the detection of Escherichia coli DNA without amplification in seawater, *Talanta*, 2013, **115**, 133–142.
73. X. Ma, W. Ding, C. Wang, H. Wu, X. Tian, M. Lyu and S. Wang, DNAzyme biosensors for the detection of pathogenic bacteria, *Sensors and Actuators B: Chemical*, 2021, **331**, 129422.

74. M. Azizi, M. Zaferani, S. H. Cheong and A. Abbaspourrad, Pathogenic bacteria detection using RNA-based loop-mediated isothermal-amplification-assisted nucleic acid amplification via droplet microfluidics, *ACS Sens.*, 2019, **4**(4), 841–848.
75. B. Huo, Y. Hu, Z. Gao and G. Li, Recent advances on functional nucleic acid-based biosensors for detection of food contaminants, *Talanta*, 2021, **222**, 121565.
76. M. Veissi, S. Maktabi, Z. Ramezani and M. Khosravi, Highly Sensitive Fluorescence Assay of Enterotoxin A in Milk Using Carbon Quantum Dots as A Fluorophore, *Food Anal. Methods*, 2021, **14**(9), 1815–1825.
77. J. L. Johnson, B. E. Rose, A. K. Sharar, G. M. Ransom, C. P. Lattuada and A. M. McNamara, Methods Used for Detection and Recovery of Escherichia coli O157:H7 Associated with a Food-Borne Disease Outbreak (†), *J. Food Prot.*, 1995, **58**(6), 597–603.
78. S.-K. Kim, H. Sung, S.-H. Hwang and M.-N. Kim, A New Quantum Dot-Based Lateral Flow Immunoassay for the Rapid Detection of Influenza Viruses, *BioChip J.*, 2022, **16**, 175–182.
79. H. Zhang, T. Xu, C. W. Li and M. Yang, A microfluidic device with microbead array for sensitive virus detection and genotyping using quantum dots as fluorescence labels, *Biosens. Bioelectron.*, 2010, **25**(11), 2402–2407.
80. L. Chen, X. Zhang, G. Zhou, X. Xiang, X. Ji, Z. Zheng, Z. He and H. Wang, Simultaneous Determination of Human Enterovirus 71 and Coxsackievirus B3 by Dual-Color Quantum Dots and Homogeneous Immunoassay, *Anal. Chem.*, 2012, **84**(7), 3200–3207.
81. O. Adegoke, M. Morita, T. Kato, M. Ito, T. Suzuki and E. Y. Park, Localized surface plasmon resonance-mediated fluorescence signals in plasmonic nanoparticle-quantum dot hybrids for ultrasensitive Zika virus RNA detection via hairpin hybridization assays, *Biosens. Bioelectron.*, 2017, **94**, 513–522.
82. A. Valipour and M. Roushani, Using silver nanoparticle and thiol graphene quantum dots nanocomposite as a substratum to load antibody for detection of hepatitis C virus core antigen: Electrochemical oxidation of riboflavin was used as redox probe, *Biosens. Bioelectron.*, 2017, **89**, 946–951.
83. A. D. Chowdhury, K. Takemura, T.-C. Li, T. Suzuki and E. Y. Park, Electrical pulse-induced electrochemical biosensor for hepatitis E virus detection, *Nat. Commun.*, 2019, **10**(1), 3737.
84. S. R. Ahmed, J. Mogus, R. Chand, E. Nagy and S. Neethirajan, Optoelectronic fowl adenovirus detection based on local electric field enhancement on graphene quantum dots and gold nanobundle hybrid, *Biosens. Bioelectron.*, 2018, **103**, 45–53.

85. X. Wang, L. Chen, X. Su and S. Ai, Electrochemical immunosensor with graphene quantum dots and apoferritin-encapsulated Cu nanoparticles double-assisted signal amplification for detection of avian leukosis virus subgroup J, *Biosens. Bioelectron.*, 2013, **47**, 171–177.
86. Q. Xiang, J. Huang, H. Huang, W. Mao and Z. Ye, A label-free electrochemical platform for the highly sensitive detection of hepatitis B virus DNA using graphene quantum dots, *RSC Adv.*, 2018, **8**(4), 1820–1825.
87. K. Ghanbari, M. Roushani and A. Azadbakht, Ultra-sensitive aptasensor based on a GQD nanocomposite for detection of hepatitis C virus core antigen, *Anal. Biochem.*, 2017, **534**, 64–69.
88. L. D. Xu, Q. Zhang, S. N. Ding, J. J. Xu and H. Y. Chen, Ultrasensitive Detection of Severe Fever with Thrombocytopenia Syndrome Virus Based on Immunofluorescent Carbon Dots/SiO(2) Nanosphere-Based Lateral Flow Assay, *ACS Omega*, 2019, **4**(25), 21431–21438.
89. L.-D. Xu, F.-L. Du, J. Zhu and S.-N. Ding, Luminous silica colloids with carbon dot incorporation for sensitive immunochromatographic assay of Zika virus, *Analyst*, 2021, **146**(2), 706–713.
90. L.-D. Xu, J. Zhu and S.-N. Ding, Immunoassay of SARS-CoV-2 nucleocapsid proteins using novel red emission-enhanced carbon dot-based silica spheres, *Analyst*, 2021, **146**(16), 5055–5060.
91. B. Sun, J. Dong, L. Cui, T. Feng, J. Zhu, X. Liu and S. Ai, A dual signal-on photoelectrochemical immunosensor for sensitively detecting target avian viruses based on AuNPs/g-C3N4 coupling with CdTe quantum dots and in situ enzymatic generation of electron donor, *Biosens. Bioelectron.*, 2019, **124–125**, 1–7.
92. S. H. Qaddare and A. Salimi, Amplified fluorescent sensing of DNA using luminescent carbon dots and AuNPs/GO as A sensing platform: A novel coupling of FRET and DNA hybridization for homogeneous HIV-1 gene detection at femtomolar level, *Biosens. Bioelectron.*, 2017, **89**, 773–780.
93. M. Shamsipur, V. Nasirian, K. Mansouri, A. Barati, A. Veisi-Raygani and S. Kashanian, A highly sensitive quantum dots-DNA nanobiosensor based on fluorescence resonance energy transfer for rapid detection of nanomolar amounts of human papillomavirus 18, *J. Pharm. Biomed. Anal.*, 2017, **136**, 140–147.
94. S. Ding, H. Hu, X. Yue, K. Feng, X. Gao, Q. Dong, M. Yang, U. Tamer, G. Huang and J. Zhang, A fluorescent biosensor based on quantum dot–labeled streptavidin and poly-l-lysine for the rapid detection of Salmonella in milk, *J. Dairy. Sci.*, 2022, **105**(4), 2895–2907.
95. L. Li, H. Zhang, D. Song, K. Xu, Y. Zheng, H. Xiao, Y. Liu, J. Li and X. Song, Simultaneous detection of three zoonotic pathogens based on phage display peptide and multicolor quantum dots, *Anal Biochem.*, 2020, **608**, 113854.

96. X. Tao, Z. Liao, Y. Zhang, F. Fu, M. Hao, Y. Song and E. Song, Aptamer-quantum dots and teicoplanin-gold nanoparticles constructed FRET sensor for sensitive detection of Staphylococcus aureus, *Chin. Chem. Lett.*, 2021, **32**(2), 791–795.
97. J. Cui, M. Zhou, Y. Li, Z. Liang, Y. Li, L. Yu, Y. Liu, Y. Liang, L. Chen and C. Yang, A New Optical Fiber Probe-Based Quantum Dots Immunofluorescence Biosensors in the Detection of Staphylococcus aureus, *Front. Cell. Infect. Microbiol*, 2021, **11**, 665241.
98. M. Zhong, L. Yang, H. Yang, C. Cheng, W. Deng, Y. Tan, Q. Xie and S. Yao, An electrochemical immunobiosensor for ultrasensitive detection of Escherichia coli O157:H7 using CdS quantum dots-encapsulated metal-organic frameworks as signal-amplifying tags, *Biosens. Bioelectron.*, 2019, **126**, 493–500.
99. D. Vijian, S. V. Chinni, L. S. Yin, B. Lertanantawong and W. Surareungchai, Non-protein coding RNA-based genosensor with quantum dots as electrochemical labels for attomolar detection of multiple pathogens, *Biosens. Bioelectron.*, 2016, **77**, 805–811.
100. P. Chen, Y. Wang, Y. Meng, Y. He, Y. Xiong, K. Ao, J. Huang, J. Chen, Y. Xie and B. Ying, Color and distance two-dimensional visual and homogeneous dual fluorescence analysis of pathogenic bacteria in clinical samples, *Sensors and Actuators B: Chemical*, 2022, **357**, 131422.
101. S. Chen, X. Chen, L. Zhang, J. Gao and Q. Ma, Electrochemiluminescence Detection of *Escherichia coli* O157:H7 Based on a Novel Polydopamine Surface Imprinted Polymer Biosensor, *ACS Appl. Mater. Interfaces*, 2017, **9**(6), 5430–5436.
102. S. Li, J. Liu, Z. Chen, Y. Lu, S. S. Low, L. Zhu, C. Cheng, Y. He, Q. Chen, B. Su and Q. Liu, Electrogenerated chemiluminescence on smartphone with graphene quantum dots nanocomposites for Escherichia Coli detection, *Sens. Actuators, B*, 2019, **297**, 126811.
103. M. Kumari and S. Chaudhary, Modulating the physicochemical and biological properties of carbon dots synthesised from plastic waste for effective sensing of E. coli, *Colloids Surf., B*, 2020, **196**, 111333.
104. L. Hao, L. Xue, F. Huang, G. Cai, W. Qi, M. Zhang, Q. Han, Z. Wang and J. Lin, A microfluidic biosensor based on magnetic nanoparticle separation, quantum dots labeling and MnO_2 nanoflower amplification for rapid and sensitive detection of salmonella typhimurium, *Micromachines*, 2020, **11**(3), 281.
105. T. N. Le, T. D. Tran and M. I. Kim, A Convenient Colorimetric Bacteria Detection Method Utilizing Chitosan-Coated Magnetic Nanoparticles, *Nanomaterials*, 2020, **10**(1), 92.
106. S.-M. You, K.-B. Jeong, K. Luo, J.-S. Park, J.-W. Park and Y.-R. Kim, Paper-based colorimetric detection of pathogenic bacteria in food through magnetic separation and enzyme-mediated signal amplification on paper disc, *Anal. Chim. Acta*, 2021, **1151**, 338252.

107. Z. Qiao, Q. Cai, Y. Fu, C. Lei and W. Yang, Visual and quantitative detection of E. coli O157:H7 by coupling immunomagnetic separation and quantum dot-based paper strip, *Anal. Bioanal. Chem.*, 2021, **413**(17), 4417–4426.
108. Y. Shang, S. Cai, Q. Ye, Q. Wu, Y. Shao, X. Qu, X. Xiang, B. Zhou, Y. Ding, M. Chen, L. Xue, H. Zhu and J. Zhang, Quantum Dot Nanobeads-Labelled Lateral Flow Immunoassay Strip for Rapid and Sensitive Detection of Salmonella Typhimurium Based on Strand Displacement Loop-Mediated Isothermal Amplification, *Engineering*, 2021.
109. S. Kuss, H. M. A. Amin and R. G. Compton, Electrochemical Detection of Pathogenic Bacteria-Recent Strategies, Advances and Challenges, *Chem. Asian J.*, 2018, **13**(19), 2758–2769.

CHAPTER 6

Application of Quantum Dots to in Vitro and in Vivo pH Detection

Z. MOHAMMADPOUR*[a] AND F. MOLAABASI[a]

[a]Biomaterials and Tissue Engineering Department, Breast Cancer Research Center, Motamed Cancer Institute, ACECR, Tehran 1517964311, Iran
*E-mail: mohammadpour@acecr.ac.ir

6.1 Introduction

pH balance in the intra- and extracellular environments is an essential factor in the maintenance of normal cell functions.[1–3] Intracellular pH significantly affects biological events, including apoptosis,[4] drug resistance,[5] ion transport,[6] growth,[7] and endocytosis.[8] Dysregulation of intracellular pH is the cause of many life-threatening afflictions, including Alzheimer's disease,[9] cancer,[10] and stroke. For example, in cancer cells, the intracellular pH is slightly alkaline, while the extracellular pH is acidic. In comparison with normal cells, the inverted pH gradient in cancer cells allows for proliferating and evading apoptosis.[11,12] pH imbalances in cellular compartments also contribute to cell malfunctions.[13–16] The ability to detect and monitor pH changes in cellular environments is, therefore, crucial to researchers' understanding of the physiological processes, pathological processes, and biological effects caused by these changes.

The vital role played by hydrogen ion concentration inside and outside cells and cell organelles inspired researchers to develop non-invasive

Detection Science Series No. 22
Quantum Dots in Bioanalytical Chemistry and Medicine
Edited by Michael Thompson and Zahra Ramezani

Published by the Royal Society of Chemistry, www.rsc.org

and precise probes for pH measurement.[17] Phosphorus nuclear magnetic resonance (P NMR), electrochemical microelectrodes, and optical probes are some of the methods used for cellular pH measurement.[17,18] P NMR spectroscopy as a tool for pH quantification operates based on the chemical shift of inorganic phosphates (ATP and ADP), which inherently exist in the cellular media, or on the shift of synthetic phosphorous-containing organic molecules.[19] The chemical shift correlates with the pH. The technique is non-invasive and fast; however, its low sensitivity and expensive instrumentation limit its practical applications. Voltammetric or potentiometric (ultra)microelectrodes as miniaturized pH meters are the other tools for pH determination at the cellular level.[20–25] Although microelectrodes provide a precise location for the measurement and real-time monitoring of pH change, their use in studies of biological processes is limited because their fabrication process is complicated and out of many researchers' reach.[24] pH-sensitive optical probes, including fluorescence and surface-enhanced Raman spectroscopy-active materials, are widely used for intra- and extracellular pH measurement. They permit sensitive pH detection in the whole cell or in the specific cell compartments. There is a broad array of fluorescent materials available for use in pH measurement, and both NMR-based and electrochemical approaches to measurement are viable; one method or the other may be better suited to a particular application.

The pH-sensitive fluorescent probes are of interest for use in cell labeling.[26–29] The pH can be measured *via* microscopy by detecting pH-dependent decreases or increases in probes' signals. pH-sensitive dyes and nanomaterials have been studied extensively, and their capacity for sensitive cellular pH detection has been verified. Fluorescent dyes, however, are subject to photobleaching. Nanomaterials, conversely, are more photostable, but to be suitable for *in vitro* or *in vivo* pH sensing, a nanomaterial must meet several criteria, including biocompatibility; pH sensitivity (high pH resolution); suitable size, shape, and composition; easy cellular uptake; and safe clearance. Fluorescent quantum dots (QDs) have drawn much recent interest in the field of pH sensing because they can be engineered to meet these criteria.

Although most literature on QDs focuses on heavy metal-based particles, the definition encompasses a wide variety of structures. QDs in solid-state physics are defined as point-like structures in which electrons are confined in three dimensions.[30] Carbon dots (CDs), Si nanocrystals (NCs), polymer dots (PDs), and heavy metal-based QDs are examples of these quantum-confined particles is true. Despite their favorable optical properties, including narrow emission and broad excitation spectra, heavy metal-based QDs' bioimaging applications are limited due to the leakage of toxic metals into cellular environments.[31] In contrast, CDs, PDs, and Si NCs are biocompatible with negligible cytotoxicity. In the following sections, we will discuss the applications of quantum-confined particles for *in vitro* and *in vivo* pH sensing.

6.2 Carbon Dots

CDs were discovered accidentally during the purification of carbon nanotubes in 2004.[17] Since then, a plethora of synthesis procedures have been developed. Although both top-down and bottom-up approaches exist, the latter is more popular in the research community due to the wide variety of carbon sources used as precursors. CDs are less than 10 nm in size and have crystalline to amorphous structures. Their most prominent characteristic is their fluorescence. Surface passivation has led to a significant reduction in QDs' non-radiative electron–hole recombination and an enhancement of their photoluminescence quantum yield (QY).[32] Depending on the synthesis conditions, their emission wavelength is tunable from blue to green, orange, yellow, and red. CDs have been widely employed in the design of optical assays for the quantification of many important biomolecules and hazardous chemicals.[33–38] CDs' biocompatibility, small size, and low toxicity are the driving forces behind their use in biology. In particular, a great deal of effort has been devoted to employing CDs as fluorescent labels in bioimaging.[39] Graphene-based QDs, considered a subtype of CDs, have also been employed in cellular bioimaging.[40]

Most of the CDs used for bioimaging are down-conversion particles that emit blue or green light.[41] They absorb high-energy photons to excite. In the short-wavelength region, the tissues' strong autofluorescence signals interfere with fluorescence imaging. Additionally, high-energy photons induce damage to the tissues, and their penetration depth is low. It is, therefore, important to prepare CDs that excite at longer wavelengths for bioimaging applications.[42,43] Kong *et al.* explored CDs for two-photon imaging and pH sensing inside cells and tissues (see Figure 6.1A).[43] To this end, they covalently attached a molecular pH sensing probe (4'-(aminomethyl phenyl)-2,2':6',2''-terpyridine (AE-TPY)) to the surface of CDs through carbodiimide chemistry. In the absence of hydronium ions, a weak emission at 498 nm was observed under two-photon excitation at 800 nm, while in the presence of hydronium ions, the reverse was observed. The exact mechanism of fluorescence change was not discussed in the original article. However, the pH sensitivity inside the living cells and living tissues at 65–185 μm depth was demonstrated using two-photon microscopy.

A distinct feature of most CDs is the dependence of their emission wavelength on the excitation wavelength. The origin of this behavior is not known. However, it has been suggested (among many other hypotheses) that the involvement of discrete multiple electronic states[46] or slower relaxation of the solvent molecules around CDs[47] might be responsible for the CDs' excitation-dependent photoluminescence.[48–50] This property is advantageous in multicolor sensing; for example, Zhuang *et al.* took advantage of it by synthesizing nitrogen and sulfur co-doped CDs and used them for cellular imaging under various excitation wavelengths (see Figure 6.1B).[44] CDs entered the cells due to their small size and exhibited different emission colors within the cells depending on the excitation wavelength

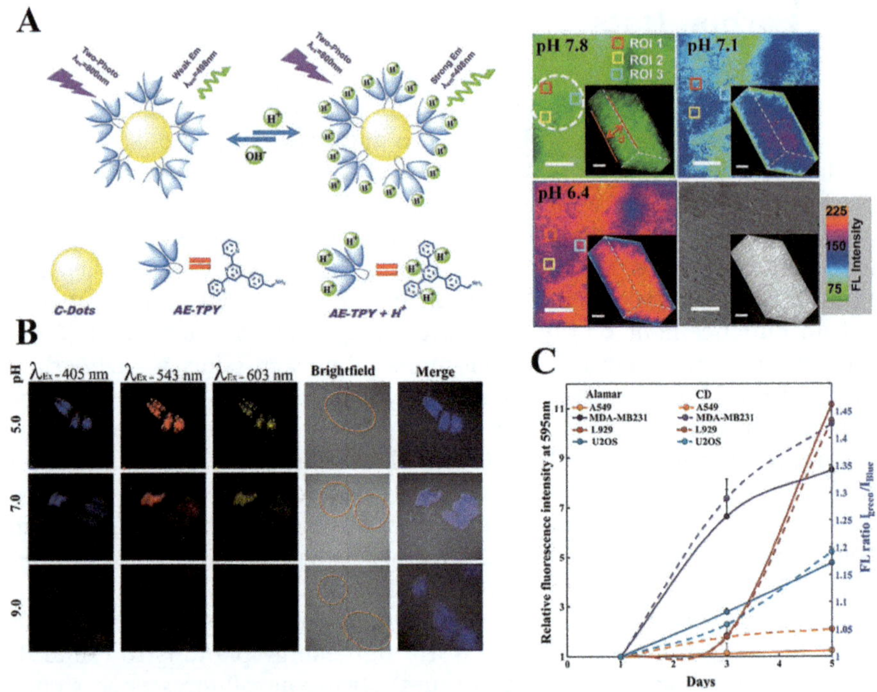

Figure 6.1 (A, left) Represents two-photon imaging for pH sensing and (A, right) represents the lung cancer tissue fluorescence imaging using the pH-sensitive CDs in various pHs. Reproduced from ref. 43 with permission from John Wiley & Sons, Copyright 2012. (B) Fluorescent images of HepG2 cells incubated with nitrogen- and sulfur-doped CDs at different pHs showing light emission at different excitations (multicolor sensing). Reproduced from ref. 44 with permission from Elsevier, Copyright 2020. (C) Growth curves of different cell lines over five days represented by the ratio of fluorescence intensity of CDs at two-emission wavelengths and standard Alamar Blue probe. The growth of cancer cells correlates with the pH reduction in the tumor microenvironment. Reproduced from ref. 45 with permission from Elsevier, Copyright 2021.

used. The CDs' emission intensity at all wavelengths decreased visibly with an increase in the pH value by two units. In another study conducted by Safavi *et al.*, the sensitivity of as-synthesized CDs to pH changes was considerably higher.[51] The fluorescence signal showed a linear response over the pH range of 6.1–7.8 with 0.1-unit resolution. The unmodified CDs exhibited a negligible response toward various ions and biomolecules that are normally found in the intracellular medium. Additionally, the signal was reversible during several cycles of pH change.

Fluorescent CDs have also been used for the imaging of cellular compartments. Lysosomes are subcellular organelles that take part in macromolecule digestion during endocytosis, phagocytosis, and autophagy.[52,53] The pH of lysosomes is acidic. Abnormal lysosome pH variations are associated with many diseases, including neurodegenerative diseases.[13,14] Monitoring lysosome pH, therefore, provides crucial insights into physiological processes. Zhang *et al.* synthesized green fluorescent CDs in which the abundant amino groups on the CDs' surfaces enabled lysosome targeting.[54] The hydrophilic functional groups, including carboxyl and amino, may be responsible for the CDs' sensitive response to pH. As the pH value decreased, the fluorescence signal diminished gradually. Co-localization of the CDs with Lyso-Tracker Red verified the successful targeting of lysosomes. Furthermore, the researchers were able to monitor real-time changes in lysosomal pH during apoptosis using the as-prepared CDs. During cell apoptosis, the fluorescence signal of the whole cell was reduced, suggesting the acidification of the cytosol. Although reports have shown that the pH value of the lysosome increases while that of the cytosol decreases during apoptosis,[55] the whole signal of the cells, including the signal associated with CDs localized in the lysosomes, was also reduced. This observation was not further discussed by the authors. Gao *et al.* also demonstrated CDs' lysosome targeting ability.[56] They observed that the CDs maintained a higher degree of photostability upon long laser exposure compared to a commercial lyso-tracker kit. It is notable that none of the reports mentioned above[55,56] conducted CDs' surface functionalization for lysosome targeting. Other studies that demonstrate successful cell organelle targeting using CDs have been reported.[21,57]

In addition to cellular and *in vivo* imaging, CDs have also been employed for cancer therapy. Phuong *et al.* investigated the applications of CD-based nanocomposites for pH-responsive tumor detection and photothermal therapy.[57] They prepared two CDs: one functionalized with a hydrophobic functional group (D-CDs) and the other functionalized with a zwitterionic compound (Z-CDs). The nanoparticles were covalently linked through a boronate ester bond. The combined CDs (C-CDs) were then functionalized with TiO_2 nanoparticles to form a nanocomposite. The fluorescence signal of C-CDs/TiO_2 was in a quenched state due to Förster resonance energy transfer (FRET). In the acidic medium of the tumor, the boronate ester bond cleaved and the CDs separated. As a result, the fluorescence signal recovered. The D-CDs and Z-CDs penetrated the membrane and nucleus, respectively. Under ultraviolet (UV) or visible irradiation, the hydrophobic D-CDs became partially hydrophilic, and the D-CDs, therefore, translocated and entered the cytosol. Upon near-infrared (NIR) irradiation, the D-CDs converted light into heat and induced a photothermal effect. No heat was generated at physiological pH even under NIR irradiation, possibly due to the absence of free D-CDs. *In vitro* and *in vivo* toxicity assessments revealed that under illumination by UV/Vis and NIR light, the toxicity of C-CDs/

TiO_2 at pH 6.8 was considerably higher than at pH 7.4. Other instances of theranostic application of CDs have been reported.[58,59]

Alongside imaging and therapy, CDs have played other roles in bio-based systems. In a study conducted by Bhattacharyya *et al.*, CDs were used as pH-sensitive probes and as cross-linkers for the formation of gelatin-based hydrogels.[60] The nanocomposite hydrogel was proposed as a candidate for intestine drug delivery, as the entrapped drug was released mostly at neutral pH values. The pH-dependent swelling ratio of the hydrogel accounted for the controlled drug release behavior of the nanocomposite hydrogel. The use of CDs as cross-linkers preserved the integrity of the hydrogel network and enabled the sustained release of the drug. However, the CDs were excited at short-wavelength regions, which makes the applicability of the proposed system for *in vivo* imaging questionable. CDs as pH sensing elements have also been employed for real-time monitoring of pH changes during cells' growth in a 3D-printed alginate scaffold (see Figure 6.1C).[45] A microfluidic system was used for the preparation of a heterogeneous population of cancer cells mimicking the tumor microenvironment. The cells were pretreated with CDs. Cancer cell proliferation in 3D tumor micromodels caused the pH to decrease, which in turn changed the CDs' fluorescence signal. Unlike end-point assays such as MTT, the coexistence of CDs and cancer cells in the tumor microenvironment enabled the authors to monitor pH changes inside the model for a long time.

pH changes also occur during the wound-healing process. At the beginning of an injury (*i.e.*, in the acute phase), the pH is acidic; over time, as the wound enters the chronic phase, the pH becomes alkaline. Omidi *et al.* designed a wound dressing consisting of chitosan and pH-responsive CDs. The sensitivity of CDs to pH was observed between 4 and 9, which covers the pH changes during the healing process. *In vivo* experiments on rat models showed that the wound that was dressed with the chitosan/CDs completely healed after 18 days. In contrast, there was a delay in the healing of the undressed wound (control) and the one dressed in chitosan. The authors proposed that the antibacterial properties of the chitosan and the CDs synergistically contributed to the accelerated healing. Moreover, the pH change during the healing process was observable through changes in the dressing's fluorescence under UV light. On day 0, the pH was acidic and the fluorescence emission was low; on day 9, the emission intensity increased visibly.

Unlike most of the above-mentioned examples, CDs can also be used as carriers for pH-sensitive fluorescent probes.[61,62] Various oxygenated functional groups on the CDs' surfaces facilitate their surface functionalization with other molecules. Shi *et al.* developed a ratiometric fluorescent pH sensor for measuring the whole intracellular pH.[61] Ratiometric measurements are free of errors originating from many variables, including optical path length and concentration; they can, therefore, be used to quantify the pH values more accurately. Shi *et al.* labeled CDs with two fluorescent dyes: fluorescein isothiocyanate (FITC) and rhodamine B isothiocyanate (RBITC). An increase in the pH value led to an enhancement of the fluorescence ratio of FITC to RBITC. pH fluctuations induced by oxidative stress were examined in the presence of redox species including H_2O_2, ClO^-, reduced glutathione (GSH), GSH inhibitor (*N*-ethyl maleimide), and GSH precursor (*N*-acetylcysteine). They found that both elevated H_2O_2 and GSH depletion contributed to intracellular acidification. In contrast, abnormally high levels of GSH and ClO^- had no effect on the intracellular pH levels.

6.2.1 Mechanism of CDs' Fluorescence Change with pH

Changes in pH alter CDs' fluorescence and even their UV-Vis absorption. In the case of the absorption spectrum, it has been suggested that the vibrational coupling between QDs' surface hydroxyl groups and the resulting H-bonding can lead to energy level broadening. Therefore, a broad UV-Vis absorption peak is observed. In alkaline pHs, however, the intra- and intermolecular H-bonding between the particles decreases, making the energy levels discrete. Thus, sharper peaks evolve. Changes in CDs' fluorescence at different pH levels do not follow a specific trend. There are CDs whose fluorescence signal decreases with an increase in the pH value and vice versa.[63–65] Other CDs do not exhibit pH-dependent changes in their fluorescence spectrum.[66] A redshift or blueshift in a CD's emission peak position may also correlate with the pH.[67–69] It is, therefore, rarely straightforward to establish a certain mechanism of pH dependency for CDs that are prepared from different precursors or through diverse procedures. One suggested mechanism of spectral redshift involves the pH-dependent generation of surface states with lower band gaps that absorb or emit light at a longer wavelength.[70] In another scenario, it has been proposed that charging or doping effects exerted by protonation and deprotonation of oxygenated surface functional groups shift CDs' Fermi level and change their optical properties.[51,65,71] pH-induced aggregation of CDs[72] and energy transfer[73] are the other suggested pH-based mechanisms for altering CDs' fluorescence.

6.3 Graphdiyne QDs

Graphdiyne is an allotrope of carbon. It is a two-dimensional nanomaterial with lateral sizes of hundreds of nanometers. Its architecture is composed of carbon–carbon triple bonds and benzene rings. Sp- and Sp^2-hybridized carbon form a large π-conjugated system. Min *et al.* cut this nanomaterial into smaller pieces *via* hydrothermal treatment and synthesized graphdiyne QDs. Similar to most CDs, the fluorescence signal of these graphdiyne QDs depended on the excitation wavelength. Also, the signal changed with the pH. Inside cells, however, no pH sensitivity was observed. The article did not discuss the contradictory pH sensitivity of graphdiyne QDs inside and outside cells. This study further found that green-emitting graphdiyne QDs were mostly localized in the lysosomes. *In vivo* experiments verified that the hemolysis of red blood cells was negligible at concentrations less than 100 μg mL^{-1}. Moreover, healthy mice were injected with graphdiyne QDs. After 14 days, their major organs were analyzed using hematoxylin and eosin staining, and no morphological differences were observed with respect to the control.

6.4 Silicon Nanocrystals

Like CDs, Si NCs are more biocompatible with living organisms compared to heavy metal-based QDs. Si NCs' wavelength emissions depend on their surface chemistry and are, therefore, tunable.[74] Surface modification of Si NCs is robust, as functionalization proceeds through covalent binding to surface-located Si atoms.[75] Renal clearance of decomposed Si NCs as silicic acid is a viable mechanism.[76] As stated in the previous sections, red-emitting nanomaterials are more suitable for imaging applications as the tissues emit less autofluorescence in the red to NIR wavelength region and light penetrates them deeper. In aqueous media, the surface of red-emitting Si NCs may be passivated by hydroxide ions or water, resulting in a peak shift to lower wavelengths and blue emissions[77]; they must, therefore, be protected from surface oxidization to retain their red emissions. Chiu *et al.* used Si as a surfactant (CTAB) to transfer hydrophobic Si NCs from the organic phase to the aqueous phase (see Figure 6.2A).[78] CTAB molecules not only protected Si NCs from surface oxidizing but also provided high stability due to their polar head groups. In acidic and neutral pHs, the Si NCs' emissions remained red and stable for more than six months. However, in alkaline conditions, the emissions turned blue. In most of the studies described above, fluorescence intensity either decreased or increased with pH. In contrast, in the study by Chiu *et al.*, the emission color, rather than the intensity, changed with pH. CTAB-stabilized Si NCs retained their red emissions upon cellular uptake. pH sensitivity inside the cells, however, was not shown.

The long lifetime of Si NCs' emissions enables time-gated data collection with a lower background signal compared to steady-state measurements. In time-gated measurement, the emission signal is recorded on a delay after excitation. In such circumstances, short-lived autofluorescence signals diminish, and the signal-to-noise ratio improves. Romano *et al.* prepared PEGylated Si NCs using a low-temperature thiolene click reaction.[79] The particles were colloidally stable in water and exhibited bright red and NIR photoluminescence with long emission lifetimes. The maximum emission wavelength, QY, and lifetime of PEGylated Si NCs (5 nm) in water were 945 nm, 0.18, and 120 μs, respectively. The pH sensitivity was verified by lifetime measurement. At acidic and neutral pH values, the lifetime increased and decreased, respectively (see Figure 6.2B). The

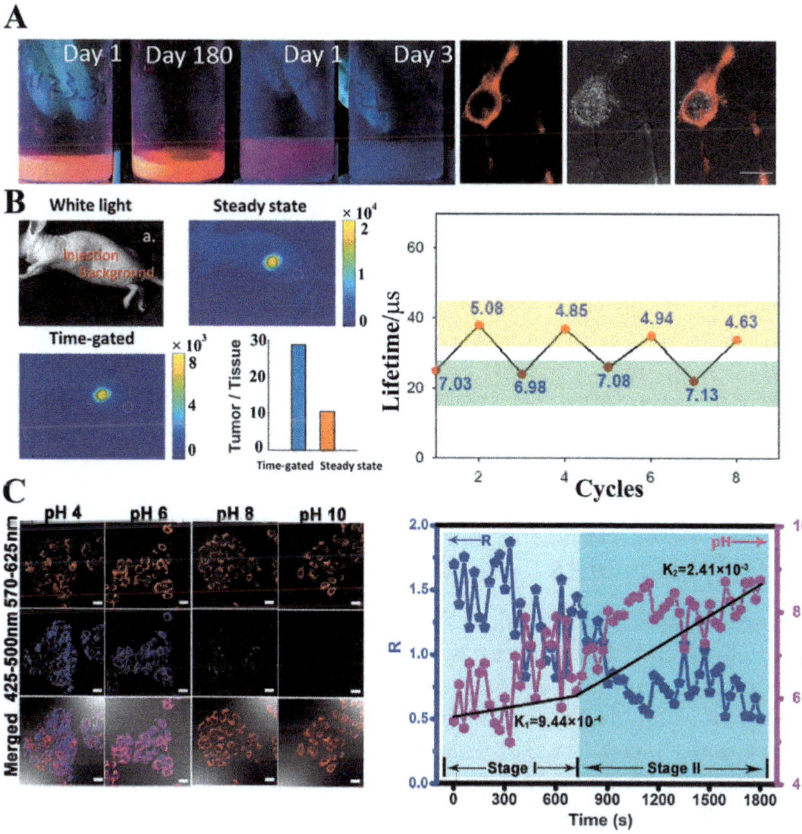

Figure 6.2 (A, left) Photostability of red-emitting Si NCs after 180 days in the acidic media (Day 1 and Day 180), and color change of Si NCs from red to pink (Day 1) and blue (Day 3) in alkaline conditions; (A, right) fluorescence microscopy, differential interference contrast (DIC), and overlaid images of N2a cells into which the red-emitting Si NCs are internalized. Reproduced from ref. 78 with permission from Springer Nature, Copyright 2013; (B, right) successive change of pH accompanied with the change in luminescent lifetime of pegylated Si NCs; (B, left) white light, steady state, and time-gated imaging of pegylated Si NCs subcutaneously injected into athymic nude mice and comparison of the signal-to-noise ratio in steady state and time-gated imaging (delay time in time-gated imaging was 10 μs). Reproduced from ref. 79 with permission from the Royal Society of Chemistry; (C, right) fluorescence intensity ratio of Si NPs at 465 nm to RBITC at 575 nm ($R = I_{465}/I_{575}$) and lysosomal pH change of HeLa cells. pH growth rate (ΔpH/Δtime) is divided into two regimens ascribed to alkalization lag phase (Stage I, $K1 = 9.44 \times 10^{-4}$) and logarithmic growth phase (Stage II, $K2 = 2.41 \times 10^{-3}$); (C, left) intracellular pH calibration. Reproduced from ref. 80 with permission from the American Chemical Society, Copyright 2016.

observed behavior in acidic and neutral conditions was reversible; it was not reversible in basic pHs. PEGylated Si NCs were intravenously injected into tumor-bearing mice; *in vivo* imaging revealed that the NCs accumulated in the liver and tumor, based on the fluorescence signal. Furthermore, a comparison of the *in vivo* imaging results verified a higher signal-to-noise ratio by an order of 3 in time-gated imaging compared to steady-state imaging (see Figure 6.2B). In a study by Chu *et al.*, fluorescent Si nanoparticles were covalently bound to pH-sensitive dopamine and pH-insensitive RBITC.[80] The fluorescence change in response to pH was established based on Si nanoparticles as mediators that accept or donate electrons to the attached molecules at various pH conditions. Using the dual-functionalized Si nanoparticles, lysosomal pH changes were monitored live, and fluctuations attributed to different stages of cell biology were demonstrated through analysis of intracellular fluorescence responses (see Figure 6.2C). Other examples of pH sensing based on Si NCs have been reported.[81,82]

6.5 Polymer Dots

Semiconducting PDs are considered attractive optical nanoplatforms due to their excellent optical characteristics, including a large absorption cross-section, tunable luminescence, high stability, brightness (2–3 orders higher than that of a single molecule dye), and biocompatibility. Of particular interest are their fast emission rate, high QY, superior photostability, and low cytotoxicity. Their surfaces can be easily modified to improve solubility, which allows for more efficient intra-particle energy transfers between excitons and dopants. All these qualities enable a broad range of biological applications, including detection (*e.g.*, metal ions, reactive oxygen species, types of biomolecules, temperatures, and pH values), fluorescence and chemiluminescence imaging, photothermal therapy, photodynamic therapy, and photoacoustic imaging. PDs are commonly composed of hydrophobic semiconducting polymers with a weight or volume fraction of 50%. They form a stable spherical morphology with a less than 30 nm diameter surrounded by solvent molecules. The efficient exciton migration along the π-conjugated backbone, which affects the polymer's bandgap and promotes energy transfer, can be altered through the selection of various monomers during the polymerization process.

PDs for biological applications can be prepared using nanoprecipitation or mini-emulsion techniques. In a typical nanoprecipitation method, semiconducting polymers and amphiphilic polymers are first dissolved in water-miscible tetrahydrofuran and, after that, are rapidly injected into water under ultrasonication. Finally, the organic solvent in the obtained mixture is evaporated by N_2 purging. In this method, the PDs are formed mainly through hydrophobic interactions associated with the significant decrease in polymer solubility that occurs when the polymers are injected into water. The PDs vary in size from 5 to 30 nm, depending on the solubility and concentration of semiconducting polymers, the strength of the ultrasonication, and the temperature of the solvent. Landfester and coworkers developed a mini-emulsion method for use in organic optoelectronic devices.[83] In this method, semiconducting polymers were dissolved in a water-immiscible organic solvent (*e.g.*, $CHCl_3$) and mixed with an aqueous solution containing surfactants (*e.g.*, sodium dodecyl sulfate) to avoid undesirable coalescence. After vigorous stirring or ultrasonication, stable oil-in-water droplets formed in the mixture, which was followed by suspension in water and evaporation of the organic solvent. Depending on the concentration of the surfactants and polymers, the PDs synthesized using the mini-emulsion method show diameters of 40 to 500 nm.[84] Both nanoprecipitation and mini-emulsion are viable techniques for PD preparation. To date, however, most studies on PDs for *in vivo* and *in vitro* pH detection report the use of the nanoprecipitation method.

A typical pH-sensitive PD design incorporates a dopant molecule with a pH-sensing site. For example, trichromatic-emission PDs (TE PDs, triple-emitting PDs with blue, green, and red emission regions) were prepared using the nanoprecipitation method.[85] The PDs had three components: poly(9,9-dioctylfluorenyl-2,7-diyl) (PFO, blue emission at 440 nm), naphthalimide-poly(styrene-co-maleic anhydride) (NA-PSMA, green emission at 530 nm), and tetraphenylporphyrin (TPP, red emission at 655 nm). The morpholine structure in NA-PSMA acted as the pH sensing site and as a lysosome targeting agent. Moreover, NH_2-NA anchored on PSMA provided the amphiphilic grafted polymer and, thus, improved the solubility of NA-PSMA through the reaction between the amino group of NH_2-NA and the anhydride. In alkaline pH values, the photo-induced electron transfer (PET) from the lone-pair electrons of morpholine's nitrogen to NA quenched the fluorescence of NA. However, in acidic conditions, the PET process stopped, and the NA's fluorescence recovered. A key feature of pH-sensitive multicomponent PDs is their ability to simultaneously detect other targets along with pH. For example, simultaneous monitoring of lysosomal pH and Cu^{2+} can elucidate the pathological processes behind neurodegenerative diseases. The designed TE PDs were used for simultaneous ratiometric detection of pH and Cu^{2+} in the HeLa cell line and zebrafishes without foreign or cross-interfering. In this work, the ratiometric multifunctional tracking was modulated by PFO, which served as the fluorescent donor and internal reference signal; the tracking was also modulated by TPP as the Cu^{2+} selective ligand (see Figure 6.3A).

Chen *et al.* proposed another strategy for developing pH-sensitive PDs using pH-sensitive dye (FITC), which they bonded with pH-insensitive dye (PFO) to design a ratiometric pH sensor with increased sensitivity.[86] The FRET mechanism between PFO (λ_{em} = 439 nm) as a donor and FITC (λ_{em} = 517 nm) as an acceptor accounted for the signal change. To amplify the FITC signal and produce robust pH-sensitive PDs, poly[(9,9-dioctyl-fluorenyl-2,7-diyl)-*co*-(1,4-benzo-{2,1',3}-thiadiazole)] (PFBT) was incorporated

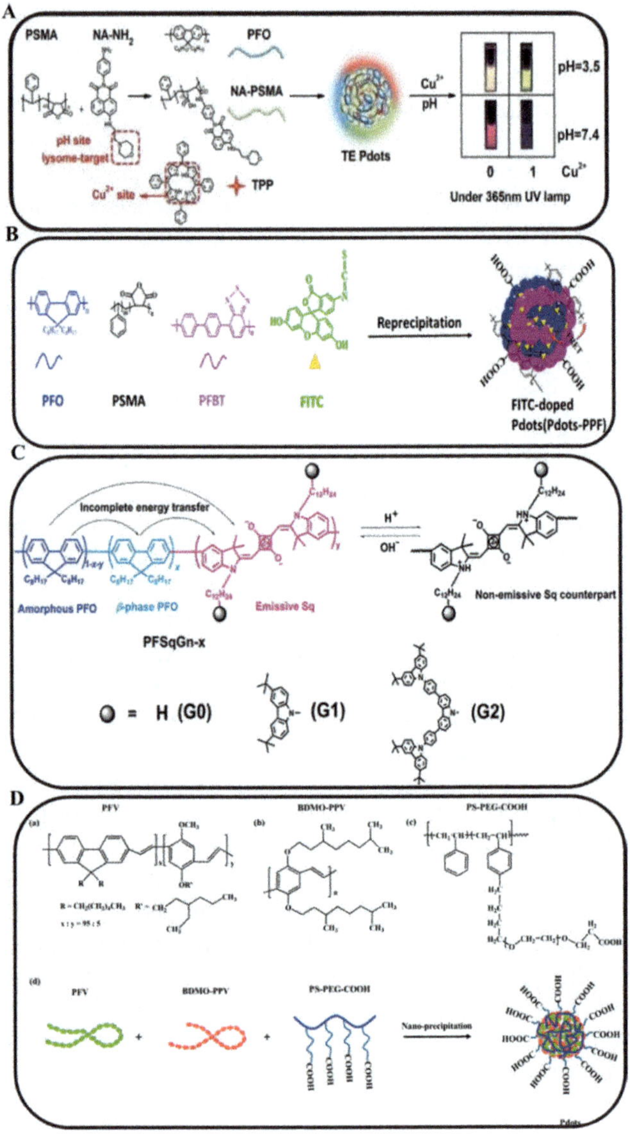

Figure 6.3 Schematic illustration of the chemical structure and nanoprecipitation procedure for preparation of (A) trichromatic-emission PDs (TE PDs) for visually detecting and simultaneously sensing pH and Cu^{2+},[85] (B) PDs-PPF,[86] (C) PFSqG2-5% PDs,[87] and (D) carboxylated PDs.[88] (A) was reproduced from ref. 85 with permission from the Royal Society of Chemistry; (B) was reproduced from ref. 86, https://doi.org/10.3390/molecules24162923, under the terms of the CC BY 4.0 license, http://creativecommons.org/licenses/by/4.0/; (C) was reproduced from ref. 87 with permission from the Royal Society of Chemistry; (D) was reproduced from ref. 88, https://doi.org/10.3390/s19061455, under the terms of the CC BY 4.0 license, http://creativecommons.org/licenses/by/4.0/.

in the designed PDs. The resulting pH-sensitive PDs, named PDs-PPF, showed stability and desirable reversibility within the range of pH 3.0 to 8.0. They were used successfully for HeLa cell imaging (see Figure 6.3B).[86] In another PD-based ratiometric pH sensor, NIR-emitting squaraine dye and PFO were used as the pH-sensitive and pH-insensitive molecular dyes, respectively, in the PDs' structures.[87] The two fluorescent motifs were dendronized using carbazoles as the bulky steric group. The large steric carbazole dendron suppressed PDs' photobleaching and squaraine dye's self-quenching at high concentrations. The squaraine dye was responsive to wide-dynamic-range pH sensing. In the developed NIR-emitting PDs (named PFSqG2-5%), which were prepared by nanoprecipitation, the energy transfer from the amorphous PFO to the squaraine dye and β-phase PFO enabled intracellular pH mapping in MCF-7 cells (see Figure 6.3C).

Carboxylated block co-polymers are another common (and potentially important) component of pH-sensitive PDs. For example, polystyrene-polyethylene glycol-carboxyl (PS-PEG-COOH) can be used to alter polymers' conformational structures and can affect the amount of energy transferred in a FRET-based system. In a design by Ou *et al.*, poly[(9,9-dihexyl-9*H*-fluorene-2,7-vinylene)-*co*-(1-methoxy-4-(2-ethylhexyloxy)-2,5phenylenevinylene)] (PFV) as the donor and poly[2,5-bis(30,70-dimethyloctyloxy)-1,4-phenylenevinylene] (BDMO-PPV) as the acceptor were mixed with PS-PEG-COOH to form a FRET-based pH-sensitive probe.[88] The energy transfer between PFV and BDMO-PPV significantly decreased in highly acidic or alkaline conditions. The authors proposed that at neutral to slightly alkaline pH values, the carboxylic acid groups would convert to carboxylates. As a result of electrostatic repulsions between the negatively charged carboxylate groups, the conformation of PS-PEG-COO$^-$ became extended, which facilitated energy transfer between the donor–acceptor pairs and led to fluorescence quenching. At pHs beyond 8, more sodium ions were added to the solution, resulting in reduced electrostatic repulsion due to the common ion effect. The resulting curled conformation of the polymer decreased the energy transfer. Similarly, at acidic pH values, the carboxyl groups were protonated and repulsion was minimized, leading to curled conformation and decreased energy transfer (see Figure 6.3D).

6.6 Quantum Dots

QDs are semiconductor NCs with a size range of 1.5–10 nm and excellent optical properties from near-ultraviolet to infrared due to the quantum confinement effect, which typically occurs in ultrasmall sizes. With the excitation and absorption of light, an electron–hole pair (known as an exciton) is produced. After the recombination of electron and hole, energy is released as a photon (a process referred to as radiative recombination). Notably, with decreasing particle size under the Bohr radius, the required energy increases to create an exciton, allowing blueshifting for emitting photons.[89] The broad absorption spectrum, high surface-to-volume ratio, narrow tunable emission, high QY, and enhanced photostability compared to organic dyes and the possibility of designing with multiple inorganic shell layers (*e.g.*, CdSe/CdS/ZnS) to minimize the lattice mismatch, decrease the cytotoxicity and improve QY, and make them interesting materials for biomedical applications.[89–91] Moreover, QDs' surfaces can be modified and functionalized with coating materials such as hydrophilic thiols, silica, amphiphilic polymers, or PEG, which make them useful candidates for *in vivo* molecular and cellular imaging.[89–93] QDs are synthesized using elements from groups II–VI (CdS, CdSe, ZnS, ZnO), III–V (GaP, InP, GaN), and IV–VI (PbS, PbSe) of the periodic table because of their semiconductive properties. QDs with longer excitation lifetimes can be easily synthesized through doping with transition metals such as Fe^{2+}, Cu^{2+}, Mn^{2+}, Cr^{2+}, and Co^{2+}. Doping improves QY while minimizing background fluorescence. Techniques for QD synthesis include hot injection synthesis, water-in-oil (W/O) microemulsion, sol–gel (specific to metal sulfide or metal oxide-based QDs), pyrolysis, and the hydrothermal method; among these techniques, W/O microemulsion is expensive and not green.[89]

QDs that are to be used as pH sensor components for longer-term studies at the cellular level (*e.g.*, studies of endosomes) must exhibit good photostability.[89] In this way, the binding of ligands with pH-dependent redox properties to the surface of QD is one of the important strategies for developing cellular pH sensors. In this regard, dopamine isothiocyanate was introduced as a candidate; it exists in two forms due to the reduced catechol at low pH and the oxidized quinone at high pH.[90] Fluorescence quenching occurred at high pH due to electron transfer from excited QDs to quinone as the electron acceptor; thus, the enhancement of the fluorescence was observed with decreasing endosome pH in COS-1 fibroblast cells (see Figure 6.4A). The size of the QDs was an important factor in fluorescence quenching efficiency. Quenching was most pronounced for smaller QDs, attributed to the increased spatial overlapping of carriers, which allows for a greater driving force and better electron transfer. Another QD–dopamine fluorescent probe sensitive to pH ranges from 4 to 10 has been reported.[91] Ratiometric sensors are preferred over redox-based sensors, as the former allow for self-calibration, leading to enhanced sensitivity and selectivity. Jin *et al.* prepared a FITC-conjugated CdSe/CdZnS QD as a ratiometric QD-based pH sensor.[93] The attachment of FITC to the GSH-stabilized CdTe/ZnS QDs created a ratiometric pH-sensitive probe in which FRET from fluorescein to QDs took place. A decrease in fluorescence intensity was observed for QDs and FITC. The ratio of QDs' fluorescence intensity at 600 nm to that of FITC at 515 nm decreased as the pH value increased. The ratiometric probe was used to measure the pH of a turbid solution. In such conditions, the fluorescence intensity fluctuates due to excitation and emission light scattering. The variation of excitation and emission intensity due to light scattering is a serious issue for pH measurements using the single intensity method. Ratiometric probes are free from such errors. In Jin *et al.*'s study, the pH that was measured by the FITC-conjugated CdTe/ZnS QDs was linearly correlated with the actual pH.[93] It is believed that for pH sensing in the lysosome and endosome, the probe should be designed with high sensitivity in the pH range of 4 to 6. This goal was achieved using a thiol-modified dye (2-(dimethylamino)ethyl)aminonaphthalimide (ANI) on the surface of the CdSe/ZnS QDs.[95] The dye is known for its pH-sensitive PET process. It was suggested that at an acidic pH, the tertiary amine group on the dye protonated, which stopped the PET in the dye so that the energy transfer occurred from the dye to the QD. This led to a shift in the fluorescence spectra accompanied by fluorescence enhancement of the QD-ANI conjugate.

There are a few reports on the simple design of pH-sensitive QDs without covalent binding of the pH-responsive polypeptide, peptide, or dye. For example, pH-responsive QDs with NIR emissions (720 nm), a long lifetime (~1 μs), and an ultrasmall size (3.5 nm) were prepared by the combination of doping and a gradient-alloyed nanostructure, resulting in Cu-doped gradient-alloyed CdZnS QDs.[96] The Cu-doped CdZnS QDs were obtained by mixing $CuCl_2$, $Zn(OAc)_2$, and $CdCl_2$ in the presence of GSH at 95 °C. The QDs' lifetime exhibited a linear response to pH in the range of 5.5 to 7.0. The as-prepared QDs were used for *in vivo* pH imaging using fluorescence lifetime imaging microscopy (FLIM). They were subcutaneously injected at two sites with different pH values in a nude mouse. In the FLIM images, the long-lifetime QDs' distinct colors were observable above the short-lived background signal. QDs have also been used as internal references for pH measurement.[92] Hu *et al.* synthesized ultrastable CdSe/ZnS QDs and coated their surfaces with an amphiphilic polymer (PE-PEG).[92] The fluorescence signal of the QDs did not vary with pH, thus serving as an internal reference. Covalent bonding of the as-prepared QDs to a pH-sensitive fluorophore, 4-aminomethyl fluorescein hydrochloride, allowed for ratiometric pH sensing in the range of 4 to 9 without any sensitivity to environmental changes (see Figure 6.4B).

Recently, a dual-functional QD-based nanocarrier with a pH-activated charge convertible property was fabricated for real-time cell imaging and controlled protein drug delivery in the tumor microenvironment. The probe consisted of a core/multishell CdSe@ZnS/ZnS QD, the surface of which was capped by a multifunctional polypeptide ligand (mPEG-*block*-poly(ethylenediamine-dihydrolipoic acid 2,3-dimethylmaleic anhydride)-L-glutamate, PEG-P(ED-DLA-DMA)LG, using the ligand-exchange strategy. Intracellular protein delivery was enabled through the loading of polypeptide ligands with cytochrome C (CC) *via* electrostatic interactions. The designed QD-PEG-P(ED-DLA-DMA)LG exhibited favorable internalization and rapid endo/lysosomal escape ability, leading to improved real-time lung carcinoma cell imaging. Furthermore, the QD-PEG-P(ED-DLA-DMA)LG-CC was effective in killing cancer cells by pH-triggered charge reversibility. Notably,

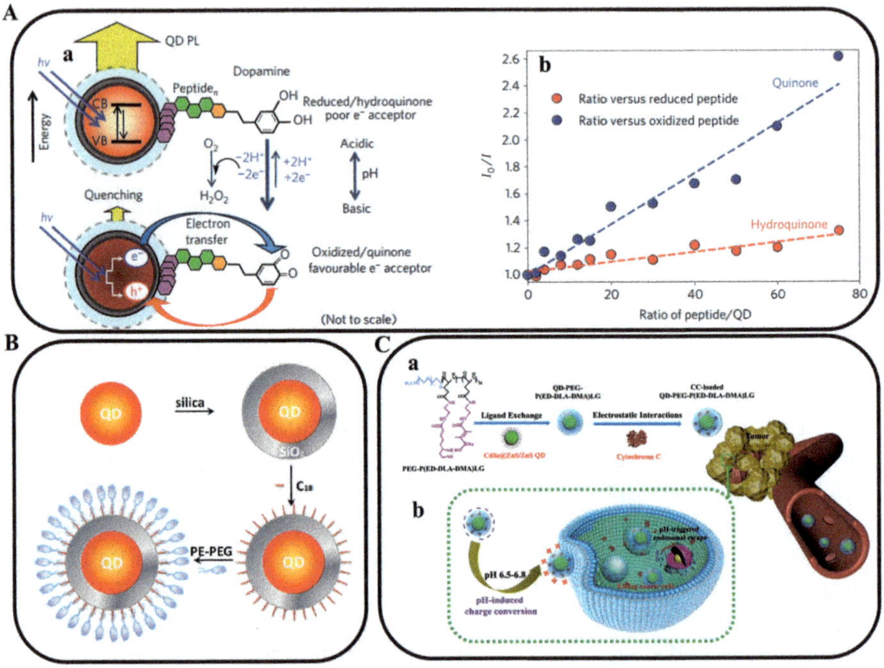

Figure 6.4 (A) At first, peptides covalently bonded to the hydroquinone as the dopamine-reduced form are self-assembled on the surface of QDs. At low pH, hydroquinone acts as a poor electron acceptor resulting in insignificant PL quenching. As the pH increases, oxidation of dopamine occurs at ambient O_2 of the buffer by which H_2O_2 species are produced. Therefore, quinone concentration as an electron acceptor increases in close proximity to the QD. This produces higher quenching efficiency with the amount of quinone. VB is the valence band and CB is the conduction band. (b) I_0=I vs dopamine/QD ratio represents quenching data at 550 nm. The experiment was done at pH 6 to keep the pre-reduced state of dopamine as hydroquinone. Reproduced from ref. 90 with permission from Springer Nature, Copyright 2010; (B) Schematic illustration of a double-phase transfer process for the synthesis of QD@SiO_2@PE-PEG. Hydrophobic QDs are anchored within the hydrophilic silica layer, followed by surface modification with OTMS, a hydrophobic silane. Then, the hydrophobic QD@SiO_2C18 is solubilized with PE-PEG polymer. Reproduced from ref. 92 with permission from the American Chemical Society, Copyright 2010; (C) Schematic illustration of designing QD-PEG-P(ED-DLA-DMA)LG by easy ligand-exchange strategy accompanied with high cytochrome c (CC) loading and (b) its work for enhanced protein drug delivery, cellular uptake, endo/lysosomal escape ability, and efficient CC releasing by affinity switching. Reproduced from ref. 94 with permission from Elsevier, Copyright 2021.

in the designed QD probe, the key point for pH-induced charge-convertible ability related to the β-carboxylic amides in the PEG-P(ED-DLA-DMA)LG polymer, which acted as both an acid-labile linker in a slightly acidic medium and an anionic binding site for loading positively charged drugs[94] (see Figure 6.4C).

6.7 Conclusion and Future Trends

This chapter has reviewed various applications of quantum-confined particles for *in vitro* and *in vivo* pH sensing. Of all the particles reviewed, CDs are most frequently used in this context. Their band gap tuning is useful for multicolor sensing, and their small size enables fast cellular penetration and clearance. CDs are inherently pH-sensitive; their surfaces can also be functionalized with pH-sensitive fluorescent dyes due to the availability of oxygenated functional groups. CDs are commonly used to measure cellular pH; additionally, their pH sensitivity has been leveraged for use in tumor detection and monitoring wound healing.

Although most quantum-confined fluorescent dots are negligibly toxic, some exhibit low stability and low fluorescence signals, which limits their practical applications for *in vitro* and *in vivo* pH sensing. Efforts have, thus, been made to develop composites with other nanomaterials that restrict QDs' motions and enhance their fluorescence. Such interactions restrict the QD's motion and play a vital role in fluorescence enhancement. Additionally, the confinement of dots in a matrix prevents the agglomeration of particles and protects them from environmental factors that damage their surfaces, which directly affects their light emission properties. The entrapment of nanodots in suitable polymer matrix would also lead to a similar effect. The current research on pH-sensitive fluorescent dots is, thus, focused on the development of robust nanocomposites with higher pH resolution, stability, and sensitivity.

Abbreviations

ANI	(2-(dimethylamino)ethyl)aminonaphthalimide
BDMO-PPV	Poly[2,5-bis(30,70-dimethyloctyloxy)-1,4-phenylenevinylene]
CDs	Carbon dots
CC	Cytochrome C
FITC	Fluorescein isothiocyanate
FLIM	Fluorescence lifetime imaging microscopy
FRET	Förster resonance energy transfer
GSH	Glutathione
NCs	Nanocrystals
NA-PSMA	Naphthalimide-poly(styrene-co-maleic anhydride)

PDs	Polymer dots
PEG-P(ED-DLA-DMA)LG	Poly(ethylenediamine-dihydrolipoic acid-2,3-dimethylmaleic anhydride)-L-glutamate
PET	Photo-induced electron transfer
PFBT	Poly[(9,9-dioctyl-fluorenyl-2,7-diyl)-co-(1,4-benzo-[84]-thiadazole)]
PFO	Poly(9,9-dioctylfluorenyl-2,7-diyl)
PFV	Poly[(9,9-dihexyl-9H-fluorene-2,7-vinylene)-co-(1-methoxy-4-(2-ethylhexyloxy)-2,5phenylenevinylene)]
PS-PEG-COOH	Polystyrene-polyethylene glycol-carboxyl
QDs	Quantum dots
QY	Quantum yield
RBITC	Rhodamine B isothiocyanate
TPP	Tetraphenylporphyrin
TE PDs	Trichromatic-emission PDs

References

1. R. T. Kennedy, L. Huang and C. A. Aspinwall, *J. Am. Chem. Soc.*, 1996, **118**, 1795–1796.
2. A. Ishaque and M. Al-Rubeai, *J. Immunol. Methods*, 1998, **221**, 43–57.
3. A. Lardner, *J. Leukoc. Biol.*, 2001, **69**, 522–530.
4. T. F. Sergeeva, M. V. Shirmanova, O. A. Zlobovskaya, A. I. Gavrina, V. V. Dudenkova, M. M. Lukina, K. A. Lukyanov and E. V. Zagaynova, *Biochim. Biophys. Acta Mol. Cell Res.*, 2017, **1864**, 604–611.
5. J. W. Wojtkowiak, D. Verduzco, K. J. Schramm and R. J. Gillies, *Mol. Pharm.*, 2011, **8**, 2032–2038.
6. B. J. Harvey, S. R. Thomas and J. Ehrenfeld, *J. Gen. Physiol.*, 1988, **92**, 767–791.
7. J. Chiche, K. Ilc, J. Laferrière, E. Trottier, F. Dayan, N. M. Mazure, M. C. Brahimi-Horn and J. Pouysségur, *Cancer Res.*, 2009, **69**, 358–368.
8. H. Sakai, G. Li, Y. Hino, Y. Moriura, J. Kawawaki, M. Sawada and M. Kuno, *J. Physiol.*, 2013, **591**, 5851–5866.
9. B. Fang, D. Wang, M. Huang, G. Yu and H. Li, *Int. J. Neurosci.*, 2010, **120**, 591–595.
10. P. Swietach, R. D. Vaughan-Jones, A. L. Harris and A. Hulikova, Philosophical transactions of the Royal Society of London Series B, *Biological sciences*, 2014, **369**, 20130099.
11. M. Sharma, M. Astekar, S. Soi, B. S. Manjunatha, D. C. Shetty and R. Radhakrishnan, *Recent Patents on Anti-Cancer Drug Discovery*, 2015, **10**, 244–258.

12. D. Wijesinghe, M. C. Arachchige, A. Lu, Y. K. Reshetnyak and O. A. Andreev, *Sci. Rep.*, 2013, **3**, 3560.
13. T. Fukuda, L. Ewan, M. Bauer, R. J. Mattaliano, K. Zaal, E. Ralston, P. H. Plotz and N. Raben, *Ann. Neurol.*, 2006, **59**, 700–708.
14. D. J. Colacurcio and R. A. Nixon, *Ageing Res. Rev.*, 2016, **32**, 75–88.
15. J.-H. Lee, W. H. Yu, A. Kumar, S. Lee, P. S. Mohan, C. M. Peterhoff, D. M. Wolfe, M. Martinez-Vicente, A. C. Massey, G. Sovak, Y. Uchiyama, D. Westaway, A. M. Cuervo and R. A. Nixon, *Cell*, 2010, **141**, 1146–1158.
16. C. Wang, T. Zhao, Y. Li, G. Huang, M. A. White and J. Gao, *Adv. Drug Delivery Rev.*, 2017, **113**, 87–96.
17. X. Xu, R. Ray, Y. Gu, H. J. Ploehn, L. Gearheart, K. Raker and W. A. Scrivens, *J. Am. Chem. Soc.*, 2004, **126**, 12736–12737.
18. D. Malhotra and J. R. Casey, *Intracellular pH Measurement*, eLS, 2015, pp. 1–7, DOI: 10.1002/9780470015902.a0002643.pub3.
19. S. Martel, J.-L. Clément, A. Muller, M. Culcasi and S. Pietri, *Bioorg. Med. Chem.*, 2002, **10**, 1451–1458.
20. R. Song, Q. Xiong, T. Wu, X. Ning, F. Zhang, Q. Wang and P. He, *Anal. Bioanal. Chem.*, 2020, **412**, 3737–3743.
21. R.-E. Munteanu, L. Stănică, M. Gheorghiu and S. Gáspár, *Anal. Chem.*, 2018, **90**, 6899–6905.
22. J. Zhou, L. Zhang and Y. Tian, *Anal. Chem.*, 2016, **88**, 2113–2118.
23. R.-E. Munteanu, L. Stănică, M. Gheorghiu and S. Gáspár, *ChemElectroChem.*, 2019, **6**, 1800–1807.
24. S. R. Ng and D. O'Hare, *Analyst*, 2015, **140**, 4224–4231.
25. J. J. García-Guzmán, C. Pérez-Ràfols, M. Cuartero and G. A. Crespo, *ACS Sens.*, 2021, **6**, 1129–1137.
26. W. Shi, X. Li and H. Ma, *Methods Appl. Fluoresc.*, 2014, **2**, 042001.
27. C. E. Rowland, C. W. Brown, I. L. Medintz and J. B. Delehanty, *Methods Appl. Fluoresc.*, 2015, **3**, 042006.
28. A. Jaworska, K. Malek and A. Kudelski, *Spectrochim. Acta, Part A*, 2021, **251**, 119410.
29. J.-T. Hou, W. X. Ren, K. Li, J. Seo, A. Sharma, X.-Q. Yu and J. S. Kim, *Chem. Soc. Rev.*, 2017, **46**, 2076–2090.
30. M. A. Reed, *Sci. Am.*, 1993, **268**, 118–123.
31. C. Zhu, Z. Chen, S. Gao, B. L. Goh, I. B. Samsudin, K. W. Lwe, Y. Wu, C. Wu and X. Su, *Prog. Nat. Sci.: Mater. Int.*, 2019, **29**, 628–640.
32. A. M. Anthony, R. Murugan, R. Subramanian, G. K. Selvarangan, P. Pandurangan, A. Dhanasekaran and A. Sohrab, *Colloids Surf., A*, 2020, **586**, 124266.
33. M. Shamsipur, A. Safavi, Z. Mohammadpour and A. R. Zolghadr, *Sens. Actuators, B*, 2015, **221**, 1554–1560.

34. Z. Mohammadpour, A. Safavi, A. Omidvar, A. Mohajeri, N. Mobaraki and M. Shamsipur, *J. Fluorine Chem.*, 2016, **190**, 12–22.
35. Z. Mohammadpour, A. Safavi and S. H. Abdollahi, *Anal. Chim. Acta*, 2017, **959**, 74–82.
36. M. Shamsipur, A. Safavi and Z. Mohammadpour, *Sens. Actuators, B*, 2014, **199**, 463–469.
37. Z. Mohammadpour, A. Safavi and M. Shamsipur, *Chem. Eng. J.*, 2014, **255**, 1–7.
38. M. Shamsipur, A. Safavi, Z. Mohammadpour and R. Ahmadi, *Microchim. Acta*, 2016, **183**, 2327–2335.
39. Y. Song, S. Zhu and B. Yang, *RSC Adv.*, 2014, **4**, 27184–27200.
40. Z. L. Wu, M. X. Gao, T. T. Wang, X. Y. Wan, L. L. Zheng and C. Z. Huang, *Nanoscale*, 2014, **6**, 3868–3874.
41. Y. Z. Yang, N. Xiao, S. G. Liu, L. Han, N. B. Li and H. Q. Luo, *Mater. Sci. Eng. C Mater. Biol. Appl.*, 2020, **108**, 110401.
42. S. Song, J. Hu, M. Li, X. Gong, C. Dong and S. Shuang, *Mater. Sci. Eng. C*, 2021, **118**, 111478.
43. B.-S. Kong, A. Zhu, C. Ding, X. Zhao, B. Li and Y. Tian, *Adv. Mater.*, 2012, **24**, 5844–5848.
44. Q. Zhuang, X. Si, L. Li, H. Zeng and Y. Ding, *Talanta*, 2020, **208**, 119780.
45. S. De, A. Joshi, D. M. Tripathi, S. Kaur and N. Singh, *Mater. Sci. Eng. C Mater. Biol. Appl.*, 2021, **128**, 112344.
46. A. Sharma, T. Gadly, A. Gupta, A. Ballal, S. K. Ghosh and M. Kumbhakar, *J. Phys. Chem. Lett.*, 2016, **7**, 3695–3702.
47. S. Khan, A. Gupta, N. C. Verma and C. K. Nandi, *Nano Lett.*, 2015, **15**, 8300–8305.
48. Y. P. Sun, B. Zhou, Y. Lin, W. Wang, K. A. Fernando, P. Pathak, M. J. Meziani, B. A. Harruff, X. Wang, H. Wang, P. G. Luo, H. Yang, M. E. Kose, B. Chen, L. M. Veca and S.-Y. Xie, *J. Am. Chem. Soc.*, 2006, **128**, 7756–7757.
49. X. Wen, P. Yu, Y.-R. Toh, X. Hao and J. Tang, *Adv. Opt. Mater.*, 2013, **1**, 173–178.
50. L. Wang, S.-J. Zhu, H.-Y. Wang, S.-N. Qu, Y.-L. Zhang, J.-H. Zhang, Q.-D. Chen, H.-L. Xu, W. Han, B. Yang and H.-B. Sun, *ACS Nano.*, 2014, **8**, 2541–2547.
51. A. Safavi, R. Ahmadi, Z. Mohammadpour and J. Zhou, *RSC Adv.*, 2016, **6**, 104657–104664.
52. L. Yu, C. K. McPhee, L. Zheng, G. A. Mardones, Y. Rong, J. Peng, N. Mi, Y. Zhao, Z. Liu, F. Wan, D. W. Hailey, V. Oorschot, J. Klumperman, E. H. Baehrecke and M. J. Lenardo, *Nature*, 2010, **465**, 942–946.
53. P. Saftig and J. Klumperman, *Nat. Rev. Mol. Cell. Biol.*, 2009, **10**, 623–635.

54. Q. Q. Zhang, T. Yang, R. S. Li, H. Y. Zou, Y. F. Li, J. Guo, X. D. Liu and C. Z. Huang, *Nanoscale*, 2018, **10**, 14705–14711.
55. C. Nilsson, K. Kågedal, U. Johansson and K. Ollinger, *Methods in cell science: an official journal of the Society for In Vitro Biology*, 2003, **25**, 185–194.
56. P. Gao, J. Wang, M. Zheng and Z. Xie, *Chem. Eng. J.*, 2020, **381**, 122665.
57. P. T. M. Phuong, H. J. Won, A. I. Robby, S. G. Kim, G. B. Im, S. H. Bhang, G. Lee and S. Y. Park, *ACS Appl. Mater. Interfaces*, 2020, **12**, 37929–37942.
58. H. Liu, Q. Wang, G. Shen, C. Zhang, C. Li, W. Ji, C. Wang and D. Cui, *Nanoscale Res. Lett.*, 2014, **9**, 397.
59. C. A. Choi, J. E. Lee, Z. A. I. Mazrad, Y. K. Kim, I. In, J. H. Jeong and S. Y. Park, *ChemMedChem.*, 2018, **13**, 1459–1468.
60. S. K. Bhattacharyya, M. Dule, R. Paul, J. Dash, M. Anas, T. K. Mandal, P. Das, N. C. Das and S. Banerjee, *ACS Biomater. Sci. Eng.*, 2020, **6**, 5662–5674.
61. W. Shi, X. Li and H. Ma, *Angew. Chem. Int. Ed. Engl.*, 2012, **51**, 6432–6435.
62. J. Espina-Casado, T. Fontanil, A. Fernández-González, S. Cal, Á. J. Obaya, M. E. Díaz-García and R. Badía-Laíño, *Sens. Actuators, B*, 2021, **346**, 130555.
63. H. Liu, T. Ye and C. Mao, *Angew. Chem., Int. Ed.*, 2007, **46**, 6473–6475.
64. Y. Guo, Z. Wang, H. Shao and X. Jiang, *Carbon*, 2013, **52**, 583–589.
65. X. Jia, J. Li and E. Wang, *Nanoscale*, 2012, **4**, 5572–5575.
66. P. Saheeda, K. Sabira, M. Dhaneesha and S. Jayaleksmi, *Luminescence*, 2018, **33**, 22–28.
67. C. Zhang, Y. Cui, L. Song, X. Liu and Z. Hu, *Talanta*, 2016, **150**, 54–60.
68. F. Yuan, L. Ding, Y. Li, X. Li, L. Fan, S. Zhou, D. Fang and S. Yang, *Nanoscale*, 2015, **7**, 11727–11733.
69. Z. Fan, S. Zhou, C. Garcia, L. Fan and J. Zhou, *Nanoscale*, 2017, **9**, 4928–4933.
70. S. Dutta Choudhury, J. M. Chethodil, P. M. Gharat, P. P K and H. Pal, *J. Phys. Chem. Lett.*, 2017, **8**, 1389–1395.
71. X. Liu, C. Yang, B. Zheng, J. Dai, L. Yan, Z. Zhuang, J. Du, Y. Guo and D. Xiao, *Sens. Actuators, B*, 2018, **255**, 572–579.
72. Y. Z. Yang, N. Xiao, S. G. Liu, L. Han, N. B. Li and H. Q. Luo, *Mater. Sci. Eng.*, 2020, **108**, 110401.
73. W. Liu, C. Li, X. Sun, W. Pan and J. Wang, *Sens. Actuators, B*, 2017, **244**, 441–449.
74. M. Dasog, G. B. De los Reyes, L. V. Titova, F. A. Hegmann and J. G. C. Veinot, *ACS Nano*, 2014, **8**, 9636–9648.
75. J. M. Buriak, *Chem. Commun.*, 1999, 1051–1060.

76. J.-H. Park, L. Gu, G. von Maltzahn, E. Ruoslahti, S. N. Bhatia and M. J. Sailor, *Nat. Mater.*, 2009, **8**, 331–336.
77. S.-W. Lin and D.-H. Chen, *Small*, 2009, **5**, 72–76.
78. S.-K. Chiu, B. A. Manhat, W. J. I. DeBenedetti, A. L. Brown, K. Fichter, T. Vu, M. Eastman, J. Jiao and A. M. Goforth, *J. Mater. Res.*, 2013, **28**, 216–230.
79. F. Romano, S. Angeloni, G. Morselli, R. Mazzaro, V. Morandi, J. R. Shell, X. Cao, B. W. Pogue and P. Ceroni, *Nanoscale*, 2020, **12**, 7921–7926.
80. B. Chu, H. Wang, B. Song, F. Peng, Y. Su and Y. He, *Anal Chem.*, 2016, **88**, 9235–9242.
81. M. Na, Y. Han, Y. Chen, S. Ma, J. Liu and X. Chen, *Anal Chem.*, 2021, **93**, 5185–5193.
82. S. Nsanzamahoro, L. Li, F. P. Mutuyimana, J. Liu, C. Ren, H. Qin and H. Chen, *New J. Chem.*, 2020, **44**, 19294–19299.
83. K. Landfester, R. Montenegro, U. Scherf, R. Güntner, U. Asawapirom, S. Patil, D. Neher and T. Kietzke, *Adv. Mater.*, 2002, **14**, 651–655.
84. Y. Yuan, W. Hou, W. Qin and C. Wu, *Biomater. Sci.*, 2021, **9**, 328–346.
85. Q. Zhang, J. Sun, R. Zhang, X. Chen, N. Chen and F. Gao, *Chem. Commun.*, 2020, **56**, 8647–8650.
86. P. Chen, I. Ilyas, S. He, Y. Xing, Z. Jin and C. Huang, *Molecules*, 2019, **24**, 2923.
87. L. Chen, L. Wu, J. Yu, C. T. Kuo, T. Jian, I. C. Wu, Y. Rong and D. T. Chiu, *Chem. Sci.*, 2017, **8**, 7236–7245.
88. J. Ou, H. Tan, Z. Chen and X. Chen, *Sensors*, 2019, **19**, 1455.
89. T. Maxwell , M. G. Nogueira Campos, S. Smith, M. Doomra, Z. Thwin and S. Santra, Nanoparticles for Biomedical Applications, in, ed. E. J. Chung, L. Leon and C. Rinaldi, Elsevier, 2020, pp. 243–265.
90. I. L. Medintz, M. H. Stewart, S. A. Trammell, K. Susumu, J. B. Delehanty, B. C. Mei, J. S. Melinger, J. B. Blanco-Canosa, P. E. Dawson and H. Mattoussi, *Nat. Mater.*, 2010, **9**, 676–684.
91. X. Ji, G. Palui, T. Avellini, H. B. Na, C. Yi, K. L. Knappenberger and H. Mattoussi, *J. Am. Chem. Soc.*, 2012, **134**, 6006–6017.
92. X. Hu and X. Gao, *ACS Nano*, 2010, **4**, 6080–6086.
93. T. Jin, A. Sasaki, M. Kinjo and J. Miyazaki, *Chem. Commun.*, 2010, **46**, 2408–2410.
94. T. Zhao, Y. Fu, M.-S. Jang, X. S. Sun, T. Wu, J. H. Lee, Y. Li, D. S. Lee and H. Y. Yang, *Mater. Sci. Eng.*, 2021, **118**, 111449.
95. S. Ast, P. J. Rutledge and M. H. Todd, *Phys. Chem. Chem. Phys.*, 2014, **16**, 25255–25257.
96. C. Chen, P. Zhang, L. Zhang, D. Gao, G. Gao, Y. Yang, W. Li, P. Gong and L. Cai, *Chem. Commun.*, 2015, **51**, 11162–11165.

CHAPTER 7

Environmental Application of Quantum Dots

NAHID POURREZA*[a], MATINEH GHOMI[a] AND MAHTAB YADOLLAHI[a]

[a]Department of Chemistry, Faculty of Science, Shahid Chamran University of Ahvaz, Ahvaz, Iran
*E-mail: npourreza@scu.ac.ir

7.1 Importance of Environmental Issues

Environmental pollution is of great concern for scientists as it continues to hamper sustainable development. Likewise, the monitoring of many compounds in the environment is of critical importance due to their potential negative health effects. Commonly monitored compounds include heavy metal ions, organic pollutants such as dyes, polycyclic aromatic hydrocarbons (PAHs), pesticides, drug residues, and other substances that have been found to be either carcinogenic or potentially carcinogenic. These compounds are inherently toxic, and their toxicity may be enhanced when they are metabolized and photo-oxidized into derivative compounds.[1] For example, heavy metal ions are biologically and technologically important, but they are also pollutants if consumed in excess. Moreover, in environmental systems where they may occur as complex mixtures, their toxicity may be increased; their detection is therefore critical. Environmental pollution can also affect agricultural products that are consumed by humans, such as foods and herbal medicines. As these compounds typically occur at low concentrations in

Detection Science Series No. 22
Quantum Dots in Bioanalytical Chemistry and Medicine
Edited by Michael Thompson and Zahra Ramezani

Published by the Royal Society of Chemistry, www.rsc.org

environmental samples, sensitive and selective methods are required for their determination.

The quantum confinement effect was discovered in the early 1980s; since then, many potential environmental applications have been suggested for quantum dots (QDs). Carbon-based QDs (CQDs), specifically, have garnered significant attention recently due to their exceptional properties.[2] Among various light emitters, QDs have received much attention in recent decades because of their easy and economical synthesis process.[3] Furthermore, modifying QDs' surfaces with various compounds and functional groups can enhance their selectivity for the determination of specific pollutants. QDs' fascinating characteristics, such as their small size, low toxicity, and chemical inertness, make them versatile materials; for instance, they have been used in wastewater treatment. Efforts are ongoing to develop new materials and synthetic methods in order to create QDs for use in the removal and determination of environmental pollutants.[4]

7.2 Quantum Dots for Heavy Metal Detection in Environment

Heavy metals are elements with a density higher than 5 g cm^{-3}. They can be found in multiple species (electronic or oxidation state, complex or molecular structure, and inorganic/organic structure as free cations or alkylated forms) depending on their environmental circumstances and their interactions with other substances. Moreover, heavy metals can be classified based on their toxicity into essential and non-essential compounds. Some of them, such as cadmium, lead, mercury, and arsenic, have been identified as carcinogenic and mutagenic according to the World Health Organization (WHO).[5,6] When heavy metals are dumped into water sources, they bioaccumulate in living organisms and eventually impair human health when they pass through the food chain.[7] Without sufficient control and monitoring, the electronic, industrial, and mining industries discharge heavy metal-containing effluent and solid waste on a regular basis into water bodies.[7,8] In addition to heavy metals, pesticides and fertilizers are also present in our food chain, as their use is unavoidable in agribusiness.[5,8] These toxic heavy metals cannot be metabolized or biodegraded in human and animal bodies, which results in their progressive accumulation. Hence, they pose a serious threat to the environment and human health. Some heavy metals, such as copper, zinc, and chromium, are less hazardous. At low doses, they are needed as micronutrients but at greater amounts, they become poisonous.[9] Heavy metal pollution occurs because of both natural and human processes and appears in complex environmental, industrial, and biological sample matrices. It is, therefore, critical that we be able to determine heavy metals' concentrations and their species, analyze the health hazards they pose, and monitor the pollution index. However, detection and analysis of these heavy metals require hazardous chemicals and costly equipment, which slows down the regulatory and monitoring procedures.

While various compounds have been used to detect heavy metals, CQDs have emerged as a subject of particular interest recently. The development of CQD-based heavy metal sensors is ongoing, with potential applications in pollution monitoring at industrial waste-contaminated sites.

CQDs are luminescent compounds that are less than 10 nm in size. Their fluorescent features were discovered coincidentally during the purification of single-walled carbon nanotubes.[2] When CQDs interact with specific compounds, their fluorescence changes. Their photoluminescent emissions occur at different wavelengths due to changes in the excitation wavelength. This phenomenon may be due to the size of CQDs, which leads to changing luminescence features. The surface functional groups can also be effective in causing this phenomenon (altering the fluorescence band gaps).[10] Several studies have examined their interactions and quenching effects. These non-toxic CQDs are very stable, soluble in water, and biocompatible. More importantly, they can be easily synthesized and functionalized.[11,12] Various structures, including diamond-like, graphite structures, graphene oxide (GO) structures, amorphous carbon structures, and amorphous carbon structures with graphitic clusters, have been considered for CQDs.[13] Graphene-based QDs, CQDs, and carbon nanodots exhibit structural differences. CQDs possess a spherical carbon-based shape with a crystal structure, whereas GQDs consist of a single layer of graphitic carbon. Carbon nanodots are amorphous and can be formed from monomers or polymers; for this reason, they are sometimes known as polymer dots.[14] Due to their luminescence and ease with which they can be prepared from readily available and eco-friendly sources, QDs are a good choice for use in sensors that measure metal ions.[15] Methods for using QDs to measure heavy metals such as Hg^{2+}, Fe^{3+}, Cu^{2+}, and Pb^{2+} have been investigated in this chapter.

7.2.1 Mercuric Ions

Mercury can be released into the environment *via* a variety of sources, including paper-pulp preservatives, chlorine, and caustic soda manufacturing.[16] The WHO recommends that mercury levels in drinking water be less than 0.001 mg L^{-1}.[17] Mercury's accumulation and toxicity in biological systems is known to cause acrodynia (also known as pink illness). Mercury poisoning alters brain functions and produces symptoms such as shyness, memory issues, irritability, and changes in vision and hearing. Lung damage, vomiting, diarrhea, nausea, and an elevated heart rate are also symptoms of metallic mercury exposure.

CQD-based fluorescence-sensing systems possess various advantages (such as excellent selectivity, sensitivity, quick sensing response time, and non-destructive nature) that make them especially useful for mercury ion

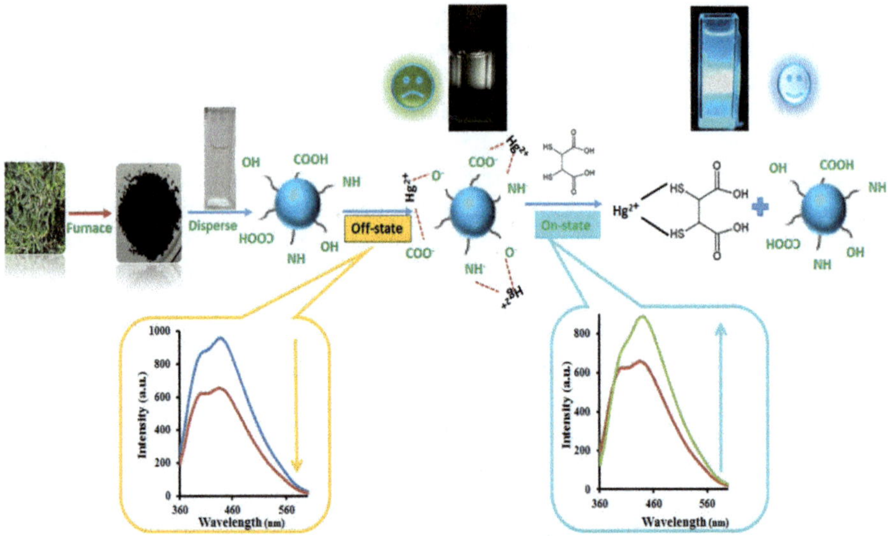

Figure 7.1 A dual off–on fluorescence probe for sensing mercury(II) and Chemet drug by Green synthesized CQDs from *Prosopis juliflora* leaves. Reproduced from ref. 22 with permission from Elsevier, Copyright 2022.

detection. Hence, CQDs have been thoroughly developed as a sensor for detecting heavy metals in water samples. Meanwhile, mercury is extremely dangerous as its poisoning manifests itself in a wide range of ways, depending on the kind of mercury involved, the method of entry, the amount of exposure, and the length of exposure. It can also cause harm to the central nervous system, damage to the liver, the kidney, and the intestine, and even death because of overdose.[18,19] Several studies have used CQD-based methods to detect mercury ions.[19–21] For example, a simple green dual fluorescence sensor was developed for mercury detection based on CQDs' "off–on" fluorescence. This sensor was prepared from a green source (*Prosopis juliflora* leaves) *via* a simple one-pot synthesis protocol. The off fluorescent was observed in the presence of various mercury ion concentrations with a luminescence quantum yield (QY) of 5%. Linear calibrations were obtained in the range of 5–500 ng mL^{-1} with a 1.26 ng mL^{-1} limit of detection (LOD). This sensor was successfully used to detect mercury in water and human serum samples. The sensing strategy is illustrated in Figure 7.1.[22]

There are several advantages to green CQD synthesis: it uses low-cost and non-toxic raw materials, involves renewable resources, is easy to perform, and is environmentally friendly. Green-synthesized CQDs can be purified and size-controlled through post-synthesis techniques such as centrifugation, dialysis, filtration, chromatography, and gel electrophoresis. This has led to a demand for research on facile and environmentally friendly green methods. Unlike conventional physico-chemical techniques, the stable synthesis of CQDs is produced at ambient temperature and neutral pH, with additional advantages such as improved stability, scalability, sustainability, and biocompatibility. While plant-mediated green synthesis presents many advantages, it does come with one obvious disadvantage: it is challenging to completely separate CQDs from biomass.[23]

In one study, green CQDs were synthesized from pomelo peels and used for mercury ion detection. In the presence of mercury ions, the fluorescence of CQDs was quenched, and the fluorescence was restored using cysteine as a chelator. The sensing mechanism can be explained as follows: the addition of mercury causes a significant drop in fluorescence intensity (*i.e.*, fluorescence is quenched) *via* electron or energy transfer; when a powerful mercury chelator (in this case, cysteine) is added to the sensing system, mercury is removed from the surface of the CQDs through the interaction between mercury ions and the S functional group of cysteine, resulting in fluorescence recovery. This approach was shown to have a detection limit of 0.23 nM; it was used to successfully detect Hg^{2+} in lake water samples.[24]

In another study, a nanohybrid ratiometric fluorescence sensor was prepared using fluorescent carbon nanoparticles and rhodamine B. The fluorescence of this sensor system was also quenched in the presence of mercury ions due to an electron transfer mechanism, which resulted in a noticeable fluorescence color fluctuation from violet to orange. This sensor was used to detect mercury ions in aqueous media with a detection limit of 42 nM.[25]

Various studies have used real-world samples to test the viability of CQD-based mercury sensors. For instance, these sensors have been used on river, lake, tap, and mineral water, and have even been employed to test for mercury in cosmetics.

The majority of the reported recovery was good and acceptable (greater than 95%).[18,26] Due to the presence of interfering species in the wastewater sample, the mercury recoveries were less than 50%.[26] In most CQD-based sensors, selective quenching in the presence of mercury ions occurs due to the existence of thiol functional groups on the surface of the CQDs. These thiol functional groups have a higher affinity with mercury ions, and, therefore, more readily form a thiol mercury complex compared to other heavy metals due to the larger ionic radius of mercury and its greater polarization. Furthermore, the S atoms can alter the local electrical density of CQDs' surfaces to promote the oxygen–mercury interaction.[26,27] While thiol is the most popular choice, other functional groups such as phosphorus, nitrogen, and oxygen on the surface of CQDs can interact with mercury ions as donor sites.[28]

7.2.2 Ferric Ions

Iron is the second most plentiful element in the earth's crust. It is a crucial part of biological systems because it is a cofactor for numerous important enzymes and proteins. Moreover, it is required for cellular metabolism, transporting oxygen in hemoglobin, and enzyme catalysis.[29] Iron deficiency can cause a variety of disorders, such as anemia, liver and renal malfunction, heart failure, diabetes, and cancer. Ferric ion monitoring *via* precise sensor systems is therefore a worthwhile subject of research.[30]

CQDs can be used as sensors for ferric ions. Their photostability, biocompatibility, low toxicity, and high sensitivity are all outstanding. In one study, a CQD-based sensor made from gelatin (as a precursor) using a one-step hydrothermal technique was reported for ferric ion detection. The detection mechanism was based on fluorescence quenching with a concentration range of 0–50 μM and a detection limit of 0.2 μM. This sensor was applied to detect ferric ions in real water samples.[31]

In another report, the maximum luminescent intensity of the CQDs illustrated good selectivity toward ferric ions in the presence of metal mixtures. This selective detection was possible due to the improved charge-transfer action and the ion-selective nature of the oxygenated functional groups, which promoted the creation of CQD–ferric ion complexes. This sensor's lowest detection limit was 0.0076 μM.[10]

In another study, *Vigna radiata* sprouts were used as carbon precursors to synthesize N-doped CQDs with high QY (about 58%). The linear range of this sensing protocol was 100–2000 μM with an LOD of 140 nM.[32] N-doped or F-doped CQDs were investigated in real samples containing ferric ions, such as tap and lake waters. The CQDs' ion detection ability was attributed to the unique coordination interaction between ferric ions and the N-doped and F-doped CQDs' surfaces, which incorporated nitrogen and fluorine functional groups, respectively. Both of these sensors illustrated good recovery (>92%); the N-doped CQDs exhibited the best recovery (100.00 ± 1.00%). Even more striking, the detection results from N-doped CQDs were nearly identical to those from atomic absorption spectrometer (AAS) analysis, with a difference of just 0.002 mg L^{-1}.[33,34] These selective responses toward ferric ions may be attributed to the enhanced charge transfer effect *via* the complex formation between the oxygen or hydroxyl functional groups of the CQDs' surfaces.[12,33]

7.2.3 Cupric Ions

Cupric ions are transition metal ions that can cause damage to biological components (such as DNA and proteins) as well as to the neurological and gastrointestinal systems.[35] In one study, cupric ions were detected by CQDs produced from biomass sources. Biomass (such as hemicelluloses, lignin, cellulose, and chitosan) can be hydrothermally treated to produce CQDs. In this method, the biopolymers are hydrolyzed and dehydrated into tiny molecules (such as furfural aldehydes, ketones, or formic acid). Next, these chemicals are polymerized and condensed to form a black solid residue. The resulting hydrothermal carbons (HTC) are then rapidly decomposed into CQDs, which exhibit intense fluorescence after treatment with a $NaOH/O_2$ solution. Based on the weight of the HTC, the conversion rate of CQDs is quite high; treating the HTC with $NaOH/O_2$ is a highly efficient technique for obtaining CQDs. The detection mechanism of these green CQDs can be attributed to the substantial interaction between the cupric ions and the surface functional groups (hydroxyl or carboxyl groups) on the CQDs. The study's findings demonstrated a suitable linear range for cupric ion detection, ranging from 0 to 30 μM with an LOD of 85 nM.[36]

In another study, hydroquinone, *o*-phenylenediamine, and terephthalic acid were employed as reaction precursors in a one-step solvothermal technique to manufacture CQDs. The findings showed that the as-prepared functionalized CQDs exhibited not only remarkable optical features but also intriguing chemical and physical features. With increasing water content in the organic solvent, the emission peak of the produced CQDs changed gradually toward the red area. This was shown graphically as a color shift. This phenomenon may be due to the presence of water, which affected the surface state of the h-CDs in acetone. This sensor was very sensitive to Cu(II); it was used to measure Cu(II) concentration in water with a detection limit of 1.8×10^{-4} mmol. The surface structures and chemical bonds of CQDs were disrupted in the presence of Cu(II), which interacted with the surfaces of the CQDs to generate non-fluorescent ground state complexes containing Cu(II).[37]

CQDs can also be combined with other materials to form cupric sensors. For instance, a novel ratiometric fluorescence-based sensor was synthesized *via* encapsulating CQDs and CdTe/CdS/ZnS QDs into a metal organic framework of ZIF-8. The synthesized composite combined the advantages of a ratiometric fluorescent sensor with those of a metal organic framework (*i.e.*, its adsorption ability). The composite demonstrated high stability and dispersibility in aqueous solution. This CQD-based composite exhibited excellent sensitivity and selectivity toward cupric ions in water samples, with a linear range of 5–100 nM and a LOD of 1.53.[38] Some functional groups, such as hydroxyl and carboxyl groups, on the surface of CQDs can selectively form chelates with cupric ions. However, the presence of sulfhydryl functional groups in biothiols leads to the dissolution of the copper–CQD complex and the restoration of the CQDs (Figure 7.2).[39] Generally, it is hypothesized that selective quenching occurs as a result of (i) the change in structure of the copper–CQD complex from octahedral to square planar and (ii) non-radiative photo-induced electron transfer between the lowest unoccupied molecular orbitals of CQDs and the d-orbitals of copper.[40]

7.2.4 Lead Ions

Lead ions are poisonous to humans: they can cause anemia, neurological and reproductive system failure, and even death. Furthermore, excessive lead concentrations in the blood can cause disorders (such as neurological,

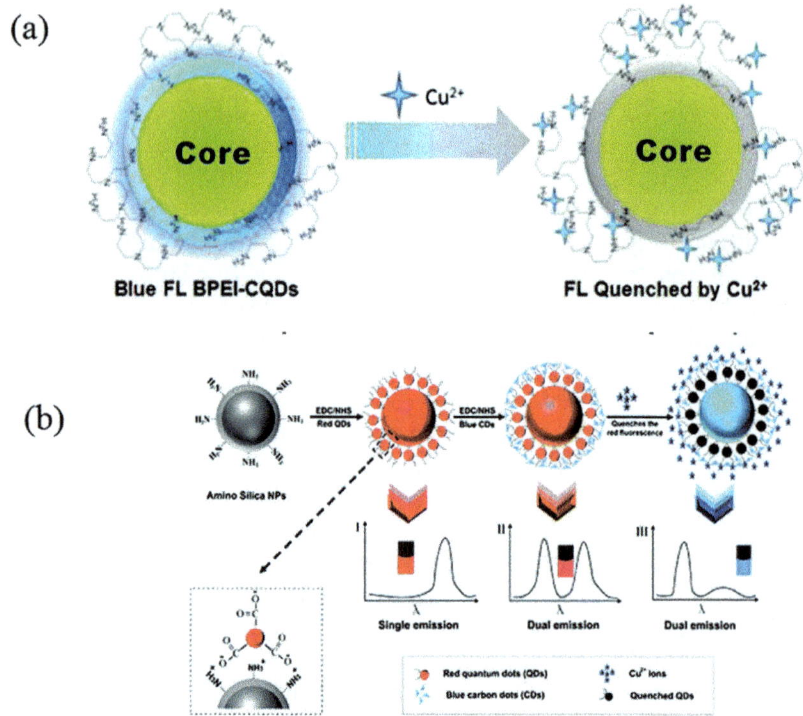

Figure 7.2 (a) Schematic illustration of a ratiometric nanoprobe structure and the visual recognition principle for copper ions and (b) fluorescence of branched polyethylenimine CQDs quenched by copper ions. Reproduced from ref. 39 with permission from the Royal Society of Chemistry.

reproductive, and cardiovascular) that may persist indefinitely. According to US Environmental Protection Agency (EPA) rules, the maximum allowable concentration of lead in water is 72 nM. As a result, detecting lead ions in water is crucial.[41]

CQDs have been used to detect lead ions in aqueous samples. Dual emission (blue and red) CQDs were prepared in a one-step hydrothermal process employing citric acid (CA) and ethylenediamine (EDA) as precursors in a water–formamide binary system. The as-prepared CQDs were then used directly to make ion-imprinted fluorescent polymers without additional separation. Two emission peaks (440 nm and 580 nm) were measured at excitation wavelengths of 365 and 520 nm, respectively. Then, ion-imprinted fluorescence polymers were synthesized by templates of chrome(III), lead ions, and the CQDs. This design reduces mass transfer resistance and improves detection sensitivity. The mesoporous ion imprinted on fluorescence polymers displayed high sensitivity and specificity for the above metals in the blue and red channels. This protocol, thus, combined the high selectivity of the molecular imprinting method with the high sensitivity of fluorescence detection. The lead ions quenched the fluorescence of red CQDs using ion imprinted polymers. The quenching effect $[(F_0 - F)/F_0]$ at 580 nm for the red emission against the concentration of lead ions was linear from 0.1 to 5.0 μM.[42]

The great affinity and capacity of lead ions to bind with CQDs is typically due to their interaction with the hydroxyl groups on CQDs' surfaces. Another possibility is related to the interaction of the unoccupied d-orbitals of lead ions with the amine group on the surface of CQDs.[43,44]

Table sugar can be used as a simple and cost-effective precursor for CQD preparation. This method of synthesis requires no chemical modifications or post-chemical treatments. Sugar-derived CQDs visibly aggregate in the presence of lead ions; using this method, this harmful ion can be detected with the naked eye. The reason for this phenomenon is the accumulation of CQDs in the presence of Pb(II) ions. A simple turbidimeter can identify the presence of lead ions with exceptional selectivity: in this study, the sensor detected lead ions with an LOD as low as 14 ng ml^{-1}; it is quite impressive relative to the permitted level in drinking water (15 ng ml^{-1}) as recommended by the US EPA.[45]

7.2.5 Other Metal Ions

CQDs and doped CQDs have been used as sensors for other dangerous heavy metals, such as Cr(VI), Co(II), and Au(III). Metalloids such as As(III) can be detected by glutathione-modified CQDs; in one study, selective quenching occurred *via* the interaction of excited As(III) with glutathione's oxygen functional groups.[46]

Arsenic is a toxic element and a significant environmental (including water) pollutant. Green multicolor luminescent CQDs were synthesized using quince fruit (*Cydonia oblonga*) for use as fluorimetric sensors for As(III) in water samples. The prepared CQDs showed a maximum emission intensity at 450 nm after excitation at 350 nm with a quantum yield of 8.55%. As(III) and MnO_4 were mixed and then added to the CQDs. This mixture exhibited more intense photoluminescence compared to permanganate alone, as the permanganate reacted with As(III) and was converted to Mn(II) ions. This increase showed a linear relation with As(III) in the range of 0.1–2 μg mL^{-1} at pH 5 using citrate buffer. The method was utilized for As(III) detection in tap water samples.[47]

Excess palladium in drinking water can cause serious health problems. In recent years, fluorescent CQD-based probes have received much attention for their potential for palladium sensing. In one study, palladium detection was based on a "fluorescence turn-off" protocol. The synthesized nanoprobes used nitrogen-doped red-emitting carbon dots (NRCDs). These biocompatible and photostable probes were prepared *via* the hydrothermal reaction of *p*-phenylenediamine with amino benzoic acid in the presence of nitric acid at 180 °C. The as-synthesized NRCDs showed high photostability with excitation-independent near-infrared emission and good water solubility, with a QY of 18%. The NRCDs were able to selectively detect Pd^{2+} and In^{3+} in water among other metal ions; they exhibited a turn-on fluorescence response in the presence of In^{3+} with a red shift, while their fluorescence was quenched in the presence of Pd^{2+}. The interactions of the –OH groups on the NRCDs' surfaces with Pd^{2+} and In^{3+} ions were also analyzed *via* their NMR and FTIR spectra. Encapsulation of NRCDs in biocompatible vesicle membranes was performed with or without Pd^{2+} and In^{3+} metal ions and investigated in sensing, confocal fluorescence imaging.[48]

In another study, a simple one-step hydrothermal treatment was used to prepare CQDs from fresh *Momordica charantia* as the only precursor. This fluorescent sensor speedily detected palladium ions *via* the static quenching effect: the sensor's fluorescence was quenched in the presence of palladium ions with linear ranges of 0–325 μM. The device exhibited good selectivity and sensitivity toward palladium ions, with a detection limit of 0.348 μM. This is due to the static quenching effect obtained after studying the results of fluorescence lifetime, quenching constant, quenching rate constant, and ultraviolet-visible (UV-Vis) absorption. The sensing system was used to identify palladium ions in tap and environmental water with a good recovery range (96.25–107.3%). These results highlight the system's significant promise as a viable platform for palladium ion sensing in aquatic matrices.[49] The comparison of some CQDs for other ions is summarized in Table 7.1.

7.3 Quantum Dots for Organic Pollutants

Environmental pollution caused by industrial expansion poses a significant threat to the environment. Concerns about toxic organic pollutants in water resources have attracted the WHO's attention. The disposal of organic pollutants (such as dyes and pesticides) has led to the pollution of aquatic environments such as groundwater aquifers, rivers, lakes, and oceans. Today, much research is focused on the determination and removal of toxic organic pollutants from water, as well as on reducing these pollutants' toxic effects on aquatic ecosystems and, hence, reducing harm to humans and animals.[57] Various methods have been used to remove toxic organic water pollutants, including adsorption, reverse osmosis, filtration, coagulation, sedimentation, and biological methods. However, these methods are less suitable for certain organic pollutants such as pesticides and dyes.[58] QDs have shown great promise in the field of pollution assays.

7.3.1 Polycyclic Aromatic Hydrocarbons

PAHs are among the most common harmful environmental organic pollutants that affect water resources. These pollutants may be introduced

Table 7.1 Some CQD-based sensors for heavy metal ions detection. to

CQDs precursor and synthetic route	Analyte type	LOD	Linear range	Applications	Ref.
Natural deep eutectic solvents/solvothermal method	Co(II)	1.2269 μM	5–250 μM	Real water samples	50
Bamboo leaves/ hydrothermal method	Cu(II)	0.115 μM	0.333–66.6 μM	Environmental water samples	51
Coconut shell/ chemical oxidation procedure	As(III)	500 pM	0.1–11 nM	Bore-well water and industrial effluents	52
CA EDA/hydrothermal	Cr(VI)	0.26 μM	0–10 μM	River water	53
Edible prickly pear cactus fruit/ hydrothermal	Cr(VI)	0.50 mg L^{-1}	0.10–0.60 mg L^{-1}	Tap/river/pond water/ industrial wastewater	54
Coal washery rejects/wet chemicaloxidationme	Cr(VI) Hg(II)	0.145 μM 0.24 μM	0–1.2 μM 0–1 μM	Water	55
Graphene quantum dots	Cd(II)	26.8 nM	0.5–100 μM	Lake water	56

water sources *via* industrial wastewater or oil spills; they may also be introduced naturally through the incomplete combustion of coal, oil, gas, and wood. Most of these compounds are carcinogens; the development of inexpensive, highly sensitive, and selective sensors for these pollutants is therefore of great importance. In one study, an electrochemical sensing method was used to detect PAHs (*e.g.*, phenanthrene, naphthalene, anthracene, benzo[*a*]anthracene, chrysene, and fluorene) in water solutions using thiolated calix[4]arenes immobilized on Fe_3O_4 magnetic nanoparticles and CdSe QDs with a size of about 10 nm as electrochemical label for PAHs. CdSe QD/PAHs interacted with Fe_3O_4/calix[4]arene nanoparticles through the cavities of calix[4]arenes, leading to the formation of calix[4]arenes supramolecules. The sensor's electrochemical signal changed significantly based on the size of the cavities on calyx[4]arenes, leading to selective sensing for various PAHs. The sensor's signal decreased in proportion to the increase in PAH concentration and vice versa (Figure 7.3). The square wave voltammetry response of this sensor therefore decreased linearly for naphthalene and anthracene in the range of 1.5×10^{-6} to 2.5×10^{-5}, and 2.1×10^{-7} to 1.4×10^{-5} M, respectively. The calculated LODs for naphthalene and anthracene were 105.5 and 20.1 ng mL^{-1}, respectively. The performance of

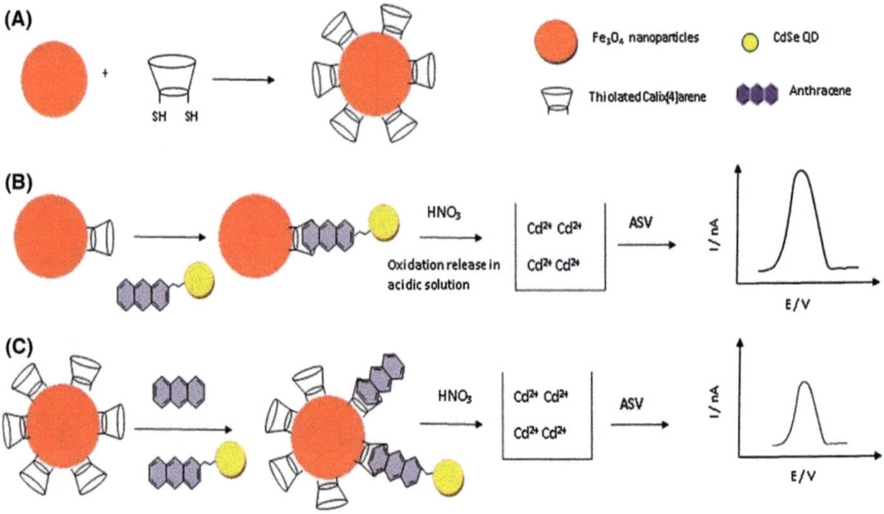

Figure 7.3 The schematic illustration of the stepwise sensor fabrication and determination process. Reproduced from ref. 59 with permission from Springer Nature, Copyright 2014.

the method was evaluated for spiked tap water samples; average recoveries of 92–113% were obtained.[59]

In another study, graphene carbon dots (GCQDs) and CQDs were prepared from CA and GO precursors using bottom-up and top-down approaches, respectively. They were then used as sensors to detect pyrene in environmental water samples based on the fluorescence emission turn-off/on method. Initially, the interaction of GCQDs with Fe^{3+} turned off the fluorescence emission; after the interaction of GCQDs/Fe^{3+} with pyrene, fluorescence emission was turned on again (Figure 7.4). For both GCQDs (CA)/Fe^{3+} and GCQDs (GO)/Fe^{3+}, fluorescence emission intensity increased linearly with the addition of different pyrene concentrations (2×10^{-6} to 10×10^{-6} mol L^{-1}). For GCQDs (GO)/Fe^{3+}, the limits of quantitation and detection were 1.08×10^{-6} mol L^{-1} and 0.325×10^{-6} mol L^{-1}, respectively. For GQDs (CA)/Fe^{3+}, the limits of quantitation and detection were 0.807×10^{-6} mol L^{-1} and 0.242×10^{-6} mol L^{-1}, respectively. The GCQD sensor also demonstrated excellent recoveries in the range of 97–107% for environmental water samples.[60]

A detection method was developed for PAHs using an L-cysteine-capped CdSeTe/ZnSe/ZnS core/shell/shell QD–GO nanocomposite. The fluorescence intensity of the QD–GO nanocomposite was enhanced compared to the unconjugated QDs in the presence of PAHs. Researchers thoroughly investigated the interaction mechanism by examining the fluorescence emission spectra of the QD–GO nanocomposite in the absence and presence of phenanthrene. They concluded that the

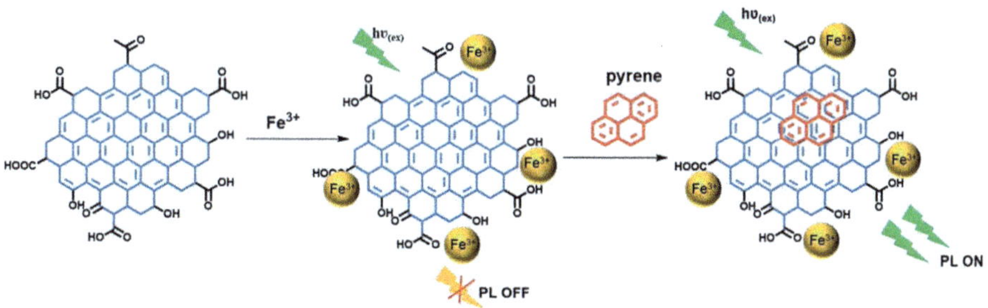

Figure 7.4 Schematic illustration of the interaction of GQDs with Fe^{3+} which results in fluorescence quenching. Upon interaction with pyrene, the fluorescence is turned on. Reproduced from ref. 60 with permission from the Royal Society of Chemistry.

fluorescence enhancement effect of the PAHs could be due to the electron-rich π system of the PAHs, and, hence, the strong π–π interaction induced by the hexagonal array of the carbon atoms of the GO nanosheets in the QDs–GO probe. An LOD of 0.19 mg L^{-1} was obtained for phenanthrene under optimal conditions; the LOD of anthracene, pyrene, and naphthalene was estimated to be about 0.26 mg L^{-1}. The probe, therefore, shows great potential for use as a sensing tool for PAH monitoring in environmental water samples.[61]

7.3.2 Photocatalytic Degradation of Organic Pollutants and Dyes

Recently, QDs have been used as photocatalytic materials to remove hazardous organic matter. Titanium dioxide (TiO_2) QDs are commonly used for the removal and decomposition of organic pollutants due to their chemical and thermal stability, non-toxicity, low cost, and excellent photocatalytic properties.[62] The most important feature of these nanoparticles is their large specific surface area and quantum confinement; the preparation of TiO_2 QDs with narrow size distributions and specific morphologies is highly desirable.[60] Tin oxide (SnO_2), an N-type semiconductor with a band gap of ~3.64 eV, has been used to treat water samples, both in the photocatalytic degradation of organic pollutants and in the removal of heavy metals.[64,65]

In recent years, researchers have also used SnO_2 QDs for the photocatalytic degradation of organic pollutants.[66] SnO_2 QDs with high photocatalytic efficiency and sizes ranging from 3 to 5 nm were prepared using a microwave method. They exhibited a methylene blue (MB) photocatalytic reaction with 90% degradation efficiency in the presence of visible light irradiation. The SnO_2 QDs' transmission electron microscopy (TEM) images and X-ray diffraction (XRD) pattern indicated that they were composed of ultra-fine

QD nanoparticles with agglomerated morphology and a cassiterite tetragonal crystal structure.[67]

The most common method for preparing TiO_2 QDs is the sol–gel reflux method. When produced using this method, these QDs can exhibit photocatalytic performance in the UV range, which allows them to remove dye pollutants such as MB, indigo carmine (IC), methyl orange (MO), and rhodamine B.[62,68,69] TiO_2 QDs with a size of about 5–10 nm and an anatase morphology were prepared using the sol–gel method without using any chelating compounds. These QDs were used for the decomposition of IC dye. The synthesized TiO_2 QDs exhibited excellent photocatalytic behavior: when introduced to water samples containing 25 ppm of IC dye, they achieved 95% IC degradation within 60 min in the presence of UV light.

ZnO QDs have also been used extensively for photocatalytic degradation of organic pollutants in water under UV light. In one study, ZnO QDs with excellent photocatalytic properties and a size of 7.1 nm were prepared using a simple sol–gel method. High-resolution transmission electron microscopy (HRTEM) and XRD images of these ZnO QDs showed that they were ultra-fine QD nanoparticles with an elongated ellipsoidal morphology and high single-phase crystallinity. Reducing the size of ZnO QD increases their surface area and also increases the excitons on their surfaces, which improves the performance of the photocatalyst. For this reason, the ZnO QDs with smaller particle sizes showed the highest percentage of MO optical degradation at room temperature.[70]

ZnS QDs can be doped with various metals to improve photocatalytic reactions. In a study by Amani-Ghadim *et al.*, for instance, these QDs were doped with Nd, Dy, and Gd using a simple reflux method to improve the photocatalytic degradation of acid azo red dye 14 (AR14). The study found that about 90% of the AR14 degraded in the presence of Gd-doped ZnS QDs under UV irradiation. The influence of g-C_3N_4 and GO on the photocatalytic activity of ZnS QDs was also investigated by the degradation of bisphenol-A (BA) and AR14. g-C_3N_4 is a 2D N-type nanomaterial with a band gap of ~2.7 eV. It exhibits good thermal and chemical stability, is non-toxic, and is inexpensive and easy to synthesize for use in the photocatalytic degradation of organic pollutants. The photocatalytic reaction efficiencies for degradation of BA and AR14 in the presence of visible light irradiation at 180 min were 99.1 and 80.5%, respectively, using Gd-doped ZnS QDs/g-C_3N_4. The improved activity of the Gd-doped ZnS QDs/g-C_3N_4 was related to the compounds' high specific surface area, their efficient separation of photo-induced electron–hole pairs, and their low electron transfer resistance and photosensitization mechanism was suggested for this activity.[71,72] The mechanism of the reactions is presented in Figure 7.5.

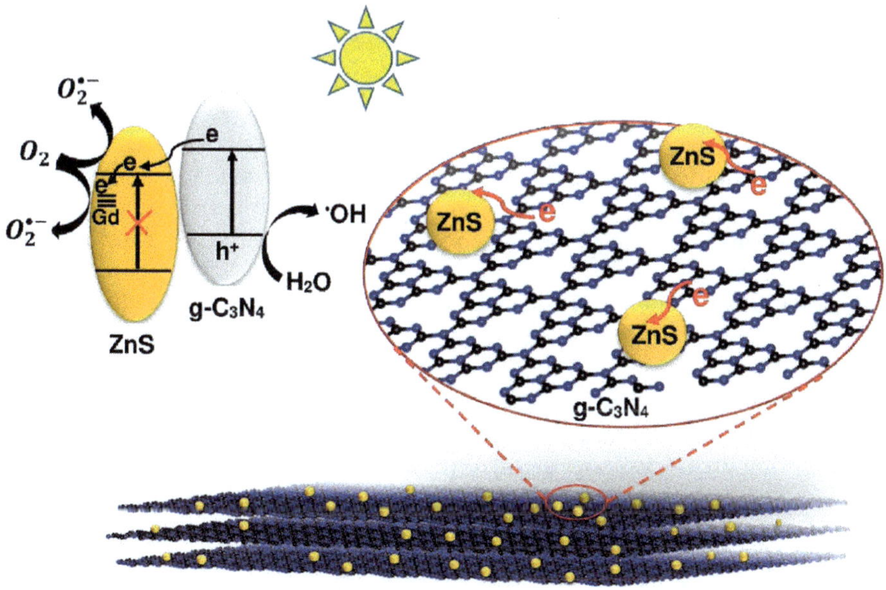

Figure 7.5 Suggested mechanism for photocatalytic activity of the Gd-doped ZnS QDs/g-C_3N_4. Reproduced from ref. 72 with permission from Elsevier, Copyright 2022.

7.4 Pesticide Detection

Worldwide developments in agriculture have led to the pollution of water resources by pesticides. Pesticides are chemical compounds (*e.g.*, organochlorine, carbamates, and organophosphates) that are widely used in agriculture to prevent crop damage by insects. However, these compounds are carcinogenic and harmful to human and animal health even at low concentrations (ppb levels) in water. It is, therefore, critical that scientists be able to detect them in water sources using efficient methods. Attempts have been made to develop a sensor for pesticide detection that leverages these chemicals' quenching effect on the photoluminescence of ZnCdSe QD-deposited film. In one study, a wet-chemical process was used to synthesize ZnCdSe QDs. The QDs were placed on the tip of a fiber optic probe as a sensing element for pesticides with the trade names DiPel, Sevin 85%, Sevin 85% wettable powder (WP), and water-dispersible granules (WG). The sensor was used to detect pesticides in water with a concentration in a linear range of 2.5–2500 μg L^{-1}.[73]

Triticonazole is a broad-spectrum systemic triazole fungicide that is widely used for seed treatment against foliar diseases. However, it is highly toxic; its residue can cause environmental pollution and it is hazardous to human health. Boron-doped CDs (B-CDs) and phosphorous-doped CDs (P-CDs) were prepared and encapsulated into a zeolitic imidazolate framework-8 (ZIF-8) and used as a dual-emissive MOF ratiometric fluorescence probe for triticonazole. The B-CDs/P-CDs@ZIF-8 composite was prepared by encapsulating B-CDs and P-CDs into zZIF-8 and used as a ratiometric fluorescence sensor for triticonazole. Two emissions were observed at 440 and 510 nm for BCDs and PCDs, respectively, after excitation using a single wavelength of 385 nm. The fluorescence intensity of the B-CDs decreased with the addition of triticonazole, but that of the P-CDs remained constant. The linear range of the sensor was 80–300 nM and its LOD was 32 nM.[74]

Nicosulfuron is the most common pesticide applied to maize; its residue can enter groundwater and cause serious harm to the environment and human health. Mn-doped ZnS QDs with a coating of molecularly imprinted polymers (MIPs) can be used to detect this pesticide. In one study, MIPs@Mn-doped ZnS QDs were prepared using tetraethoxy silane (TEOS) cross-linking and 3-aminopropyl triethoxy silane (APTES) monomer with a size of ~27 nm by the sol–gel method. The synthesized MIPs@Mn-doped ZnS QDs are excellent fluorescence probes for nicosulfuron. To evaluate the fluorescence potential of these MIPs@Mn-doped ZnS QDs, nicosulfuron was added to 30 mL water samples in concentrations of 0.024, 0.24, and 2.4 μmol L^{-1}, and the fluorescent response of the nicosulfuron was measured. The quantitative recoveries ranged from 89.6% to 96.5% with relative standard deviations (RSDs) of 2.5% to 5.7%. The MIPs@Mn-doped ZnS QDs, thus, showed great potential for detecting nicosulfuron pesticide in river water samples.[75]

7.5 Quantum Dots for Drug Residue Assay

Pharmaceuticals are biologically active molecules that are designed to target specific biological processes in human or animal bodies. They are produced in large quantities for use in medical and veterinary contexts, and they can enter ecosystems through a variety of pathways, including disposal of unused or expired medication; human, animal, and hospital waste; animal carcasses; chemical fabrication; and effluents discharged from medical manufacturing industries. Additionally, pharmaceuticals that enter the environment through one pathway may be further dispersed through food chains. Most pharmaceuticals are active at low concentrations, which makes them particularly potent contaminants. The European Union (EU) has, thus, established maximum drug residue limits in livestock to protect human health.[76,77]

The primary concern regarding pharmaceutical pollution is antibiotic resistance. Studies have demonstrated that measured environmental concentrations of antibiotics are high enough to exert a selective pressure on clinically relevant bacteria, which may lead to an increase in the prevalence of antibiotic-resistant bacterial strains. Scientists have, therefore, focused on antibiotics over other pharmaceutical residues in an effort to curb the human and animal health risks associated with the formation of antibiotic-resistant bacteria.[78] Various methods have been used to detect veterinary residues, including high-performance liquid chromatography (HPLC), mass spectrometry (MS), and capillary electrophoresis (CE). All these methods are time-consuming and require expensive equipment. The detection of antibiotics is especially important as they are commonly used in medical fields and can affect both humans and their environment.

QDs are widely used as sensing probes for the detection and analysis of drug residues due to their interesting fluorescence properties. When CQDs bind to a particular antibiotic, their fluorescence may be enhanced or quenched, which can be used to distinguish antibiotics and monitor their concentration. QDs' selectivity can also be enhanced by modifying their surfaces with various compounds and special functional groups.[79,80]

Tetracycline (TC) is a broad-spectrum antibiotic that is used extensively in human and animal disease control and agricultural feed additives because of its excellent therapeutic effect and low cost; it is, thus, a common pharmaceutical pollutant. TC has a long life in the environment, and, as a result, its residue can be present in soil, animal products, surface water, drinking water, and groundwater. As the release of TC into the surrounding environment can cause bacterial drug resistance, its determination in various samples is of great importance.

One source of drug residue contamination is veterinary drugs that are used to promote weight gain and improve breeding efficiency. Furthermore, drug residues can also be present in milk, eggs, and other animal products. These drug residues can be toxic to humans and threaten their health if present above certain levels. CD/QD-MIPs have become popular tools for drug residue detection. Most are used to determine drug residues in biological samples, such as bovine serum albumin and fetal bovine tissue; however, it is important to develop sensors that are effective at determining residues in other contexts, as biological samples can enter the environment *via* various waste products. A selective and sensitive fluorescent sensor based on GQDs and iron nanoparticles embedded in MIP (GQDs@Fe_3O_4/MIP) was used for the enrichment and determination of ceftazidime (a broad-spectrum antibiotic) in milk samples. The synthesis route and TEM images for GQDs@Fe_3O_4/MIP are shown in Figure 7.6. In optimal conditions, linearity was in the range of 0.10–10.0 μg L^{-1} and LOD was 0.05 μg L^{-1}.[81]

Carbon nitride QDs with enhanced fluorescence and good stability were synthesized by a simple solvothermal method and used as a selective and sensitive sensor for TC determination. The fluorescent behavior of C_3NQDs toward TC was found to be governed by the static quenching mechanism and the inner filter effect. This sensor was successfully used to determine TC in the range of 0.23 to 202.70 mM in water and milk powder samples.[82]

In another study, europium-doped CQDs (Eu-CQDs) with high stability were prepared using a simple and rapid carbonization method. The

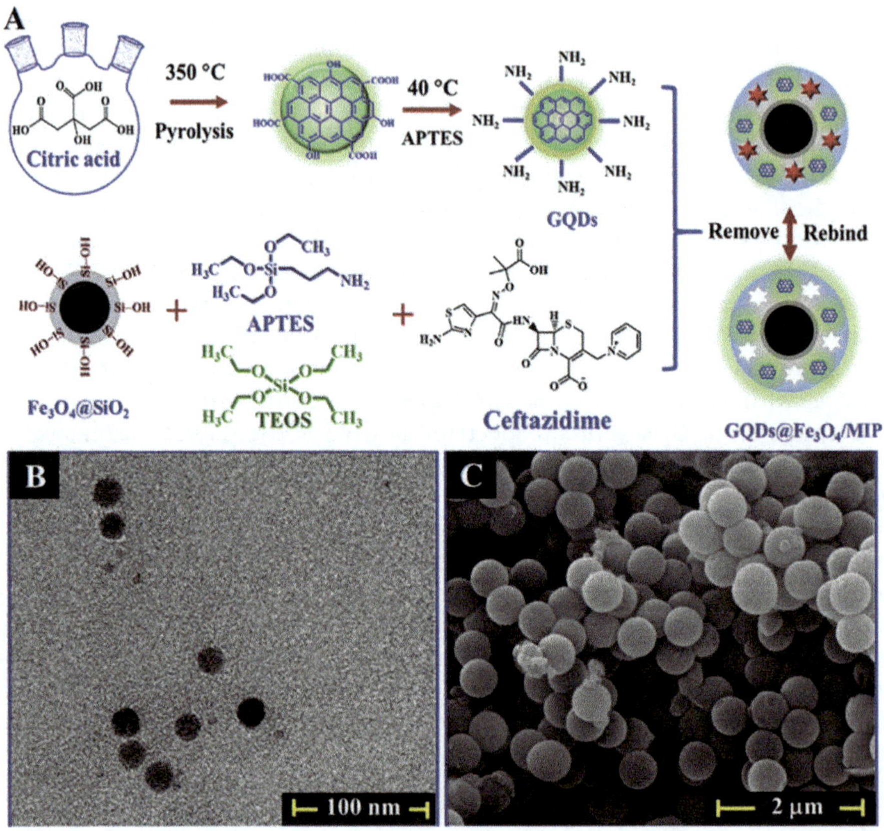

Figure 7.6 The synthesis of the GQDs@Fe GQDs@Fe_3O_4/MIP fluorescent probe for ceftazidime detection (A), TEM image of GQD nanoparticles (B), and SEM image of nanocomposite (C). Reproduced from ref. 81 with permission from Elsevier, Copyright 2020.

Eu-CQDs were used as a selective sensor for the detection of TC. TC quenched the Eu-CQDs' blue fluorescence, which enabled the determination of TC in complex water samples with a LOD of 0.3 mM.[83]

N-doped carbon dots (N-CDs) were synthesized and functionalized with bipyridine and copper as a fluorescence turn off–on sensor for the detection of Cu^{2+} and ciprofloxacin (CIP), one of the most commonly used antibiotics for disease control. N-CDs were fabricated *via* the hydrothermal method using CA and EDA as carbon and nitrogen sources, respectively. Their average size was 5.7 nm, and they had a high QY (84%). The fluorescence sensor showed good recoveries in spiked lake water samples.[84]

Photoluminescent thioglycolic acid-modified QDs (TGA-CdTe QDs) were synthesized and characterized by monitoring their optical properties, such as UV-Vis absorption and fluorescence spectra. TGA-CdTe QDs with an average diameter of approximately 2.2 nm exhibited maximum emissions at 518 nm upon excitation at 350 nm. These QDs were used to detect kanamycin, an aminoglycoside-class antibiotic that is used to treat a myriad of infections. In the presence of this antibiotic, the synthesized TGA-CdTe QDs' photoluminescence was enhanced following the Langmuir-binding isotherm model. This effect was used to monitor the concentration of kanamycin in various samples, such as milk and stream water. The signal enhancement caused by the presence of kanamycin was linearized (using C/P *vs.* C) to cover two orders of magnitude under the adjusted robust experimental conditions. Kanamycin was successfully detected in milk, yellow fever vaccine, and stream water samples at ng mL^{-1} levels. The quantification was made using TGA-CdTe QDs dispersions as an optical probe. The same experimental conditions were also used to examine the effect of other aminoglycosides (such as neomicin, streptomycin, gentamycin, amikacin, and tobramycin) on the photoluminescence of the TGA-CdTe QDs. With the exception of streptomycin, these other aminoglycosides also enhanced the photoluminescence as they interacted with the QDs. However, the efficiency of the photoluminescence enhancement varied depending on the specific type of aminoglycoside used.[85]

AMX is another common type of antibiotic used for bacterial infection treatments. The excessive use and improper application of AMX, especially in domestic animals and marine life, can lead to its accumulation in water and wastewater, where it poses a pollution threat and contributes to the formation of drug-resistant bacteria. Detecting AMX in water is a crucial step toward its removal. In one study, CdTe QDs were coated with MIP-QDs for use as a highly selective and sensitive fluorescence probe for the detection of trace AMX residue in milk, eggs, and honey. The MIP-QDs' fluorescence was quenched by AMX; under optimal conditions, the intensity of the MIP-QDs' photoluminescence showed a linear relation with AMX concentration in the range of 0.20–50.0 μg L^{-1} with a detection limit of 0.14 μg L^{-1}.[86]

B-doped CQDs (B-CQDs) were prepared using CA as the carbon source and boric acid as the modifier and then used as a fluorescent probe for the selective and sensitive detection of AMX in water. The fluorescence signal of the B-CQDs-AMX system was enhanced by increasing AMX concentration in the range of 1.43–429.12 μmol L^{-1} with an LOD of 0.825 μmol L^{-1}. The detection mechanism was found to be *via* a primary hydrogen bond between AMX and B-CQDs.[87]

7.6 Conclusion

QDs have a vast array of applications in the detection of environmental pollution. They have been shown to be sensitive and selective sensors for various analytical and bioanalytical purposes. This chapter has provided an overview of QDs' uses in the detection of heavy metals, organic pollutants, drug residues, and pesticides; we have also discussed QDs' applications in photocatalysis. Green-synthesized QDs have emerged as a relatively non-toxic, low-cost, water-soluble, and physiochemically and photochemically stable solution for pollutant detection. Future research in the field of environmental pollutant detection should therefore look to expand on the synthetic methods and scientific applications for these eco-friendly and low-cost QDs. Despite all the efforts made by researchers in this field, several challenges remain. For instance, more work must be done to better control QDs' sizes and shapes, size distribution, and surface/bulk elemental composition for use in various contexts.

Abbreviations

2-DNMs	Two-dimensional nanomaterials
AAS	Atomic absorption spectrometer
APTES	3-Aminopropyl triethoxy silane
APTES	Aminopropyl triethoxy silane
AR14	Acid azo red dye 14
BA	Bisphenol-A
BN	Hexagonal boron nitride
CA	Citric acid
CE	Capillary electrophoresis
CIP	Ciprofloxacin
CQDs	Carbon-based quantum dots
EDA	Ethylenediamine
EPA	Environmental Protection Agency
EU	European Union
$g\text{-}C_3N_4$	Graphitic carbon nitride
GO	Graphene oxide
GQDs	Graphene-based quantum dots
HPLC	High-performance liquid chromatography
HRTEM	High-resolution transmission electron microscopy
IC	Indigo carmine
LDHs	Layered double hydroxides
LOD	Limit of detection
MB	Methylene blue
MIPs	Molecularly imprinted polymers
MO	Methyl orange
MS	Mass spectrometry
N-CDs	N-doped carbon dots
NRCDs	Nitrogen-doped red-emitting carbon dots
PAHs	Polycyclic aromatic hydrocarbons
QDs	Quantum dots
QY	Quantum yield
RSDs	Relative standard deviations
SEM	Scanning electron microscopy

TC	Tetracycline
TEM	Transmission electron microscopy
TEOS	Tetraethoxy silane
TGA	Thioglycolic acid
UV	Ultraviolet
WG	Water-dispersal granules
WP	Wettable powder
XRD	X-ray diffraction

References

1. B. Gidwani, V. Sahu, S. S. Shukla, R. Pandey, V. Joshi, V. K. Jain and A. Vyas, *J. Drug Delivery Sci. Technol.*, 2021, **61**, 102308.
2. X. Xu, R. Ray, Y. Gu, H. J. Ploehn, L. Gearheart, K. Raker and W. A. Scrivens, *J. Am. Chem. Soc.*, 2004, **126**, 12736.
3. U. Badıllı, F. Mollarasouli, N. K. Bakirhan, Y. Ozkan and S. A. Ozkan, *TrAC, Trends Anal. Chem.*, 2020, **131**, 116013.
4. M. A. Farzin and H. Abdoos, *Talanta*, 2021, **224**, 121828.
5. J.-J. Kim, Y.-S. Kim and V. Kumar, *J. Trace Elem. Med. Biol.*, 2019, **54**, 226.
6. M. Hemmati, M. Rajabi and A. Asghari, *Microchim. Acta*, 2018, **185**, 1.
7. Z. Rahman, *J. Hazard. Mater.*, 2020, **396**, 122682.
8. L. Xia, Y. Lu, H. Meng and C. Li, *J. Hazard. Mater.*, 2020, **393**, 122487.
9. S. Clemens and J. F. Ma, *Annu. Rev. Plant Biol.*, 2016, **67**, 489.
10. Z. Sun, X. Li, Y. Wu, C. Wei and H. Zeng, *New J. Chem.*, 2018, **42**, 4603.
11. G. S. Das, J. P. Shim, A. Bhatnagar, K. M. Tripathi and T. Kim, *Sci. Rep.*, 2019, **9**, 1.
12. Z. M. Khan, R. S. Rahman, S. Islam and M. Zulfequar, *Opt. Mater.*, 2019, **91**, 386.
13. Y. Park, J. Yoo, B. Lim, W. Kwon and S.-W. Rhee, *J. Mater. Chem. A*, 2016, **4**, 11582.
14. S. Zhu, Y. Song, X. Zhao, J. Shao, J. Zhang and B. Yang, *Nano Res.*, 2015, **8**, 355.
15. V. Sharma, P. Tiwari, N. Kaur and S. M. Mobin, *Environ. Chem. Lett.*, 2021, **19**, 3229.
16. S. Morais, F. G. Costa and M. d. L. Pereira, *Environmental health–emerging issues and practice*, 2012, **10**, 227.
17. J. E. Gray, P. M. Theodorakos, D. L. Fey and D. P. Krabbenhoft, *Environ. Geochem. Health*, 2015, **37**, 35.
18. X. Li, C. Chai, Y. Zhang, Y. Wang, J. Lv, W. Bian and M. M. F. Choi, *Opt. Mater.*, 2020, **99**, 109559.

19. N. Pourreza and M. Ghomi, *Sens. Actuators, B*, 2017, **251**, 609.
20. H. Qi, X. Sun, T. Jing, J. Li and J. Li, *RSC Adv.*, 2022, **12**, 1989.
21. Q. Zhou, Y. Liu, Y. Wu, Z. Li, Y. Li, M. Liu, T. Qu and C. Chen, *Chemosphere*, 2021, 274, 129959.
22. N. Pourreza and M. Ghomi, *Mater. Sci. Eng. C*, 2019, **98**, 887.
23. V. Manikandan and N. Y. Lee, *Environ. Res.*, 2022, **212**, 113283.
24. W. Lu, X. Qin, S. Liu, G. Chang, Y. Zhang, Y. Luo, A. M. Asiri, A. O. Al-Youbi and X. Sun, *Anal. Chem.*, 2012, **84**, 5351.
25. M. Lan, J. Zhang, Y.-S. Chui, P. Wang, X. Chen, C.-S. Lee, H.-L. Kwong and W. Zhang, *ACS Appl. Mater. Interfaces*, 2014, **6**, 21270.
26. G. Liu, H. Jia, N. Li, X. Li, Z. Yu, J. Wang and Y. Song, *Microchem. J.*, 2019, **145**, 718.
27. R. Tabaraki and N. Sadeghinejad, *Ecotoxicol. Environ. Saf.*, 2018, **153**, 101.
28. M. Pajewska-Szmyt, B. Buszewski and R. Gadzała-Kopciuch, *Mater. Chem. Phys.*, 2020, **242**, 122484.
29. M. W. Hentze, M. U. Muckenthaler, B. Galy and C. Camaschella, *Cell*, 2010, **142**, 24.
30. N. Pourreza and M. Ghomi, *Talanta*, 2018, 179, 92.
31. U. Latief, S. Ul Islam, Z. M. S. H. Khan and M. S. Khan, *Spectrochim. Acta A*, 2021, **262**, 120132.
32. N. Kaur, V. Sharma, P. Tiwari, A. K. Saini and S. M. Mobin, *Sens. Actuators B*, 2019, **291**, 275.
33. Y. Liang, Y. Liu, S. Li, B. Lu, C. Liu, H. Yang, X. Ren and Y. Hou, *Opt. Mater.*, 2019, **89**, 92.
34. D. Hong, X. Deng, J. Liang, J. Li, Y. Tao and K. Tan, *Microchem. J.*, 2019, **151**, 104217.
35. T. Luo, L. Bu, S. Peng, Y. Zhang, Z. Zhou, G. Li and J. Huang, *Talanta*, 2019, **205**, 120117.
36. Y. Zhao, S. Jing, X. Peng, Z. Chen, Y. Hu, H. Zhuo, R. Sun and L. Zhong, *Cellulose*, 2020, **27**, 415.
37. C. Yu, D. Zhang, Q. Zhu, D. Chao, H. Liu and L. Zhou, *Dyes Pigm.*, 2022, **198**, 110008.
38. Y. Ma, G. Xu, F. Wei, Y. Cen, Y. Ma, Y. Song, X. Xu, M. Shi, S. Muhammad and Q. Hu, *J. Mater. Chem. C*, 2017, **5**, 8566.
39. S. Huang, W. Li, P. Han, X. Zhou, J. Cheng, H. Wen and W. Xue, *Anal. Methods*, 2019, **11**, 2240.
40. J. Praneerad, N. Thongsai, P. Supchocksoonthorn, S. Kladsomboon and P. Paoprasert, *Spectrochim. Acta, Part A*, 2019, **211**, 59.
41. R. Brochin, S. Leone, D. Phillips, N. Shepard, D. Zisa and A. Angerio, *GU J. Health. Sci.*, 2008, **5**, 1.

42. H. Lu, S. Xu and J. Liu, *ACS Sens.*, 2019, **4**, 1917.
43. Y. Liu, Q. Zhou, J. Li, M. Lei and X. Yan, *Sens. Actuators, B*, 2016, **237**, 597.
44. Y. Jiang, Y. Wang, F. Meng, B. Wang, Y. Cheng and C. Zhu, *New J. Chem.*, 2015, **39**, 3357.
45. V. A. Ansi and N. K. Renuka, *Sens. Actuators, B*, 2018, **264**, 67.
46. K. Radhakrishnan and P. Panneerselvam, *RSC Adv.*, 2018, **8**, 30455.
47. Z. Ramezani, M. Qorbanpour and N. Rahbar, *Colloids Surf., A*, 2018, **549**, 58.
48. S. Pawar, S. Kaja and A. Nag, *ACS Omega*, 2020, **5**, 8362.
49. Y. Dong, Y. Zhang, S. Zhi, X. Yang and C. Yao, *ChemistrySelect*, 2021, **6**, 123.
50. H. Ren, M. Li, Y. Liu, T. Zhao, R. Zhang and E. Duan, *Sci. Total Environ.*, 2022, **811**, 152389.
51. Y. Liu, Y. Zhao and Y. Zhang, *Sens. Actuators, B*, 2014, **196**, 647.
52. B. RasulKhan, S. K. Ponnaiah, P. Periakaruppan, G. Venkatachalam and J. Balasubramanian, *New J. Chem.*, 2020, **44**, 18149.
53. S. Liu, J. Cui, J. Huang, B. Tian, F. Jia and Z. Wang, *Spectrochim. Acta, Part A*, 2019, **206**, 65.
54. Z. Lua, J. Lia, L. Shena and H. Wanga, *Dig. J. Nanomater. Biostructures*, 2022, **17**, 221.
55. A. Boruah and B. K. Saikia, *ChemistrySelect*, 2022, 7, e202104567.
56. Y. Liu, Z. Gao, X. Shao, J. Yang, X. Tang, J. Wang, W. Chen, H. Lin, M. Deng and T. Zhu, *J. Mater. Sci.*, 2022, **57**, 17336.
57. Y. Vasseghian, M. Hosseinzadeh, A. Khataee and E.-N. Dragoi, *Sci. Total Environ.*, 2021, **796**, 149000.
58. M. S. Al Ja'farawy, A. Purwanto and H. Widiyandari, *Environ. Nanotechnol., Monit. Manage.*, 2022, **18**, 100681.
59. B. Sehatnia, R. E. Sabzi, F. Kheiri and A. Nikoo, *J. Appl. Electrochem.*, 2014, **44**, 727.
60. S. A. Nsibande and P. B. Forbes, *RSC Adv.*, 2020, **10**, 12119.
61. O. A. Adegoke and P. B. Forbes, *Talanta*, 2016, **146**, 780.
62. S. Javed, M. Islam and M. Mujahid, *Ceram. Int.*, 2019, **45**, 2676.
63. X. Xu, Z. Gao, Z. Cui, Y. Liang, Z. Li, S. Zhu, X. Yang and J. Ma, *ACS Appl. Mater. Interfaces*, 2016, **8**, 91.
64. B. Babu, A. N. Kadam, G. T. Rao, S.-W. Lee, C. Byon and J. Shim, *J. Lumin.*, 2018, **195**, 283.
65. A. Priyadharsan, V. Vasanthakumar, S. Karthikeyan, V. Raj, S. Shanavas and P. Anbarasan, *J. Photochem. Photobiol., A*, 2017, **346**, 32.
66. L. Chu, X. Han, J. Zhang, S. Jiang, R. Xue, S. Dong and J. Sun, *Colloids Surf., A*, 2021, **612**, 125915.

67. X. Liu, L. Pan, T. Chen, J. Li, K. Yu, Z. Sun and C. Sun, *Catal. Sci. Technol.*, 2013, **3**, 1805.
68. S. Sood, S. Kumar, A. Umar, A. Kaur, S. K. Mehta and S. K. Kansal, *J. Alloys Compd.*, 2015, **650**, 193.
69. L. Gnanasekaran, R. Hemamalini and K. Ravichandran, *J. Saudi Chem. Soc.*, 2015, **19**, 589.
70. W. A. Mohamed, I. A. Ibrahem, A. M. El-Sayed, H. R. Galal, H. Handal, H. A. Mousa and A. A. Labib, *Adv. Powder Technol.*, 2020, **31**, 2555.
71. J. Shim, J. S. Lee, J. hyeon Lee, Y. J. Yun, S. K. Park, B. Angadi and D. I. Son, *Composites, Part B*, 2021, **225**, 109307.
72. A. R. Amani-Ghadim, S. Arefi-Oskoui, R. Mahmoudi, A. T. Sareshkeh, A. Khataee, F. Khodam and M. S. Seyed Dorraji, *Chemosphere*, 2022, 295, 133917.
73. N. Bakar, M. Salleh, A. Umar and M. Yahaya, *Adv. Nat. Sci.: Nanosci. Nanotechnol.*, 2011, **2**, 025011.
74. R. Shokri and M. Amjadi, *Spectrochim. Acta, Part A*, 2021, **246**, 118951.
75. X. Ren and L. Chen, *Anal. Bioanal. Chem.*, 2015, **407**, 8087.
76. C. L. Chitescu, M. Lupoae and A. M. Elisei, *Rev. Chim.*, 2016, **67**, 1008.
77. K. E. Arnold, A. B. Boxall, A. R. Brown, R. J. Cuthbert, S. Gaw, T. H. Hutchinson, S. Jobling, J. C. Madden, C. D. Metcalfe, V. Naidoo, R. F. Shore, J. E. Smits, M. A. Taggart and H. M. Thompson, Assessing the exposure risk and impacts of pharmaceuticals in the environment on individuals and ecosystems, *Biol. Lett.*, 2013, 20130492.
78. P. Bottoni, S. Caroli and A. B. Caracciolo, *Toxicol. Environ. Chem.*, 2010, **92**, 549.
79. L. n. Qiao, S. Qian, Y. Wang, S. Yan and H. Lin, *Eur. J. Chem.*, 2018, **24**, 4703.
80. Y. Fu, S. Zhao, S. Wu, L. Huang, T. Xu, X. Xing, M. Lan and X. Song, *Dyes Pigm.*, 2020, **172**, 107846.
81. O. Bunkoed, P. Raksawong, R. Chaowana and P. Nurerk, *Talanta*, 2020, **218**, 121168.
82. R. Bai, H. Sun, P. Jin, J. Li, A. Peng and J. He, *RSC Adv.*, 2021, **11**, 24892.
83. M. L. Liu, B. B. Chen, T. Yang, J. Wang, X. D. Liu and C. Z. Huang, *Methods Appl. Fluoresc.*, 2017, **5**, 015003.
84. V. D. Dang, A. B. Ganganboina and R.-A. Doong, *ACS Appl. Mater. Interfaces*, 2020, **12**, 32247.
85. S. Khan, E. M. Miguel, C. F. de Souza, A. R. da Silva and R. Q. Aucelio, *Sens. Actuators, B*, 2017, **246**, 444.
86. K. Chullasat, P. Nurerk, P. Kanatharana, F. Davis and O. Bunkoed, *Sens. Actuators, B*, 2018, **254**, 255.
87. X. Zhang, Y. Ren, Z. Ji and J. Fan, *J. Mol. Liq.*, 2020, **311**, 113278.

CHAPTER 8

Quantum Dots for Toxin Detection in Foods and Beverages

ZAHRA RAMEZANI*[a], FATEMEH SEDAGHATI[b] AND ROGHAYEH HEIRAN[b]

[a]Toxicology Research Center, Medical Basic Sciences Research Institute, Ahvaz Jundishapur University of Medical Sciences, Ahvaz, Iran; [b]Estahban Higher Education Center-Shiraz University, Estahban, Iran
*E-mail: zramezani@ajums.ac.ir; zahramezani@gmail.com

8.1 Introduction

Food safety is a fundamental, worldwide public issue with far-ranging consequences: unsafe food causes diseases, which can reduce productivity and influence socio-economic activity. Food can become unsafe when contaminated with pathogens, toxins, or chemicals. Foodborne diseases are most often caused by the consumption of food contaminated with foodborne pathogens, such as bacteria, parasites, viruses, and fungi.[1] A study of significant biotoxins produced by bacteria, fungi, and algae (as well as their sources, lethal doses, and structures) was conducted by Rajput.[2] Pathogens may be consumed along with food or beverages and infect the human host's intestinal tract in a process referred to as foodborne infection. Even if they are not directly consumed, pathogens can produce toxins that are ingested by humans; this type of illness is referred to as foodborne

Detection Science Series No. 22
Quantum Dots in Bioanalytical Chemistry and Medicine
Edited by Michael Thompson and Zahra Ramezani

Published by the Royal Society of Chemistry, www.rsc.org

intoxication. The incubation period (*i.e.*, the time between exposure to the pathogen and the onset of the symptoms) of foodborne infection is usually much longer than that of foodborne intoxication.

Screening for foodborne pathogens is important from a public health perspective; it is also of concern for national security reasons, as foodborne pathogens may be used as bioterrorist weapons. To date, many biosensors based on organic dyes have been prepared for this purpose. Recently, QDs, with their superior properties, have attracted scientific attention for their potential use in biosensor design. In this chapter, we will first provide an overview of several key pathogens and their toxins, then discuss various applications of QDs for their detection. The advantages and disadvantages of QDs in relation to organic dyes will also be discussed.

8.1.1 Foodborne Bacteria and Their Toxins

Bacteria are a common type of foodborne pathogen. There are many different types of bacteria with a wide variety of shapes and properties. Some pathogenic bacteria can form spores and, hence, are highly heat-resistant (*e.g.*, *Clostridium perfringens*, *Clostridium botulinum*, *Bacillus cereus*, and *Bacillus subtilis*). Some can generate heat-resistant toxins (*e.g.*, *Clostridium botulinum* and *Staphylococcus aureus*). The optimal temperature range for the growth of most pathogens is between 20 °C and 45 °C; however, some specific foodborne pathogens (*i.e.*, psychotrophic bacteria such as *Yersinia enterocolitica* and *Listeria monocytogenes*) can grow in temperatures below 10 °C or under refrigerated conditions.[1,3,4]

A wide range of bacterial pathogens can produce toxins.[5] Bacterial toxins are usually classified into two main groups: (i) endotoxins (lipopolysaccharides or LPS), which are produced by Gram-negative bacteria; and (ii) exotoxins (peptides and proteins), which are generated by both Gram-negative and Gram-positive bacteria. Endotoxins are one of the main components in the outer membrane of the cell wall of Gram-negative bacteria. They are not secreted from a bacterium until it is killed. Exotoxins are proteins produced inside pathogenic bacteria. They are released extracellularly as the organism grows; as they are secreted into the surrounding medium, they may travel from the locus of infection to a distant part of the body and cause damage. Bacteria such as *Clostridium botulinum*, *Clostridium tetani*, *Corynebacterium diphtheria*, and *Staphylococcus aureus* can produce exotoxins.[2] A specific bacterial pathogen may produce a single toxin or even multiple toxins. Common bacterial toxins and their respective sources, main structures, and lethal doses are listed in Table 8.1.[2]

Bacterial toxins have a wide range of molecular weights and, hence, a wide variety of physico-chemical properties. They may be present at any step of the food production chain.[6] It is critical to note that although toxins cannot propagate from one host to another, they can remain in food after processing and cooking (*i.e.*, after their corresponding bacterial pathogens have been killed) and cause severe damage to tissues and organs.[7,8]

Table 8.1 An overview of major bacterial toxins and their source of origin, LD50 value, and main structural component. Reproduced from ref. 2 with permission from Elsevier, Copyright 2021.

	Source of toxin	Toxin and LD50 (lethal dose 50%)	Main structural component
1	*C. botulinum* *C. baratii* *C. butyricum*	Botulinum neurotoxin 0.0003 μg kg^{-1} (mouse lethal doses)	Single-chain protein (150 kDa)
2	*V. cholera*	Cholera toxin, 250 μg kg^{-1} (mouse lethal dose)	AB5 structure
3	*C. perfringens*	Epsilon toxin, 0.1 μg kg^{-1} (mouse lethal dose)	Single-chain protein (36 kDa)
4	*S. aureus*	Staphylococcal Enterotoxins, 20 μg kg^{-1} (monkey lethal dose)	Single-chain protein (25–30 kDa)
5	*C. tetani*	Tetanus toxin, 0.001 μg kg^{-1} (mouse lethal dose)	Single-chain protein (150 kDa)
6	Symbiotic bacteria – *Pseudoalteromonas*, *Pseudomonas*, and *Vibrio*	Tetrodotoxin (TTX), 232 μg kg^{-1} (oral LD50)	Alkaloid

These toxins can cause cytotoxicity by changing the physiological activity and integrity of the plasma membrane.[9] For example, exposure to Botulinum neurotoxins (BoNTs, category A bioterrorism agents) leads to the proteolysis of soluble *N*-ethylmaleimide-sensitive factor attachment protein receptor (SNARE) proteins and disruption of acetylcholine release in exocytosis, which leads to muscle paralysis and potentially death.[10] Listeriolysin O liberated by *Listeria monocytogenes* follows the mechanism of the production of pores for subsequent disruption of the phospholipid bilayer and finally cell lysis. In another example, the Shiga toxin produced by *E. coli* O157:H7 can prevent protein synthesis and activate apoptosis and necrosis.

Various studies have described the features of the most common foodborne bacteria, their toxins, their associated diseases, and some of their related outbreaks.[3,8,11–13]

8.1.2 Foodborne Fungi and Their Mycotoxins

The term "mycotoxin" is derived from two Greek words: "mikes" (fungi) and "toxin" (poison).[14] Mycotoxins are secondary metabolites generated in the food chain by fungi species such as *Aspergillus*, *Penicillium*, *Alternaria*, and *Fusarium*.[15] Fungal toxins from *Fusarium* and *Alternarium* are frequently categorized as "field fungi", while the species *Aspergillus* and

Penicillium are referred to as "storage fungi" due to the production of mycotoxins related to agro-industrial storage conditions.[16] Mycotoxins produced during harvesting, storage, processing, and distribution of crops can infect about 25% of the world's food production, which is a serious concern for both public health and economics.[17] Most mycotoxins are thermally and chemically stable during food processing, including boiling, cooking, baking, and pasteurization. Humans may be exposed to mycotoxins through the direct ingestion of polluted crops or *via* animal products such as milk, meat, and eggs. Fungal toxins differ from bacterial toxins in that they cannot be detected by human and animal immune systems because they are not protein molecules.[18] Many mycotoxins are highly toxic and pose a significant threat to human health, even in low doses. This quality also makes them candidates for use in acts of bioterrorism.[15]

To date, between 300 and 400 different mycotoxins have been identified. Some of them, such as aflatoxins (AFs), trichothecenes (*e.g.*, deoxynivalenol (DON) and T-2 toxin), zearalenone (ZEN), fumonisins (FS), ochratoxins (OT), and patulin (PAT), show a great variety of toxic effects and pose serious threats to human and animal health.[19] Mycotoxins can be classified in terms of the nature and degree of their toxicity. With regard to toxicity, mycotoxins may be carcinogenic (aflatoxins, ochratoxins, fumonisins), mutagenic (aflatoxins, sterigmatocystin), teratogenic (ochratoxin), estrogenic (zearalenone), hemorrhagic (trichothecenes), immunotoxic (aflatoxins and ochratoxins), nephrotoxic (ochratoxins), hepatotoxic (aflatoxins and phomopsins), dermotoxic (trichothecenes), or neurotoxic (ergotoxins, penitrems, lolitrems, and paxilline).[14]

To protect human health, the maximum acceptable limits for various mycotoxins are established in most countries. The maximum allowable limits of mycotoxins vary according to different agencies and for different crops.[19] The most common mycotoxins, their main sources, and some commonly contaminated food commodities are listed in Table 8.2, together with the United States FDA (US FDA) and the European Union (EU) regulatory limits for their levels in food and animal feed.[19]

8.1.3 Marine Biotoxins

Among the thousands of known microalga species, more than 100 produce persistent natural toxins that can threaten human and animal health, harm aquatic ecosystems, and affect the economy.[20] Marine biotoxins are mostly small molecular compounds. They are highly diverse but can be classified in terms of solubility: that is, as hydrophilic or lipophilic toxins. They are also categorized in terms of their poisoning effects: for example, paralytic shellfish poisoning (PSP), diarrhetic shellfish poisoning (DSP), amnesic shellfish poisoning (ASP), neurotoxic shellfish poisoning (NSP), and ciguatera fish poisoning (CFP).[20] Harmful algal blooms (HABs), such as red tides, blue-green algae, and cyanobacteria, are major sources of these biotoxins.[21]

Table 8.2 Major mycotoxins and the US and EU limits on food and animal feed levels. Reproduced from ref. 18. (No special permission is required).

Mycotoxin	Fungal species	Food commodity	US FDA (μg kg^{-1})	EU (EC 2006) (μg kg^{-1})
Aflatoxins B1, B2, G1, G2	*Aspergillus flavus Aspergillus parasiticus*	Maize, wheat, rice, peanut, sorghum, pistachio, almond, ground nut, tree nuts, fig, cottonseed, spices	20 for total	2–12 for B1 4–15 for total
Aflatoxin M1	Metabolite of aflatoxin B1	Milk, milk products	0.5	0.05 in milk 0.025 in infant formulae and infant milk
Ochratoxin A	*Aspergillus ochraceus Penicillium verrucosum Aspergillus carbonarius*	Cereals, dried vine fruit, wine, grapes, coffee, cocoa, cheese	Not set	2–10
Fumonisins B1, B2, B3	*Fusarium verticillioides Fusarium proliferatum*	Maize, maize products, sorghum, asparagus	2000–4000	200–1000
Zearalenone	*Fusarium graminearum Fusarium culmorum*	Cereals, cereal products, maize, wheat, barley	Not set	20–100
Deoxynivalenol	*Fusarium graminearum Fusarium culmorum*	Cereals, cereal products	1000	200–50
Patulin	*Penicillium expansum*	Apples, apple juice, and concentrate	50	10–50

There are two common forms of exposure to marine biotoxins: (i) direct contact (*e.g.*, swimming and breathing in aerosolized toxins in water droplets) and (ii) the ingestion of toxin-infected seafood (*e.g.*, shellfish, crabs, and fish), which can lead to different levels of intoxications.[20,22] Marine toxins are resistant to cooking and freezing, which makes them a serious threat to public health. Figure 8.1 schematically illustrates the spread of marine biotoxins through various species. It also classifies shellfish toxins and their chemical structures.[22]

Earlier studies have described regulatory limits, the lowest observable adverse effect level (LOAEL), the no observable adverse effect level (NOAEL), and the acute reference dose (ARfD) for various marine biotoxins.[20,22,23]

8.2 Application of QDs in Foodborne Pathogen Detection

Both conventional and novel strategies for the detection of bacterial, fungal, and marine toxins are discussed in the following sections, with a specific focus on QD-based methods.

8.2.1 Bacterial Toxin Detection

Alahi and Mukhopadhyay have described various approaches to monitoring pathogens and endotoxins.[24] The most common toxins produced by Gram-positive pathogens are the neurotoxins from *Clostridium botulinum*, enterotoxins secreted by *Bacillus cereus* and *Clostridium perfringens*, and the family of enterotoxins produced by *Staphylococcus aureus* and other *staphylococci*. As mentioned, not all bacteria pose a health threat by themselves; however, the toxins they produce can be severely harmful to humans.[5]

In the last decade, many procedures have been developed for sensing bacterial toxins, including direct and indirect approaches. Indirect assays provide valuable information, especially for unknown bacterial toxins or in cases when direct methods are unavailable.[6] Older methods of detection include biological assays (whole animal tests) (*e.g.*, mouse lethality tests, monkey and kitten emesis tests, and rabbit and guinea pig skin tests), part animal tests (*e.g.*, ileal loop tests), and cell culture systems (*e.g.*, Chinese hamster ovary (CHO) cells).[25] Various methods, including immunological assays, nucleic acid-based methods, chromatographic techniques, and mass spectrometers, have also been developed to detect bacterial toxins. Immunological assays can be classified based on the specific method used, including gel diffusion, hemagglutination, coagglutination, reverse passive latex agglutination (RPLA), enzyme-linked immunosorbent assay (ELISA), enzyme-linked immunofiltration assay (ELIFA), and radioimmunoassay (RIA). Nucleic acid-based methods involve polymerase chain reaction

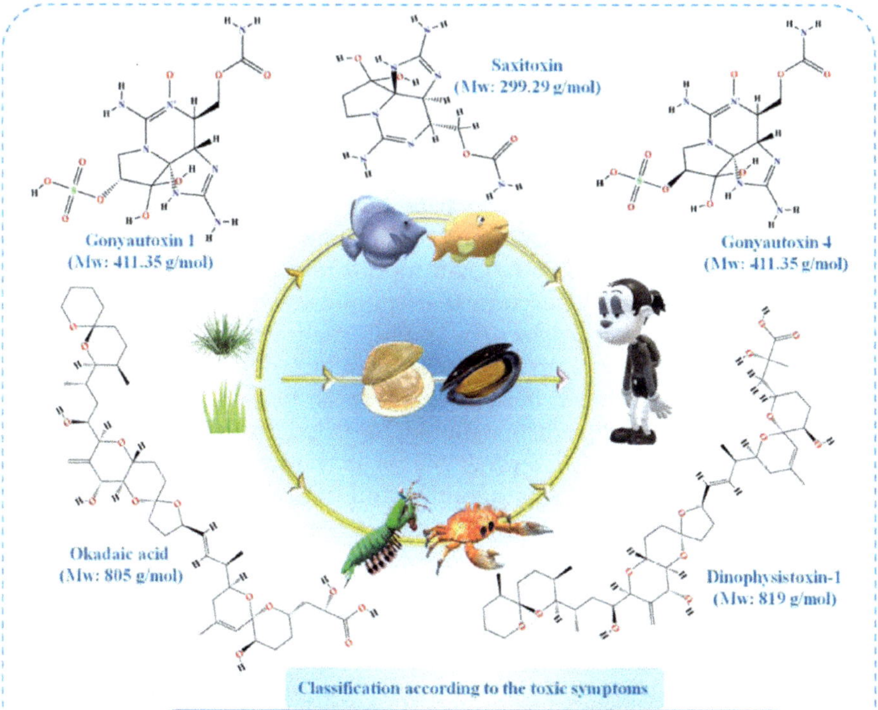

	Paralytic shellfish poisoning (PSP)	Diarrhetic shellfish poisoning (DSP)	Neurotoxic shellfish poisoning (NSP)	Amnesic shellfish poisoning (ASP)	Azaspiracid shellfish poisoning (AZP)
Main toxins	STXs, GTXs	OA, DTXs, YTXs, PTXs	BTXs	DA	AZAs
Main origin	*Alexandrium catenella*	*Dinophysis fortii*	*Karenia brevis*	*Pseudo-nitzschia pungens*	*Azadinium spinosum*
Main food source	Clams, mussels, oysters, crabs, fish	Mussels, scallops, clams	Clams, mussels	Mussels, clams, crabs, anchovies	Mussels, oyster
Symptoms	Vomiting, facial and respiratory paralysis, death	Diarrhea, nausea, vomiting, abdominal pain	Abdominal pain, paresthesias, myalgias, ataxia	Myoclonus, loss of short-term memory, coma, hemiparesis	Diarrhea, nausea, vomiting, and abdominal pain
Symptom onset time	5–30 min	30 min–3 h	30 min–3 h	15 min–>35 h	30 min–6 h

Figure 8.1 Illustration of the accumulation of shellfish toxins in the marine food chain, a threat to consumers, and the classification of the shellfish toxins and chemical structures of representatives. STX, saxitoxin; OA, okadaic acid; DTX, dinophysistoxin; YTX, yessotoxin; PTX, pectenotoxin; BTX, brevetoxins; DA, domoic acid; AZP, azaspiracid. Reproduced from ref. 22 with permission from Elsevier, Copyright 2022.

(PCR), quantitative PCR (qPCR), loop-mediated isothermal amplification (LAMP), nucleic acid sequence-based amplification (NASBA), and microarray technology.[6,25–27] Chromatography is also considered a reliable detection technique; chromatographic methods are generally categorized as gas or liquid chromatography (GC or LC, respectively) equipped with various kinds of detectors. They are highly sensitive and selective techniques for toxin analysis,[25] especially when coupled with forms of mass spectroscopy such as GC-MS, LC-MS, and LC-MS/MS.

Although all the above-mentioned methods have unique advantages, some of them have limitations (such as high analysis costs, complicated instrumentation requirements, time consumption, or a lack of portability) that restrict their use for on-site toxin monitoring. The detection speed is of particular concern in the context of public health emergencies; even procedures that are categorized as rapid, such as antibody-based and nucleic acid-based assays, may require specialized equipment and specific enrichment to achieve better sensitivity, which may increase analysis time.[28] Technological advancements have led to the introduction of immunochromatographic assays (ICAs) as new screening tools. ICA was developed to connect chromatography and immunochemical reactions for the analysis of various toxins. The most common ICA assay types are dipstick format, lateral flow device (LFA), and column flow. All these methods provide rapid results and are easy to use in the field without specialized equipment.[29] Asakura *et al.* developed an immunochromatographic test strip to detect cholera toxins.[26] Likewise, an immunochromatographic test for the rapid and simultaneous detection of *Clostridium difficile* antigens and toxins was reported.[30] In another study, affinity immunochromatography column (AICC) assays were used to identify *Clostridium botulinum* neurotoxin types A, B, and E rapidly and sensitively in food samples without pre-enrichment.[28] However, these strip tests are semi-quantitative and do not provide a complete picture of contaminants.[15] Lee *et al.* have summarized the advantages and limitations of these detection methods.[27]

Biosensors offer a promising new approach to biotoxin detection and one that meets most of the requirements for efficient analysis. The term "biosensor" (short for "biological sensor") refers to a powerful analytical device with two key components: a bioreceptor and a transducer. The bioreceptor is designed to interact with the target analyte (*e.g.*, a specific toxin) and produce a signal in proportion to the concentration of the analyte. The transducer converts the biological interactions into a measurable signal. Bioreceptors can be classified based on the materials from which they are composed:

1. Biological material (such as enzymes, antibodies, nucleic acids, and cell receptors)
2. Biologically derived material (such as aptamers and recombinant antibodies)
3. Biomimetic (such as imprinted polymers and synthetic catalysts).[27]

Bioreceptors may also be categorized as catalytic or affinity-based.[31] Transducer elements in biosensors can be categorized into electrochemical, optical, mass change, *etc.*[15] These are further classified into sub-classes. Among the main biosensor classes, optical and electrochemical biosensors have significant advantages over the others because of their simplicity, sensitivity, and specificity.

In the last decade, innovative nanotechnology-based methods of toxin detection have attracted researchers' attention worldwide.[2,24] Nanomaterial-based detection methods can overcome the limitations of conventional detection techniques, such as slow recognition time, low biocompatibility, or instability.[32] Nanomaterials are of different sizes, shapes, and functional properties. Owing to their exceptional chemical and physical features, large surface area, and lower density, they are excellent candidates for use in new detection methods and for the improvement of existing ones. Nanomaterials can improve detection by various means, such as supporting the immobilization process, producing a signal probe, or amplifying signals. They show great affinity for biomolecules, which facilitates the immobilization of antibodies, enzymes, nucleic acids, and many other substances and, thus, facilitates the development of a wide variety of sensing platforms. Nano-based platforms enable rapid, simple, sensitive, specific, and cost-effective detection and can be used in the field. Nanomaterials can be incorporated into both of a biosensor's main parts; that is, they may be used in both bioreceptors and transducers.

Various research groups have published many articles on this subject. Nanomaterials such as noble metal nanoparticles, carbon nanomaterials, QDs, magnetic nanoparticles, and silica nanomaterials have been used for detection. For example, the use of nanomaterials in antibody-based assays for the detection of bacterial toxins was discussed by Märtlbauer *et al.*[6] Recently, nanotechnology has been applied to the detection of various food toxins.[33] In one study, a cost-effective, rapid, sensitive, and selective electrochemical aptasensor was designed for monitoring low-level Staphylococcal enterotoxin B in foodstuffs.[34] An overview of the application of various nanosensors for the detection of biotoxins along with the mechanism of their interaction with toxic or microbial components was reported.[2]

QDs, as a type of nanotechnology, show great promise for use in toxin detection. This is especially the case for immunoassays and nano-biosensors, owing to QDs' unique electronic, optical, fluorescent, photoluminescent, chemiluminescent, and electro-chemiluminescent properties.[35] Moreover, their properties, such as their nanoscale size (similar to proteins), narrowband emission, broad excitation spectra for multicolor imaging, and flexibility in surface modification, make them attractive for use in multiplex immunoassays.[36] QDs also have several advantages over conventionally used fluorescent labels: their emission bands are tuneable, symmetrical, and narrow, and they exhibit high photostability.[36]

In the following sections, we discuss applications of QDs (including carbon-based QDs or CQDs, graphene-based QDs or GQDs, and semiconductor QDs) for toxin detection and provide a better perspective for developing many more QD-based methods for monitoring toxins.

8.2.1.1 Graphene-based QDs

GQDs are a new class of fluorescent nanoparticles that have attracted great attention in the field of food safety and quality assurance. Due to their outstanding properties, such as low cytotoxicity, photostability, high surface area, and adjustable band gap, GQDs have many applications in the detection and reduction of foodborne pathogens, especially bacteria, fungi, and their toxins.[37] Early detection of food pathogens is vital to controlling human poisoning and avoiding the spread of contamination.

There are few reports about the use of micromotors in food safety applications. Self-propelled micro- and nanomotors are micro- and nanoscale devices constructed to move freely in a solution by converting a specific stimulus into mechanical work.[38] A Janus micromotor is a specific type of micromaterial with an asymmetric structure whose surface has two or more physical properties. Two kinds of reactions co-occur in an asymmetric structure. Janus microsensors can be used to detect enterobacterial contamination.[39] On this subject, Jurado-Sánchez *et al.* demonstrated that GQD-functionalized Janus micromotors could be used as mobile sensors for toxins produced by enterobacteria, enabling the detection of contaminated food.[40] They developed an "ON–OFF" fluorescence sensing method for detecting *Salmonella enterica* endotoxins.

For the construction of Janus micromotors, platinum nanoparticles for enhanced bubble-propulsion and functionalized GQDs modified with 3-aminophenylboronic acid (PABA) as a receptor of the target were utilized. Lipopolysaccharides (LPS) from *Salmonella enterica* were applied as target endotoxins. When the target was added to the micromotors, their interaction with the QDs resulted in the quenching of the micromotors' fluorescence, which was readily detected using an optical microscope. A significantly enhanced mass transport was obtained owing to the bubble-ejection mechanism and continuous mixing across a contaminated sample. Hence, this phenomenon results in an increase in the reaction rates between micromotors and endotoxin. The microsensor was capable of monitoring concentrations far below the level considered toxic to humans (275 $\mu g\ mL^{-1}$). It was used to sense *Salmonella* toxins in food samples such as milk, mayonnaise, and egg yolks and whites, and produced results in 15 min; the gold standard method, in contrast, requires several hours. Figure 8.2 illustrates the principle of the "ON–OFF" fluorescence micromotor for the fast and selective sensing of enterobacterial contamination.[40]

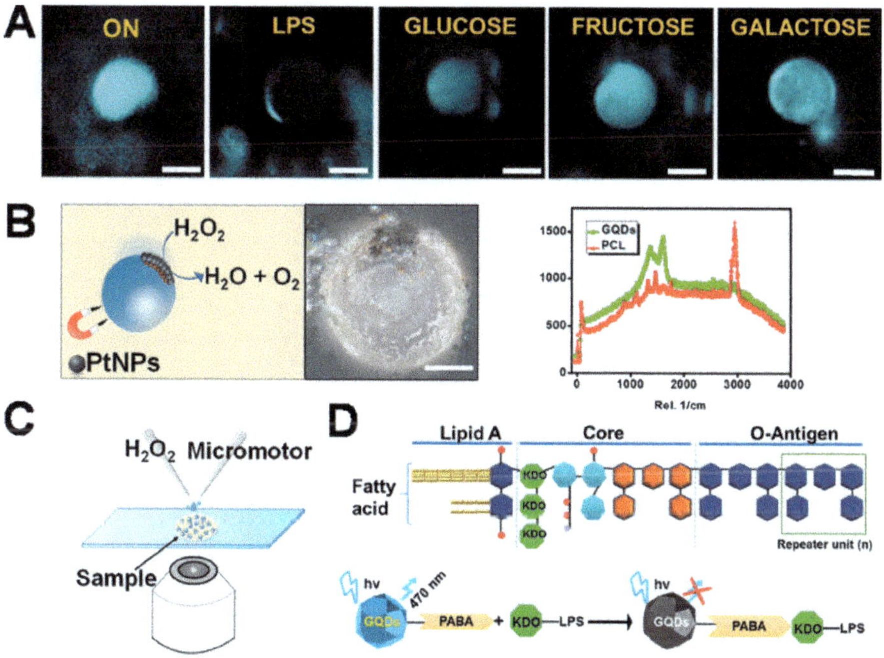

Figure 8.2 Self-propelled Janus microsensors for the detection of LPS from *Salmonella enterica* (serotype enteritidis). (A) Real-time optical visualization of the LPS recognition event: time-lapse fluorescent images of the micromotors before (ON) and after LPS addition (OFF) and selectivity of the protocol in the presence of interfering saccharides. (B) Schematic of the microsensor operation and characterization. Optical images and Raman mapping showing the distribution of nanoparticles, GQDs, and polycaprolactone (PCL) in the microsensor and corresponding Raman spectra. (C) Schematic of the setup for the Janus micromotors-based sensing protocol. (D) Schematics for the structure of the LPS from *Salmonella enterica* and mechanism of quenching by LPS union to the GQDs recognition units contained in the microsensors. Experimental conditions: 15% H_2O_2, surfactant, 5% (w/v) sodium cholate. Scale bars, 20 μm. Reproduced from ref. 40 with permission from the American Chemical Society, Copyright 2018.

8.2.1.2 Carbon-based QDs

CQDs possess excellent photochemical features, such as bright fluorescence, stable photoluminescence properties, excellent chemical stability, and good water solubility. Functional groups can be added to their surfaces easily, and they exhibit both excellent biocompatibility and low toxicity. They are prepared through simple and relatively fast procedures using readily available and inexpensive chemicals or natural carbon sources. CQDs, therefore, have many chemical and biochemical applications and

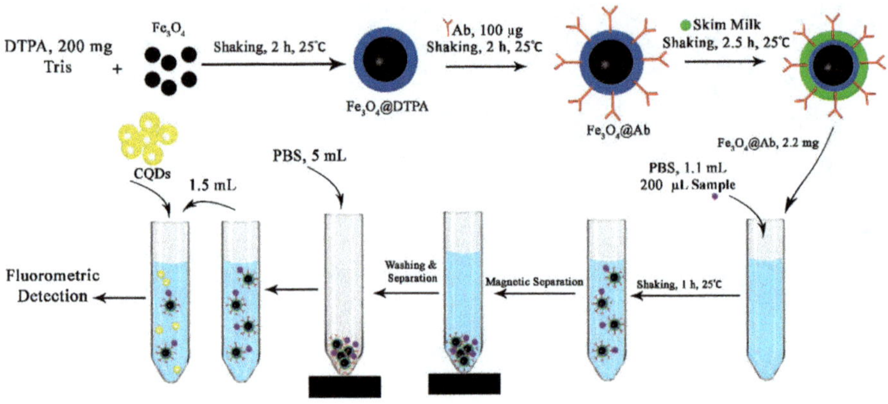

Figure 8.3 Graphical presentation of the application of CQDs in separation and detection of SEA in pasteurized and homogenized milk.

are a reliable candidate for use in biochemical sensing. Recently, CQDs were used to detect Staphylococcal enterotoxin A (SEA) in milk samples.[41] In this study, SEA was separated magnetically from homogenized and pasteurized milk using SEA antibody-loaded Fe_3O_4 (Fe_3O_4@Ab) nanoparticles and then dispersed in a phosphate buffer. A CQD solution was added (see Figure 8.3), and the fluorescence intensity of the resulting suspension was recorded.

While preparing Fe_3O_4@Ab, they realized that the kind of surface functional group determined the extent of Ab loading on the magnetic particles. Both EDC-NHS and DTPA were evaluated for the attachment of Ab to the surface of Fe_3O_4. The percentage of Ab attachment was obtained using an ELISA test, which showed that the Ab's attachment to DTPA was 95%: much higher than that of EDC-NHS.

A fast and reliable technique is required for the determination of SEA in a complex matrix such as milk. In this study, the magnetic nanoparticles collected the toxin from the complicated matrix and washed out any non-specific adsorbed constituents. Fe_3O_4@Ab@SEA was used as the signaling molecule; there was no need to elute the extracted SEA. Fe_3O_4@Ab@SEA has ultraviolet-visible (UV-Vis) absorption maxima at 440 nm, the wavelength emitted by CQDs. Upon the addition of CQDs to the Fe_3O_4@Ab@SEA suspension, the emitted light of the CQDs (as the donors) was absorbed by the signaling molecules (as the acceptors), and fluorescence quenching was observed. Fluorescence resonance energy transfer (FRET) – the spectral overlap between the donor's emission and the acceptor's absorption wavelength – was likely the mechanism behind this quenching. The UV-Vis spectra of SEA and CQDs overlapped at 350 nm as well, and the absorption of the excitation wavelength by the signaling molecules may have decreased the fluorescence intensity of CQDs, and the inner filter effect (IFE) may have been involved to a lesser extent in the quenching mechanism. Data to further evaluate the quenching mechanism was not collected.

Conventional ELISA is the method of choice for SEA quantification. However, it has several disadvantages: its detection limit is high (ng mL^{-1}), a single kit is highly expensive and each contains a limited number of sample assays, it requires two-step incubation, and milk matrices such as proteins and fat may interfere with the assay. The limit of detection and linear range of the present QD assay were in the pg mL^{-1} range – a significant improvement over all previously reported SEA quantification methods for milk quality control.

8.2.2 Mycotoxin Detection

Mycotoxin contamination is an unpredictable but unavoidable occurrence in a wide range of food matrixes, and multiple mycotoxins may be present in a single sample. The concentration of mycotoxins is often very low, which means that determining their presence in samples requires sensors with a low detection limit: parts per billion (ppb) or even parts per trillion (ppt). The development and improvement of methods for detecting various mycotoxin species is, therefore, both important and significantly challenging.[16]

As with bacterial toxin detection, there are multiple strategies for mycotoxin detection. Chromatography techniques, such as high-performance liquid chromatography coupled with mass spectrometry (HPLC-MS) or fluorescence detection (HPLC-FD), are usually considered the gold standard procedures against which the reliability and accuracy of new methods are evaluated.[42] HPLC-based techniques can detect multiple toxins at ppb levels with high sensitivity, but they have a number of drawbacks: they are difficult to handle, relatively expensive, and time-consuming. These issues have limited HPLC-based techniques' use in portable analysis. Immunoassay methods and immunochromatography systems are also used extensively for detecting mycotoxins. ELISA, as a standard biosensing method, is a common mycotoxin assay that is also used as a point of comparison for other methods.[15] Its sensitivity is relatively high (generally below ng mL^{-1}) and commercial ELISA kits are available. However, it, too, has some disadvantages: it is not very portable, labor-intensive, and time-consuming, which together limit its applications for on-site mycotoxin analysis.[16]

Lateral flow immunoassay (LFI) is an immunochromatography-based detection method whose rapidity and convenience make it a good candidate for qualitative and semi quantitative mycotoxin analysis.[43] Smartphone-based point-of-care tests (POCTs) are a potential alternative to conventional methods of mycotoxin detection; they are cost-effective, precise, and easy to implement in resource-limited areas. Colorimetric, fluorescent, microscopic imaging, electrochemical, and electrochemiluminescence techniques are common in smartphone-based applications. For example, in one OTA detection study, when a UV-LED in the range of 360–380 nm was employed as a source of excitation, an OTA sample in an alkaline solution emitted blue fluorescence; a smartphone was used to capture the fluorescence image.[44] Sivakumar and Lee have published an excellent review of the recent progress in smartphone-based techniques for food safety and mycotoxin detection.[45]

Nanoparticle-based analysis constitutes a wide area of research that has opened up many new possibilities for mycotoxin monitoring. The incorporation of nanoscale materials and biological components has led to the development of a new generation of biosensors known as nano-biosensors. These sensors increase productivity and save time through their capacity to detect multiple mycotoxins simultaneously. The detection of mycotoxins using LFI has been reviewed in several reports;[46–48] Lai has systematically described the developed nanomaterials and some novel strategies for improving LFI's performance.[43] In smartphone-based techniques, up-conversion nanoparticles (UCNPs) have been used to detect multiple mycotoxins simultaneously.[49] Most research, however, remains focused on enhancing the sensitivity and selectivity of biosensors and studying their mycotoxin detection applications in the field.[18,50] As a result of the dangers they pose to human health and their potential economic impact, several specific mycotoxin groups are overrepresented in the literature. These groups include aflatoxins, ochratoxin A (OTA), zearalenone (ZEN), fumonisins, trichothecenes such as deoxynivalenol (DON), and T-2 and HT-2 toxins.[51] There are numerous excellent reviews that focus on the detection of these mycotoxins.[15–17,42,43,50–55]

8.2.2.1 Graphene-based QDs

In this section, we focus on GQDs' applications in the design of various fluorescence, electrochemical, and electrochemiluminescence sensors for fungi and their toxins.

Huang *et al.* reported the facile synthesis of nitrogen-doped GQDs@SiO_2 nanoparticles (NGQDs@SiO_2 NPs) and their application for the detection of OTA.[56] In this study, the NGQDs@SiO_2 NPs were used in an aptasensor that detected OTA using electrochemiluminescence (ECL) and fluorescence methods. An OTA-specific thiolated aptamer was successfully tagged with core–shell Fe_3O_4@Au magnetic beads (MBs), and amino-modified capture DNA (cDNA) was attached to the surface of the NGQDs@SiO_2 NPs. Next, to fabricate a multifunctional aptasensor, the NGQDs@SiO_2 NPs were linked to the surface of the Fe_3O_4@Au MBs *via* the specific DNA hybridization between cDNA and the aptamer. With the addition of OTA, the aptamer on the surface of each Fe_3O_4@Au MB was linked to OTA to form an aptamer–OTA complex, leading to the release of NGQDs@SiO_2 NPs from the MBs' surface into the solution. As OTA concentration increased in the system, more NGQDs@SiO_2 NPs were released into the solution, and fewer NGQDs@ SiO_2 NPs were collected on the surface of the electrode.

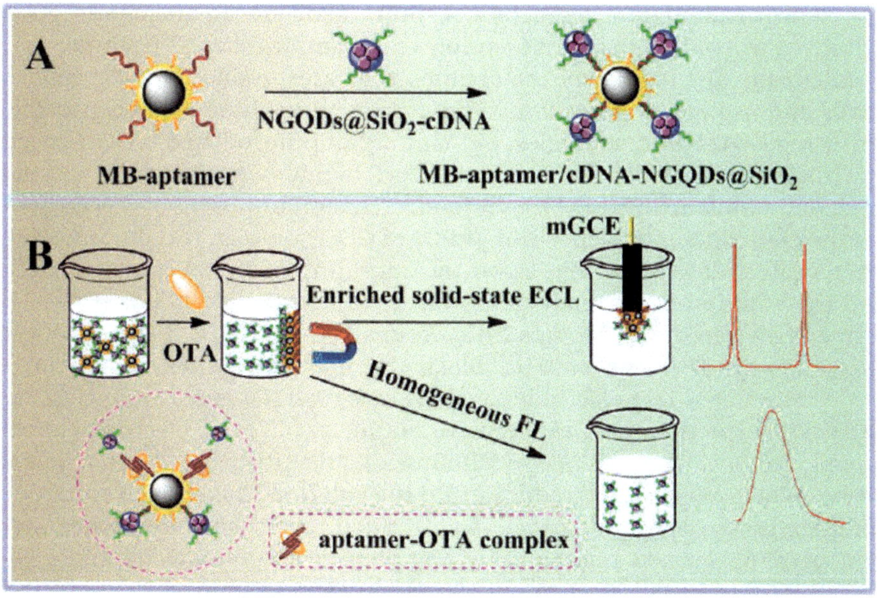

Figure 8.4 (A) Schematic illustration of the aptasensor fabrication and (B) its working principle for OTA detection by using two detection channels. Reproduced from ref. 56 with permission from the American Chemical Society, Copyright 2015.

After magnetic separation, the signals related to released and unreleased NGQDs@SiO_2 NPs were measured using fluorescence and solid-state ECL analysis, respectively, to determine the concentration of OTA. Figure 8.4 schematically illustrates the aptasensor's fabrication and detection methods.[56] This aptasensor showed a low detection limit, a wide linear range, and high specificity for OTA; it showed great promise for determining the presence and extent of OTA contamination in peanuts.

Lu *et al.* proposed a FRET-based approach for determining OTA.[57] As the absorption spectrum of GQDs and the fluorescence emission spectrum of nanoceria overlap, these nanoparticles are an excellent donor–acceptor pair to use in FRET probes. To construct the sensor, DNA1 and DNA2 were selected to complement the OTA aptamer, and then DNA1@nanoceria and DNA2@GQD probes were made. FRET occurred in the presence of DNA1@nanoceria and DNA2@GQD because of the electrostatic interaction between the two probes. The addition of the OTA aptamer resulted in an increased distance between the nanoceria and the GQD, which interrupted the FRET. Moreover, when target OTA was added to the system, the aptamer dissociated from the probes and bound to its target, which restored the FRET between the nanoceria and the GQDs. The ratio of the fluorescence intensities at 450 and 360 nm is considered as a response of the aptasensor.

The sensor exhibited a low detection limit, good linear range, and good selectivity when used to analyze complex real samples such as peanuts.

Aflatoxins are secondary metabolites that are produced by *Aspergillus flavus* and *Aspergillus parasiticus*. These metabolites have also been detected using GQD-based strategies. Sumana's group introduced a GQD-based electrochemical sensor for the label-free, highly sensitive, and specific detection of aflatoxin B1 (AFB1) in food.[58] This immunosensor was created *via* the electrophoretic deposition (EPD) of GQDs onto an indium tin oxide (ITO)-coated glass electrode. Upon the activation of the electrode surface and the addition of antibody (anti-AFB_1), covalent amide bonds (CO–NH) formed between the antibodies' NH_2 groups and the GQDs' COOH groups. Subsequently, BSA was used to block the non-active sites on the electrode. The GQDs/ITO electrode, anti-AFB1/GQDs/ITO, and BSA/anti-AFB1/GQDs/ITO were examined *via* cyclic voltammetry (CV) and electrochemical impedance spectroscopy (EIS) techniques using $[Fe(CN)_6]^{3-/4-}$ as a redox probe. When AFB_1 was introduced into the solution, the antigen–antibody complex at the sensor's surface was formed. Increasing the concentration of AFB1 caused more of the complex to accumulate on the electrode's surface, which in turn decreased the electrochemical currents and enhanced the charge transfer resistance. This immunosensor was used on maize to demonstrate its practical applications for the rapid detection of AFB_1 in food samples. In another study, Sumana *et al.* constructed two sensors to detect AFB_1 in food: (1) an electrochemical sensor based on gold nanoparticles and GQDs; and (2) a biosensor based on molybdenum disulfide (MoS_2) nanosheets decorated with GQDs. Using a method similar to the one described above, they used BSA/anti-AFB_1/GQDs-AuNPs/ITO and anti-AFB1/MoS_2@GQDs/ITO bioelectrodes to sense the toxins.[59,60]

Recently, a fluorescence aptasensor for the sensitive detection of AFB_1 was developed based on the dual signal amplification of three DNA nanostructures [DNA walker, DNA tetrahedron nanostructures (DTNs), and network hybridization chain reaction (HCR)].[61] The first step of signal amplification was *via* targets mediated with DNA walker nanostructures on the surface of AuNPs and the second step of amplification was achieved by GQDs nanotrees based on network HCR. Figure 8.5 illustrates the structure and function of this aptasensor for AFB_1 sensing.[61] The aptasensor showed a wide linear range and low detection limit for monitoring AFB_1 in complex food samples such as peanut oil.

Aflatoxin M1 (AFM_1) is another important toxin that has been detected using a GQD-based electrochemical sensor. In one study, the sensing electrode for AFM_1 detection and quantification was a GCE coated with a GQD multilayer, α-cyclodextrin (α-CD), and AgNPs (GQDs-α-CD-AgNPs). The GQDs-α-CD-AgNPs' composite film combined the α-CD's excellent electrical conductivity with the GQDs as the amplification elements and the AgNPs as the electrocatalytic agents. This sensor was used for the

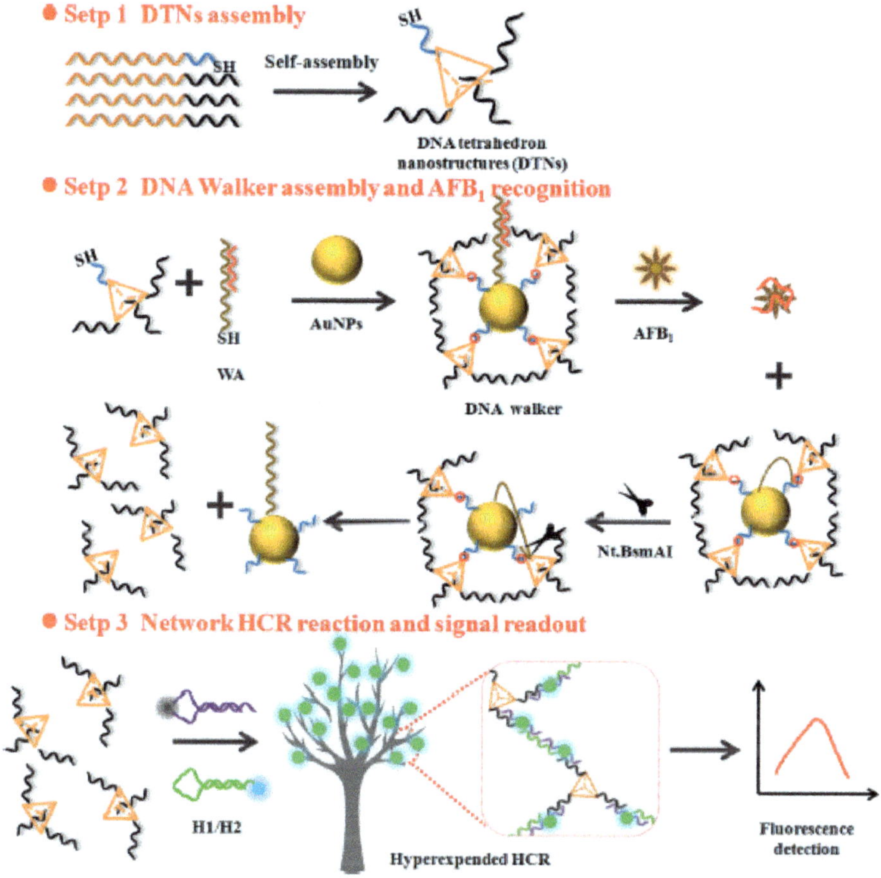

Figure 8.5 Schematic illustration of DNA nanomaterial-assisted dual amplified aptasensor for sensitive AFB1 detection based on DNA walker, DTNs, and network HCR. Step 1: the assembly of DTNs. Step 2: the assembly of DNA walker and the first signal amplification based on target-mediated DNA walker. Step 3: the second signal amplification based on network HCR. Reproduced from ref. 61 with permission from Elsevier, Copyright 2021.

electrochemical determination of AFM_1 in milk samples *via* linear sweep voltammetry (LSV).[62]

To determine the level of zearalenone (ZEN, known as F-2 toxin) in food products, You's group developed a self-enhanced ECL aptasensor.[63] In this study, the self-enhanced ECL luminophore incorporated $Ru(bpy)_3^{2+}$ and NGQDs (NGQDs) as its co-reactants to shorten the electron transfer and increase the luminous efficiency of $Ru(bpy)_3^{2+}$. When the NGQDs-NH_2-Ru@SiO_2 luminophore was decorated on the surface of the GCE, a strong ECL signal was observed due to the presence of NGQDs. After modification of the ZEN aptamer, the ECL signal is reduced. Upon the addition of ZEN, ECL quenched significantly. The proposed ECL aptasensor was used to detect ZEN contamination in corn flour, proving its suitability for use in ZEN detection in food products.

8.2.2.2 CQDs in Mycotoxin Detection

CQDs, as one of the newest classes of luminescent material, have recently attracted attention for their potential use in the ultrasensitive and rapid detection and quantification of mycotoxins. Most of the currently reported detection methods are based on fluorescence quenching of CQDs. Two recent assays have focused on ZEN detection and quantification.[64,65] One of them was based on a fluorometric lateral flow immunochromatographic assay (FLFIA), in which CQDs conjugated to ovalbumin and Ag nanoparticles loaded with ZEN antibody (ZEN-Ab) were used as donor and acceptor probes, respectively, in a FRET system. This study constituted the first reported use of CQDs as photosensitizer energy donors in a FRET-based immunochromatographic assay.

FLFIAs are rapid food safety screening tools. Traditional FLFIAs are based on turn-off fluorescence; they are less visually sensitive compared to turn-on FLFIAs. FLFIAs consist of three different pads (sample, conjugate, and absorbent pads), a nitrocellulose (NC) membrane, and a polyvinyl chloride (PVC) sheet as a substrate. An NC membrane was placed in the middle of the PVC sheet, and at the second 1/3 of its length, two lines were marked as the control line (C line) and the test line (T line). A few drops of CQD-OVA and a mixture of CQDs-OVA and Ag-ZEN-Ab were placed on the C line and T line, respectively, and left to dry. As indicated in Figure 8.6, the absorption and sample pads were connected to the NC membrane.[65] The conjugate pad connected the sample pad to the NC membrane. A few microliters of samples containing ZEN standards or a negative sample (*i.e.*, a sample without ZEN) mixed with Ag–Ab were dropped onto the sample pad. After 20 min, the turn-on pattern was observed under 365 nm portable UV light. The CQDs' emission wavelength was 458.5 nm (under the excitation of 356.5 nm), which was an approximate match with Ag–Ab's absorbance wavelength of 442 nm. The emission energy is transferred to Ag–Ab in the absence of ZEN. When ZEN was present, ZEN and CQDs-OVC competed to bind with Ag–Ab. The ZEN that bonded with the Ag–Ab moved toward the absorption pad, the CQDs' emissions (the C line) were maintained, and both the C line and the T line were illuminated under UV light. In the absence of ZEN, Ag–Ab interacted with CQD-OVA; the CQDs' fluorescence was quenched due to the FRET energy transfer, and the C line's color disappeared under UV light.

Compared to similar techniques, this immunosensor has the advantage of lower detection limits, but as with all immunosensors, it suffers from a short lifetime due to the denaturation of the Ab and quenching of the

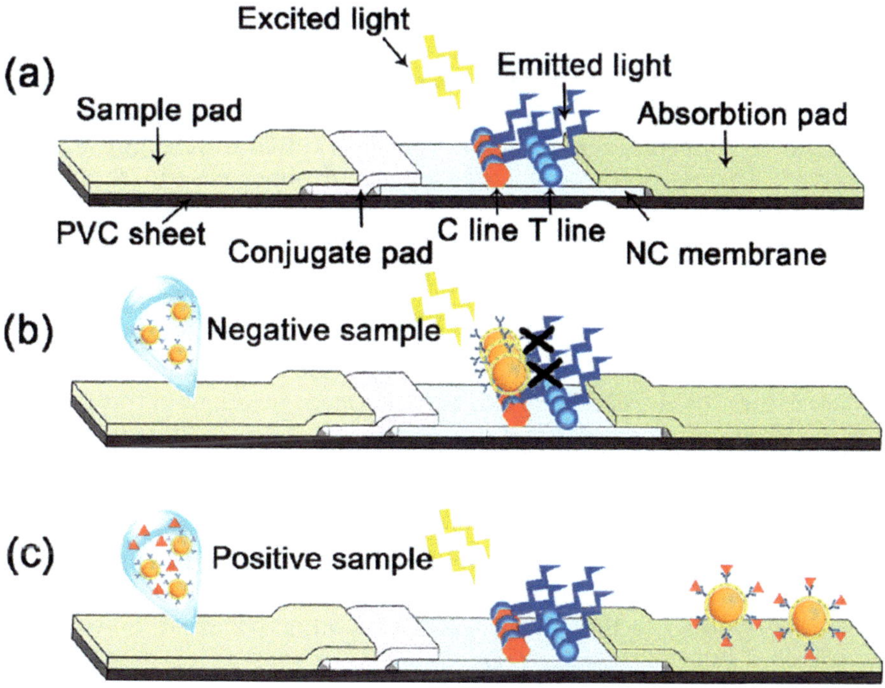

Figure 8.6 Schematic representation of the strip prepared in FLFIAs assay. Reproduced from ref. 65 with permission from Springer Nature, Copyright 2018.

donor probe that occur due to improper strip storage. An Ab-free method based on CQDs would, therefore, be the ideal choice for ZEN detection.

For this purpose, an eco-friendly fluorescence probe for ZEN determination was developed using encapsulated molecularly imprinted fluorescence quenching particles (MIFQPs).[64] The imprinted polymer had specific binding sites for ZEN and showed high, stable fluorescence of CQDs. The CQDs' fluorescence was quenched when ZEN was absorbed into the polymer. The quenching occurred due to the electron transfer from the CQDs (the emitters) to the ZEN molecules. The sensitivity of the MIFQP was about four times that of the nonimprinted polymer fluorescence quenching particle (NIFQP). Additionally, the quenching ability of ZEN was much higher than that of other tested mycotoxins due to its specific binding. This highly selective and sensitive method was used for the determination of ZEN in corn samples.

An ultrasensitive visual detection system was introduced for the point-of-care assay of the mycotoxin citrinin.[66] Yellow light-emitting nitrogen-doped CQDs (NCQDs) were prepared through a simple solvothermal method using diammonium citrate and urea as the carbon source. The NCQDs' yellow emissions were quenched in the presence of Congo red dye; they recovered when citrinin was added. UV-Vis and steady-state fluorescence evaluations revealed that the cause of this quenching was the non-covalent interaction of NCQDs with Congo red. A slight suppression in the NCQDs' emissions was observed with the addition of various concentrations of the dye to a fixed NCQD concentration; a red shift in the emitted wavelength was also observed. The red fluorescence of Congo red appeared and suppressed the yellow color when the NCQDs' surfaces were completely covered with Congo red. Upon the addition of citrinin, the NCQDs' yellow emissions recovered within 6 h. Mg^{2+} ions were added to the NCQDs–Congo red system to reduce the emission recovery time and make the sensor suitable for point-of-care use. Mycotoxins form strong chelates with Mg^{2+}, and as Mg^{2+} has an affinity for a higher coordination number, citrinin molecules attach to the CDs' surfaces through Mg^{2+}-induced chelation. In this process, the citrate/amine of the CD and the B-hydroxy/B-keto acid functionality of the citrinin from the analyte displace the Congo red molecules from the CD's surface, which initiates the recovery of the CD's fluorescence.

The fluorescence recovery time was 5 min. The technique was simple, fast, and sensitive, with a long lifetime. It was also economical, as no antibodies or aptamers were incorporated into the sensor. The detection technique worked well in the presence of other potential concomitant mycotoxins, such as AFB1 and OTA.

An ultrasensitive electrochemical aptasensor was developed using an AFB1 aptamer immobilized on the surface of a CQDs–Cu_2O nanocomposite.[67] CQDs–Cu_2O were deposited on the surface of a glassy carbon electrode (GCE) to improve the aptamer's immobilization efficiency. Cu_2O nanoparticles, as p-type semiconductors, improved the electrochemical performance of the electrode and aptamer immobilized on the CQDs-Cu_2O/GCE through the chemisorption process that occurred between the carboxylic group on the surface of the CQDs and the amine groups of the aptamers. This process increased electron transfer and led to higher sensitivity for AFB1 quantifications. The method reliably detected AFB1 in wheat flour in the range of 3 ag mL^{-1} to 1.9 μg mL^{-1}.

Recently, a new fluorescent immunosensor based on NCQDs was designed for AFM1 monitoring in milk. A simple hydrothermal procedure was used to synthesize NCQDs with a maximum emission at 450 nm upon excitation at 350 nm. The obtained CQDs were conjugated to the anti-AFM1 antibody (Ab) *via* EDC/NHS carbodiimide coupling.

The fluorescent response of the CQDs/Ab probe was investigated using different concentrations of AFM1 (0.01–1 ng mL^{-1}) under optimized experimental conditions. With the increasing AFM1 concentrations, the fluorescent intensity of the CQDs/Ab probe quenched gradually. This procedure was also used to develop a CQD-based fluorescent lateral flow assay for the sensitive, direct, on-site detection of aflatoxin M1 residue in milk and dairy products.[68]

8.2.3 Marine Toxin Detection

Early and rapid detection is one of the most effective ways to reduce the potential risks associated with marine toxins. As mentioned above, traditional detection methods have many limitations, such as low sensitivity and specificity, complicated procedures, and ethical problems. Immunoassays and biosensors are recommended alternative approaches: they are highly sensitive and specific, time saving, rapid, convenient, relatively low cost, and have a very low detection limit.[21] Some of the main available analytical techniques for marine toxin detection have been reviewed by Daguer *et al.* and Schirone *et al.*[20,23]

In an excellent report, Wang *et al.* summarized the then-latest (*i.e.*, 2016–2021) advances in research on aptasensors for rapid shellfish toxin screening in marine food chains. They also provided detailed explanations of aptamers, working principles, and analytical figures of merit. They discussed the aptasensors' applicability to real samples.[22] Various other aptamer-based biosensors have been reported.[21,69]

8.2.3.1 Graphene-based QDs

Several studies have examined GQDs' applications for marine toxin monitoring; we summarize their findings below.

Okadaic acid (OA) is a significant marine toxin whose determination was investigated by Chen's group.[70] Luminophores with good biobinding ability and excellent ECL action exhibited better performance when used in an ECL-based biosensor. Compared to pristine GQDs without passivation, heteroatom-dopped GQDs offered better ECL performance. In this study, an electrolysis method was employed to synthesize water-dispersed and uniform-sized phosphorus and sulfur co-doped GQDs (P, S-GQDs). The P, S-GQDs were used as ECL markers in the fabrication of an immune-ECL sensor for OA detection. A carboxylated multiwall carbon nanotubes-poly(diallyl dimethyl ammonium) chloride-Au nanocluster (CMCNT-PDDA-AuNC) composite was decorated on the surface of a glassy carbon electrode (GCE) to construct a CMCNT-PDDA-AuNC GCE as a sensing platform. OA was also immobilized on the modified electrode. Monoclonal antibody-labeling against OA (anti-OA-Mab) and P, S-GQDs was conjugated to make P, S-GQDs-Ab as the ECL immunoprobe. The sensor functioned based on the competition of P, S-GQDs-Ab for linking to the immobilized and free OA. When P, S-GQDs-Ab was linked with the immobilized OA, ECL emission was observed. When free OA was added to the system, P, SGQDs-Ab bound with it; as a result, less P, S-GQDs-Ab was immobilized on the electrode's surface, which led in turn to a reduction in ECL intensity. This sensor was used in mussel samples, and good recoveries were obtained. Compared to ELISA kits, it was more sensitive and exhibited a lower matrix effect. Figure 8.7 presents a schematic illustration of this immunosensor's construction and use.[70]

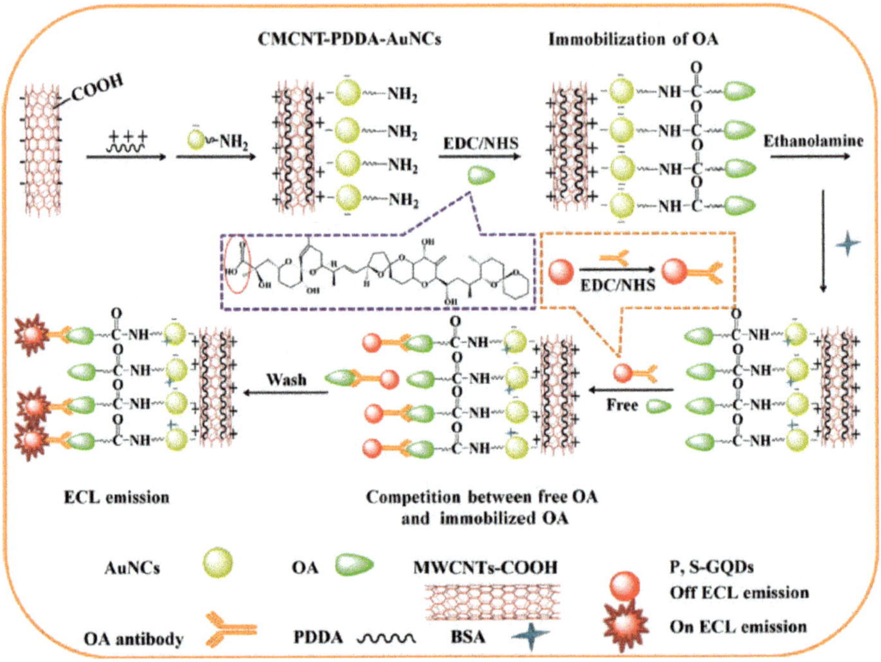

Figure 8.7 Schematic representation for the fabrication and application of the immunosensor. Reproduced from ref. 70 with permission from Elsevier, Copyright 2020.

Saxitoxin (STX) is another major marine toxin found in shellfish. A GQD-based fluorescent aptamer assay for sensing STX has been reported.[71] In this aptasensor, the aptamer STX-41, GQDs, magnetic reduced graphene oxide (MRGO), and DNase (I) were the recognition element, fluorescent signal, energy receptor, and enzymatic signal amplifier, respectively. This assay combined the advantages of FRET and nuclease-assisted target recycling signal amplification. In the absence of a target, the interaction of the GQD–aptamer probe with MRGO resulted in fluorescence quenching due to the FRET mechanism. When STX was added, it caused the GQD–aptamer to be removed from MRGO and form the GQD-aptamer/target complex instead, leading to the recovery of the fluorescence signal. Moreover, targets were released by nuclease digestion and recycled to produce more fluorescent signals, causing signal amplification. The sensor was tested on real clam samples. Figure 8.8 provides a schematic representation of the GQD fluorescent aptamer assay.[71]

8.2.3.2 *Carbon-based QDs*

CQDs, as a new alternative to traditional dyes, were used to improve FRET efficiency and sensitivity in the sensing of maitotoxin (MTX) attached

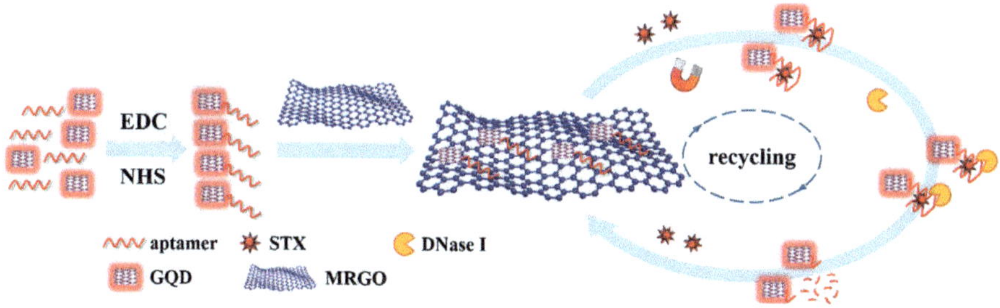

Figure 8.8 Schematic illustration of the GQD fluorescent aptamer assay based on FRET and nuclease-assisted target recycling signal amplification. Reproduced from ref. 71 with permission from Springer Nature, Copyright 2021.

to Au nanoparticles using the fluorescence technique.[72] CQDs as donors were prepared from biochar (a natural carbon source that is both biologically and environmentally safe) using a simple procedure. The maximum emission intensity of as-prepared CQDs was 527 nm. Au nanoparticles were added to CQD solutions; when various concentrations of MTX were subsequently added to the solutions, the MTX attached to the surfaces of the Au nanoparticles *via* the thiol groups and absorbed the CQDs' emissions. Au–MTX has a maximum absorption wavelength of 550 nm and acts as an acceptor in this FRET fluorescence sensing process. The sensor was specific, fast, and highly selective; it was able to quantify the toxin in the range of 1–600 pmole L^{-1} with a detection limit of 0.3 pmole L^{-1}. Currently, it is the only reported sensor to use CQD quenching to quantify marine toxins. Figure 8.9 presents a schematic representation of the FRET system between the CQDs and the Au nanoparticles attached to MTX.[72]

8.3 Applications of Other QDs in the Analysis of Food Pathogens

Tables 8.3–8.5 summarize the applications of various semiconductor QDs for the monitoring of various biotoxins. They almost show sensitive and specific characteristics.

As shown in these tables, more reports focus on mycotoxin detection compared to bacterial or marine toxin detection. According to the available literature, the most important mycotoxins (such as ZEN, OTA, AFB_1, AFM1, and FB_1) have been monitored using semiconductor QDs. Detection techniques for these toxins based on QDs' luminescence include fluorescence-linked immunosorbent assays, lateral flow fluorescent strip immunosensors, multiplexed immunochromatographic assays, and electrochemiluminescence biosensors. To better compare various

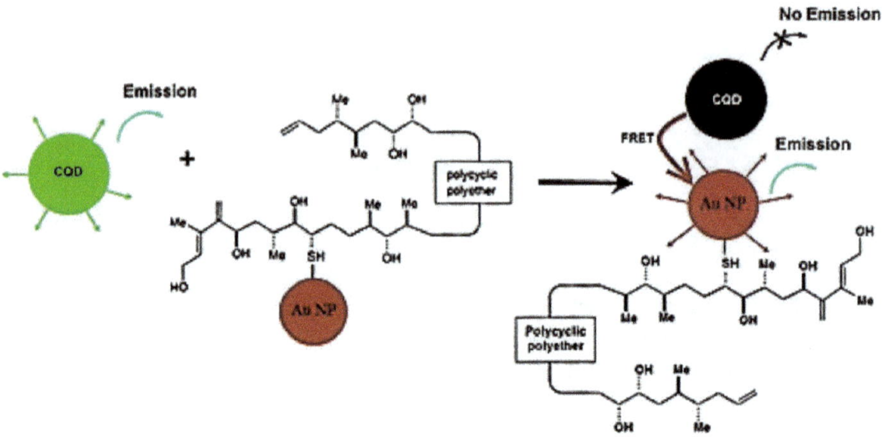

Figure 8.9 FRET between CQDs and MTX-loaded Au nanoparticles. Reproduced from ref. 72 with permission from Elsevier, Copyright 2020.

strategies, typical performance characteristics of each method (including detection limit, linear range, recovery, precision, and reliability) are also summarized in the tables. Various QDs have been used to detect toxins, including binary and ternary QDs, core–shell QDs, and QD nanobeads (QBs).

A simple and easy-to-construct CdSSe@ZnS QB-based FS immunosensor was developed for the rapid, on-site detection of AFB_1 in lotus seeds. By inserting many QDs in the polymer nanobeads, QBs possess the excellent optical properties of QDs and show 1000 times higher fluorescent intensity and stronger tolerance against solution pH and ion strength than the corresponding QDs.[78] Compared to conventional laboratory methods such as LC-MS/MS, this sensor was relatively simple, low-cost, and portable, and its detection limit was lower.[78]

Some feeds can be contaminated with more than one mycotoxin, which makes the development of multiplex assays a critical task. Table 8.4 lists a number of publications on semiconductor QD-based assays for the simultaneous determination of multiple mycotoxins. In order to achieve the highest possible sensitivity in a multiplex immunochemical assay, different labels must be applied, and the signals caused by the various toxins must be separated. When organic dyes are used as labels, statistical methods must inevitably be used due to the labels' asymmetrical unsharpened emission peaks. Alternatively, multicolor (*i.e.*, multiwavelength) QDs can be used as labels to solve this problem.[85]

Detection systems must be fast, specific, and accurate, and they must be capable of sensing the concentration of a given toxin below its maximum acceptable limit. To achieve these goals, one study developed a multiplex immunosorbent assay based on hydrophobic Cd-based QDs @SiO_2 for monitoring three mycotoxins (AFB1, DON, and ZEN) in cereal-based

Table 8.3 Applications of various semiconductor QDs for monitoring of bacterial toxin.

QDs	Toxin	Strategy	Analytical performance	Real sample	Ref.
MagQDs[a]	BoNT/A[b]	LFA[d] biosensor	**LOD[k]**	Milk and grape juice	73
Fe_3O_4@PEI-QDs	SEB[c]		2.52 and 2.86 pg mL^{-1}		
(QD: CdSe/ZnS) QD655,[e] QD525,[f] QD585[g]	BoNT/E[h]	FRET[i] between QDs and dark quencher-labeled peptide probes	LOD[k] 0.02 and 2 ng mL^{-1} for BoNT/E light chain (LcE) and holotoxin LOQ[l]: 4 pM (0.2 ng mL^{-1}) for LcE Liner range: 4 pM–120 nM (0.2 ng mL^{-1}–6 μg mL^{-1})		10
NM[j]	BoNT/A[b]	Fluorescence sandwich immunoassay	LOD[k]: 31 pM (96-well plate format), 5 pM (bead-based assay format)		74
CdTe	Simultaneous detection of *Vibrio cholera* O1 (Ogawa) and of the *Clostridium botulinum* toxin type A	Multicolor immunochromatographic test strip format	LOD[k] 0.8 ng mL^{-1} (*C. botulinum* toxin A) and 103 CFU.mL^{-1} (*V. cholera*)	The developed test system for simultaneous detection of *C. botulinum* toxin A and *V. cholera* O1 can be efficiently used in food production and food/water safety control.	75

(*continued*)

Table 8.3 (*continued*)

QDs	Toxin	Strategy	Analytical performance	Real sample	Ref.
ZnS	SEB[c]	Electrochemical and fluorescence immuno-assay methods	LOD[k]: 0.02 ng mL^{-1} (fluorescence assay) and 1.0 ng mL^{-1} (electrochemical assay) 10 ng mL^{-1} to 1 μg mL^{-1} (electrochemical assay)		76

[a]Magnetic QD nanoparticles.

[b]Botulinum neurotoxin type A.

[c]Staphylococcal enterotoxin B.

[d]Lateral flow assay.

[e]Qdot 655 ITK carboxyl quantum dots.

[f]Qdot 525 ITK carboxyl quantum dots.

[g]Qdot 585 ITK carboxyl quantum dots.

[h]Botulinum neurotoxin serotype E.

[i]Förster resonance energy transfer.

[j]Not mentioned.

[k]Limit of detection.

[l]Limit of quantification.

Table 8.4 Applications of various semiconductor QDs for monitoring of mycotoxins.

QDs	Toxin type	Strategy	Analytical performance	Real sample	Ref.
CdSe/ZnS	AFB_1[a] and ZEN[b]	QBs-FLISA[g]	LOD: 9.3 and 102.1 pg mL^{-1} (for AFB_1 and ZEN) LOD in feedstuffs 0.28 and 3.1 μg kg^{-1} (for AFB_1 and ZEN) Linear range: 15 to 500 pg mL^{-1} for AFB1 0.15 to 5 ng mL^{-1} for ZEN Recoveries: 82.50–116.21% RSD: < 11.3% Accuracy: compared with LC-MS/MS analysis	Maize, distillers dried grain, maize germ, and maize gluten	77
CdSSe@ZnS	AFB_1	QBs-FS immunosensor[i]	LOD: 1 ng mL^{-1} (2 μg kg^{-1}) Linear range: 1–19 ng mL^{-1} (2–38 μg kg^{-1}) Recovery: 94.0–116.0% RSD: <10% Accuracy: compared with LC-MS/MS analysis	Lotus seeds	78
CdSe/ZnS	AFB_1	FRET method[j]	Dynamic range: 25–2500 ppm Accuracy: compared with HPLC method	Maize grains	79
CdTe and PbS QDs	AFB_1	Ratiometric electrochemical	LOD: 4.5 pg mL^{-1} Linear range: 5 pg mL^{-1}–50 ng mL^{-1}	Peanuts	80

(*continued*)

Table 8.4 (*continued*)

QDs	Toxin type	Strategy	Analytical performance	Real sample	Ref.
		aptasensor (square wave voltammetry)	RSD: 4.7% (for 10 ng mL^{-1}) Recoveries: 90.0–93.6% (RSD < 6.7%)		
CdZnTe QDs	AFB_1	FRET-based apatsensor	LOD: 20 pg mL^{-1} Linear range: 50 pg mL^{-1}–100 ng mL^{-1} RSD: 4.8% Recovery: 95.7–106.3% (RSD < 9.3%) Accuracy: compared with HPLC method	Peanut	81
CdSe/SiO2 QBs	AFB_1, ZEN DON	LFA[k]	LOD: 10 pg mL^{-1} (AFB1) 80 pg mL^{-1} (ZEN) 500 pg mL^{-1} (DON) The LODs in cereals AFB_1 0.24 μg kg^{-1} ZEN 1.92 μg kg^{-1} DON 12 μg kg^{-1} Recovery: 85.5–119.0%, RSD < 16.4% The IC_{50} for AFB_1 0.07 ng mL^{-1} ZEN 0.45 ng mL^{-1} DON 5.23 ng mL^{-1} Accuracy: compared with LC-MS/MS analysis	Cereal (maize, wheat, distillers dried grains, peanut press pulp, shotcrete corn husk and corn germ meal samples)	82

(*continued*)

Table 8.4 (*continued*)

QDs	Toxin type	Strategy	Analytical performance	Real sample	Ref.
CdSe/ZnS	AFB_1 and ZEN	QB-based ICA[l]	LOD: 1.65 pg mL^{-1} (AFB_1) 59.15 pg mL^{-1} (ZEN) Linear range: 2–300 pg mL^{-1} (AFB_1) 0.1–15 ng mL^{-1} (ZEN) Recoveries: 81.77–119.70% (AFB_1) 94.18–111.4% (ZEN) RSD < 12% IC_{50} value: 38.98 pg mL^{-1} (AFB_1) 1.23 ng mL^{-1} (ZEN) Accuracy: compared with ELISA method (70 samples)	Maize	83
Cd/Te	AFB_1	FRET-based fluorescence aptasensor	LOD: 0.004 μg μL^{-1} (4 ng μL^{-1}) Linear range: 0.002 (6 nM)–0.2 μg μL^{-1} (600 μM) Recovery: 83.4–97.26% Standard deviation of recovery: 0.02–0.53 Accuracy: comparison with standard commercial AFB_1 kit	Wheat and maize	84
Cd-based QDs	DON, ZEN, AFB_1	TAM FLISA[m]	LOD: 6.1 μg kg^{-1} (DON) 5.3 μg kg^{-1} (ZEN) 5.4 μg kg^{-1} (AFB_1)	Maize and wheat	85

(*continued*)

Table 8.4 *(continued)*

QDs	Toxin type	Strategy	Analytical performance	Real sample	Ref.
			Linear range: 12–110 μg kg^{-1} (DON) 9–84 μg kg^{-1} (ZEN) 6.5–76 μg kg^{-1} (AFB_1) Accuracy: compared with LC–MS/MS method		
CdSe/ZnS core/shell QDs	AFM_1[f]	Frit-based immunoassay	Cut-off value: 0.02 μg kg^{-1} For test of reliability and acceptability: comparison with LC–MS/MS method The rates for false positive and false negative results: below 5%, the specificity and sensitivity rates > 95%	Milk	86
CdTe	AFM_1 or ATM1	ECLIA[n]	LOD: 0.3 pg mL^{-1} Linear range: 1.0–1.0 × 10^5 pg mL^{-1} Recovery: 95.4% (intra-assay RSD = 2.3%) RSD: 3.5% Accuracy: compared with ELISA method	Milk	87
CdSeS/ZnS QDs	DON[d]	Homogenous FRET-based fluorescent immunoassay	LOD: 28 μg kg^{-1} in spiked wheat Linear range: 1–10 ng mL^{-1}	Wheat	88
CdSe/CdS and CdSe/CdS/ZnS	ZEN and DON	Multicolor LFIA[o]	The minimal cutoff levels 20 μg kg^{-1} (ZEN)–130 μg kg^{-1} (DON)	Maize and wheat	89

(continued)

Table 8.4 *(continued)*

QDs	Toxin type	Strategy	Analytical performance	Real sample	Ref.
			The cutoff levels are typically set at 80% of the EC maximum level, corresponding to 40 μg kg^{-1} for ZEN and 400 μg kg^{-1} for DON. Sensitivities: 95.6% DON and 100% for ZEN in maize samples 96.4% for DON and 100% for ZEN in wheat samples The specificities: 96.9% for DON and 100% for ZEN in maize samples 95.2% for DON and 100% for ZEN in wheat samples Accuracy: compared with LC–MS/MS method		
CdSeS/ZnS	DON, ZEN T2 and HT2-toxin	Multiplex LFIA[p]	Cut-off limit[x]: DON: 1000 μg kg^{-1}; ZEN: 80 μg kg^{-1} T2/HT2: 80 μg kg^{-1} The average recovery ± SD: DON: 103% ± 10 ZEN: 97% ± 5.9 T2: 100% ± 8.3 HT2: 93% ± 11.4 Accuracy: compared with LC–MS/MS method	Wheat	90

(continued)

Table 8.4 (*continued*)

QDs	Toxin type	Strategy	Analytical performance	Real sample	Ref.
InP/ZnS	ZEN and DON	Multiplex LFIA	Cut-off value: 50 μg kg^{-1} (ZEN) 500 μg kg^{-1} (DON) Sensitivities: 96% DON and 97% for ZEN in maize samples 95% for DON and 97% for ZEN in wheat samples The specificities: 98% for DON and 96% for ZEN in maize samples 97% for DON and 98% for ZEN in wheat samples Accuracy: comparison of with LC–MS/MS method	Maize and wheat	91
CdSe@CdS	OTA[c]	QD-based label-free ECL aptasensor[h]	LOD: 0.89 ng mL^{-1} Linear range: 1–100 ng mL^{-1} Recovery: 98.1–105.6% (Lily samples and RSD = 4.5–6.5%) 97.3–101.5% (Rhubarb samples and RSD = 2.1–4.2%)	Real lily and rhubarb samples	92
ZnCdSe/ZnS quantum dots	OTA	Nb-FRET immunosensor[q]	LOD: 5 pg mL^{-1} Linear range: 0.005–1 ng mL^{-1} The final LODs: 0.05 μg kg^{-1} in rice	Cereals (rice, oats, barley, and wheat)	93

(*continued*)

Table 8.4 (*continued*)

QDs	Toxin type	Strategy	Analytical performance	Real sample	Ref.
			0.125 μg kg^{-1} in barley 0.2 μg kg^{-1} in wheat 0.5 μg kg^{-1} in oats The average recovery rate: 81–111% (intra-assay) 80–109% (inter-assay) RSD: 3–9.4% (intra-assay) 4.5–9% (inter-assay) Accuracy: comparison of with LC–MS/MS method		
CdTe QDs	OTA	Label-free fluorescent aptasensor	LOD: 0.16 ng mL^{-1} Linear range: 0.2–20 ng mL^{-1} Recovery: 98.9–102.2% RSD: 2%	Astragalus membranaceus	94
CdTe	OTA	ECL-RET[r]	Linear range: 0.0005–50 ng mL^{-1} LOD: 0.17 pg mL^{-1} RSD: 1.9% (intra-assay) 3.9% (inter-assay) Recovery: 96.1–100.7%	Corn	95
CdSe/ZnS QDs	ZEN, OTA, and FB1[e]	mICA[s]	LOD: 5 ng mL^{-1} (OTA) 20 ng mL^{-1} (FB1) 10 ng mL^{-1} (ZEN) Accuracy: comparison with the UPLC-FLD method (15 samples)	Maize	96

(*continued*)

Table 8.4 (*continued*)

QDs	Toxin type	Strategy	Analytical performance	Real sample	Ref.
CdTe or PbS	OTA, FB_1	Electrochemical aptasensor	LOD: 20 pg mL^{-1} (FB1) 5 pg mL^{-1} (OTA) Linear range: 10 pg mL^{-1} to 10 ng mL^{-1} (OTA) 50 pg mL^{-1} to 50 ng mL^{-1} (FB1) Recovery: 85.0–94.2% (OTA) 92.0–103.0% (FB1) RSD: 6.1% (OTA) 5.3% (FB1) Accuracy: comparison of with HPLC method	Maize	97
CdTe	OTA	Fluorescence ELISA	LOD: 2.2 pg mL^{-1} Linear range: 2.4 pg/mL–625 pg mL^{-1} Recovery: 86.6–106% (intra-assay, RSD = 4.22–13.8%) 84.4–107% (inter-assay, RSD = 7.13–14.5%) Accuracy: comparison with a commercial ELISA kit	Corn	98
ZnCdSe/ZnS	OTA	Aptasensors based on FRET mechanism	LOD: 376.3 pM (0.152 ng mL^{-1}) Linear range: 0.5–20 nM Recovery:	Coffee and red wine	99

(*continued*)

Table 8.4 (*continued*)

QDs	Toxin type	Strategy	Analytical performance	Real sample	Ref.
			93.4–124% (red wine, RSD = 8.3–10.2%) 96.1–123.2% (coffee, RSD = 9.4–12.5%)		
CdTe	OTA	Fluorescence aptasensor	LOD: 1.0 ng mL^{-1} Dynamic range: 1.0–1000 ng mL^{-1} Recovery: 92–123% RSD < 12%	Red wine	100
Not-mentioned	OTA	Fluorescent strip sensor based on aptamer–QD technology	LOD: 1.9 ng mL^{-1} Linear range: 0–10 ng mL^{-1} RSD: 4.7% Accuracy: comparison with HPLC method	Red wine	101
CdTe	FB_1[e]	cFELISA[t]	LOD: 0.33 ng mL^{-1} Linear range: 0.39–12.50 ng mL^{-1} Recovery: 80.9–87.7% (intra-assay, RSD = 5.5–8.6%) 85.8–93.1% (inter-assay, RSD = 11.5–17.4%) IC_{50}[y]: 1.95 ng mL^{-1} Accuracy: comparison with the UPLC-FLD method	Corn	102
ZnCdSe/ZnS QD	ZEN	ICAs[u]	LOD: 1 μg L^{-1} Visual LOD: 6 μg kg^{-1} (maize samples)	Cereal sample (maize, wheat, and rice)	103

(*continued*)

Table 8.4 *(continued)*

QDs	Toxin type	Strategy	Analytical performance	Real sample	Ref.
			Accuracy: comparison with the commercial ELISA kit		
MPA-QDs[w]	ZEN	Fluorescent ELISA	DOL: 4.1 pg mL^{-1} Linear range: 2.4 pg mL^{-1} to 1.25 ng mL^{-1} Recovery: 84.5–108.5 (intra-assay, RSD = 1.4–3.9%) 84.1–112.3 (inter-assay, RSD = 2.4–3.1%) IC_{50}: 75 pg mL^{-1} Accuracy: comparison with LC–MS/MS method	Corn	104
CdSe/ZnS	ZEN	QB-ICA[v]	LOD: 0.0625 ng mL^{-1} Linear range: 0.125–10 ng mL^{-1} Recovery: 89.5–108% (intra-assay, RSD = 1.5–3.0%) 93.5–106% (inter-assay, RSD = 1.9–4.4%) IC_{50}: 1.01 ± 0.09 ng mL^{-1} Accuracy: comparison with LC–MS/MS method	Corn	105

[a]Aflatoxin B_1 (AFB_1).

[b]Zearalenone (ZEN).

(continued)

Table 8.4 (continued)

QDs	Toxin type	Strategy	Analytical performance	Real sample	Ref.

[c]Ochratoxin A (OTA).

[d]Deoxynivalenol (DON).

[e]fumonisin B_1 (FB_1).

[f]Aflatoxin M_1 (AFM_1 or ATM_1).

[g]QD nanobeads-based fluorescence-linked immunosorbent assay (QBs-FLISA).

[h]QD- based label-free electrochemiluminescence aptasensor (QD-based label-free ECL aptasensor).

[i]QB-based lateral flow fluorescent strip immunosensor (QBs-FS immunosensor).

[j]Fluorescence resonance energy transfer method (FRET).

[k]Fluorometric lateral flow immunoassay (LFA).

[l]QD nanobead-based multiplexed immunochromatographic assay (QB-based ICA).

[m]Triple-analyte multiplex Fluorescent labeled immunosorbent assay (TAM FLISA).

[n]Electrochemiluminescent immunoassay (ECLIA).

[o]Multicolor lateral flow immunoassay (Multicolor LFIA).

[p]Multiplex lateral flow immunoassay (LFIA).

[q]Nanobody-mediated FRET-based immunosensor (Nb-FRET immunosensor).

[r]Electrochemiluminescence resonance energy transfer (ECL-RET).

[s]Multiplex immunochromatographic assay (mICA).

(continued)

Table 8.4 *(continued)*

QDs	Toxin type	Strategy	Analytical performance	Real sample	Ref.

[t]A competitive fluorescence enzyme-linked immunosorbent assay (cFELISA).

[u]Lateral flow immunochromatographic assays (ICAs).

[v]QD submicrobead-based immunochromatographic assay (QB-ICA).

[w]Mercaptopropionic acid capped CdTe QDs (MPA-QDs).

[x]The cutoff level was estimated as the lowest mycotoxin concentration, which resulted in no fluorescence of the test lines.(18).

[y]IC_{50} : The half maximal inhibitory concentration or the median inhibition concentration (31).

Table 8.5 Applications of various semiconductor QDs for monitoring of marine toxins.

QDs type	Toxin type	Strategy	Analytical performance	Real sample	Ref.
Qdot525 (CdSe core encapsulated in the shell of ZnS and polymer)	Microcystin-LR (cyanobacterial hepatotoxins)	Electrochemical immunoassay	Linear range: 0.227–50 μg L^{-1} LOD: 0.099 μg L^{-1}	*Microcystis aeruginosa*	106
CdSe/CdS/ZnS	Okadaic acid	Fluorescence microplate assay	Linear range: 10.0–100.0 μg kg^{-1} LOD: 0.25 μg kg^{-1} LOQ: 0.84 μg kg^{-1} RSD < 4.5% The mean recoveries: 92.5–101.0% (in extracts, RSD < 7.7%) 92.9–101.3% (in purified extracts after QuEChERs, RSD < 7.6%) For test of reliability and acceptability: comparison with HPLC-MS/MS and ELISA method	Shellfish samples (*Mytilus edulis*, *Cyclina sinensis*, and *Tegillarca granosa* samples)	107
CdSe QDs	Okadaic acid and brevetoxins	Paper-based microfluidic aptasensor (FRET mechanism)	LOD: 0.4 ng mL^{-1} (OA) 0.56 ng mL^{-1} (brevetoxins) Linear range: 0–5.5 ng mL^{-1} (OA) 0–2 ng mL^{-1} (brevetoxins) Recoveries: 89.63–105% (OA) 88–112.53% (brevetoxins)	Mussel tissue, sausage, fresh egg white	108

(continued)

Table 8.5 (*continued*)

QDs type	Toxin type	Strategy	Analytical performance	Real sample	Ref.
			For test of reliability and acceptability: comparison with ELISA method		
CdSe/ZnS	*Trichinella spiralis*	Immunochromatographic assay	LOD: 189.92 ng mL^{-1} Linear range: 250–850 ng mL^{-1} For test of reliability and acceptability: comparison with ELISA kit method	Swine sera	109

products (maize and wheat). The detection limits were lower than the maximum acceptable limits of the target mycotoxins. Liquid chromatography coupled with tandem mass spectrometry (LC-MS/MS) confirmed this method's reliability.[85]

An electrochemical magneto-controlled aptasensor was developed for the simultaneous detection of OTA and FB_1 in maize.[97] This aptasensor was fabricated using a multi-step process. First, aptamers specific for OTA and FB1 were assembled on the surface of CdTe or PbS QD-coated silica hybrid spheres, and core–shell Fe_3O_4@Au MBs were used to immobilize the thiolated complementary DNAs (cDNAs) of the corresponding aptamers. Next, a magneto-controlled aptasensor was constructed *via* the conjugation between aptamers and cDNA. In the presence of the targets, the aptamer preferred to bond with them, leading to the partial release of the labels from the MBs. The quantity of labels remaining on the MBs' surfaces was negatively correlated with the target's concentration and could be magnetically collected. After an acid-dissolution step, the electrochemical signals of Cd^{2+} and Pb^{2+} dissolved from CdTe and PbS QDs on the MBs' surfaces were utilized for simultaneous measurements of OTA and FB1 by using square wave voltammetry (SWV) stripping measurements. The sensing platform showed a wide linear range of 10 pg mL^{-1} to 10 ng mL^{-1} for OTA and 50 pg mL^{-1} to 50 ng mL^{-1} for FB1. The aptasensor's performance in this study confirmed its applicability for quality control in food processing. An HPLC technique was used as a standard for examining the accuracy of this new method.[97]

8.4 Conclusion

Foodborne pathogens and the toxins they release constitute a significant threat to human health and the environment. They can cause outbreaks of diseases and may affect regional or even global economies. Some toxins are heat-resistant and may persist in food products even after several steps of processing, and they may remain during storage. Pathogen toxin detection is, therefore, of vital importance from an agricultural and food industry point of view. For this purpose, a variety of techniques have been introduced, including chromatography, (enzymatic) immunoassays, and spectroscopy. Demand is high for methods that are simple, fast, reliable, inexpensive, high-throughput, and capable of being used in the field. Among the various approaches to toxin detection, fluorescent sensors are the method of choice, as they satisfy the above conditions; efforts are going on to improve fluorescent immunosensors' specificity for toxins of particular concern.

Semiconductor QDs work well as fluorescence donors in most fluorometric methods of toxin determination. They have highly stable luminescence, can be functionalized for specific applications, and can be incorporated into polymer matrixes as fluorophores. Their applications in sensing strategies are limited, however, by the complicated pre-treatment required to add functional groups to their surfaces and by the potential of heavy metal-based QDs, especially, to act as environmental pollutants and health hazards. CQDs have recently been introduced as a good substitute for semiconductor QDs in various assays, including those for toxins. They exhibit low toxicity, high stability, and bright fluorescence and are both biocompatible and environmentally friendly. CQDs can be synthesized from a variety of source materials using simple methods, and it is likewise easy to scale them up. CQDs' surface functional groups can also be determined easily through the selection of appropriate starting materials, with no need for additional complicated steps. They, thus, show great potential for use in biosensors for point-of-care food quality control in the food packaging, agricultural, and beverage industries. Research in biosensor design for toxin detection and quantification is still in its infancy; further studies are required to replace Ab and aptamer recognition elements in these kinds of sensors without sacrificing their selectivity and sensitivity.

Abbreviations

AFB1	Aflatoxin B
AFM1	Aflatoxin M_1
AICC	Affinity immune-chromatography column
ARfD	Acute reference dose
ASP	Amnesic shellfish poisoning
ATM1	Aflatoxin M_1
AZP	Azaspiracid
BoNT/A	Botulinum neurotoxin type A
BoNT/E	Botulinum Neurotoxin Serotype E
BoNTs	Botulinum neurotoxins
BSA	Bovine serum albumin
BTX	Brevetoxins
cDNA	Capture DNA
cFELISA	Competitive fluorescence enzyme-linked immunosorbent assay
CFP	Ciguatera fish poisoning
CHO	Chinese hamster ovary cells
CMCNT-PDDA-AuNC	Carboxylated multiwall carbon nanotubes-poly(diallyldimethylammonium) chloride-Au nanocluster
CQDs	Carbon quantum dots
CV	Cyclic voltammetry
DA	Domoic acid
DON	Deoxynivalenol

DSP	Diarrheic shellfish poisoning
DTX	Dinophysistoxin
E. coli	*Escherichia coli*
ECL	Electrochemiluminescence
ECLIA	Electrochemiluminescent immunoassay
ECL-RET	Electrochemiluminescence resonance energy transfer
EIS	Electrochemical impedance spectroscopy
ELIFA	Enzyme-linked immunofiltration assay
ELISA	Enzyme-linked immunosorbent assay
EPD	Electrophoretic deposition
EU	European
FB1	Fumonisin B_1
FDA	Food and Drug Administration
FLFIAs	Fluorometric lateral flow immunochromatographic assay
FRET	Förster resonance energy transfer
FRET	Fluorescence resonance energy transfer
GC	Gas chromatography
GCE	Glassy carbon electrode
GC-MS	Gas chromatography mass spectroscopy
GQDs	Graphene-based quantum dots
HAB	Harmful algal blooms
HCR	Hybridization chain reaction
HPLC-FD	High-performance liquid chromatography fluorescence detection
HPLC-MS	High-performance liquid chromatography-mass spectrometry
ICAs	Lateral flow immunochromatographic assays
ICAs	Immunochromatographic assays
IFE	Inner filter effect
ITO	Indium tin oxide
LAMP	Loop-mediated isothermal amplification
LC	Liquid chromatography
LC-MS	Liquid chromatography mass spectroscopy
LD50	Lethal dose 50%
LFA	Lateral flow assay
LFA	Fluorometric lateral flow immunoassay
LFIA	Lateral flow immunoassay
LOAEL	Lowest observable adverse effect level
LOD	Limit of detection
LOQ	Limit of quantification
LPS	Lipopolysaccharides
LSV	Linear sweep voltammetry
MagQDs	Magnetic quantum dot nanoparticles

MBs	Magnetic beads
MICA	Multiplex immunochromatographic assay
MIFQP	Molecularly imprinted fluorescence quenching particles
MPA-QDs	Mercaptopropionic acid capped CdTe QDs
MRGO	Magnetic reduced graphene oxide
MTX	Maitotoxin
NASBA	Nucleic acid sequence-based amplification
Nb-FRET immunosensor	Nanobody-mediated FRET-based immunosensor
NC	Nitrocellulose
NGQDs	Nitrogen-doped graphene-based quantum dots
NPs	Nitrogen-doped graphene-based quantum dots@SiO_2 nanoparticles
NM	Not mentioned
NOAEL	No observable adverse effect level
NSP	Neurotoxic shellfish poisoning
OA	Okadaic acid
OTA	Ochratoxin A
OVA	Ovalbumin
PABA	3-Aminophenylboronic acid
PAT	Patulin
PCR	Polymerase chain reaction
POCTs	Smartphone-based point-of-care tests
PSP	Paralytic shellfish poisoning
PTX	Pectenotoxin
PVC	Polyvinyl chloride
QA-based ICA	Quantum dot nanobead-based multiplexed immunochromatographic assay
QB-ICA	Quantum-dot submicrobead-based immunochromatographic assay
QBs-FLISA	Quantum dot nanobeads-based fluorescence-linked immunosorbent assay
QBs-FS immunosensor	QBs-based lateral flow fluorescent strip immunosensor
QD525	Qdot 525 ITK carboxyl quantum dot
QD585	Qdot 585 ITK carboxyl quantum dot
QD655	Qdot 655 ITK carboxyl quantum dot
QDs	Quantum dots
qPCR	Quantitative polymerase chain reaction
RIA	Radioimmunoassay
RPLA	Reverse passive latex agglutination
SEA	Staphylococcal enterotoxin A
SEB	Staphylococcal enterotoxin B
STX	Saxitoxin

TAM FLISA	Triple-analyte multiplex fluorescent labeled immunosorbent assay
UCNPs	Upconversion nanoparticles
US FDA	United States Food and Drug Administration
YTX	Yessotoxin
ZEN	Zearalenone
α-CD	α-Cyclodextrin

References

1. R. T. Bacon and J. N. Sofos, Characteristics of biological hazards in foods, *Food safety handbook*, 2003, **10**, 157–195.
2. J. K. Rajput, Nanomaterial-based sensors as potential remedy for detection of biotoxins, *Food Control*, 2021, 108686.
3. T. Bintsis, Foodborne pathogens, *AIMS Microbiol.*, 2017, **3**(3), 529.
4. R. K. Barakat and L. J. Harris, Growth of Listeria monocytogenes and Yersinia enterocolitica on cooked modified-atmosphere-packaged poultry in the presence and absence of a naturally occurring microbiota, *Appl. Environ. Microbiol.*, 1999, **65**(1), 342–345.
5. T. Martinović, U. Andjelković, M. Š. Gajdošik, D. Rešetar and D. Josić, Foodborne pathogens and their toxins, *J. Proteomics*, 2016, **147**, 226–235.
6. K. Zhu, R. Dietrich, A. Didier, D. Doyscher and E. Märtlbauer, Recent developments in antibody-based assays for the detection of bacterial toxins, *Toxins*, 2014, **6**(4), 1325–1348.
7. B. S. Inbaraj and B. H. Chen, Nanomaterial-based sensors for detection of foodborne bacterial pathogens and toxins as well as pork adulteration in meat products, *J. Food Drug Anal.*, 2016, **24**(1), 15–28.
8. J. K. Rudkin, R. M. McLoughlin, A. Preston and R. C. Massey, Bacterial toxins: Offensive, defensive, or something else altogether?, *PLoS Pathog.*, 2017, **13**(9), e1006452.
9. J. M. Williams and B. Tsai, Intracellular trafficking of bacterial toxins, *Curr. Opin. Cell Biol.*, 2016, **41**, 51–56.
10. Y. Wang, K. M. Schill, H. C. Fry and T. V. Duncan, A quantum dot nanobiosensor for rapid detection of botulinum neurotoxin serotype E, *ACS Sens.*, 2020, **5**(7), 2118–2127.
11. J. S. Henkel, M. R. Baldwin and J. T. Barbieri, Toxins from bacteria, *Molecular, Clinical and Environmental Toxicol.*, 2010, 1–29.
12. J. W. Grate, R. M. Ozanich Jr., M. G. Warner, C. J. Bruckner-Lea and J. D. Marks, Advances in assays and analytical approaches for botulinum-toxin detection, *TrAC, Trends Anal. Chem.*, 2010, **29**(10), 1137–1156.

13. E. Lemichez and J. T. Barbieri, General aspects and recent advances on bacterial protein toxins, *Cold Spring Harbor Perspect. Med.*, 2013, **3**(2), a013573.

14. D. Bueno, R. Munoz and J. L. Marty, Common Methods to Detect Mycotoxins: A Review with Particular Emphasis on Electrochemical Detection, *Sensing in Electroanalysis*, ed. K. Kalcher, R. Metelka, I. Švancara and K. Vytřas, 2013.

15. A. Sharma, R. Khan, G. Catanante, T. A. Sherazi, S. Bhand, A. Hayat and J. L. Marty, Designed strategies for fluorescence-based biosensors for the detection of mycotoxins, *Toxins*, 2018, **10**(5), 197.

16. M. M. da Silva Rodrigueiro, C. A. de Amorim, K. S. Mustafé Oliveira, F. J. de Oliveira Morais, P. Joshi and K. C. Blanco, Advances in Aflatoxins Monitoring of Stored Food, *Biointerface Res. Appl. Chem.*, 2022, **12**(4), 4990–5003.

17. A. Rhouati, G. Bulbul, U. Latif, A. Hayat, Z.-H. Li and J. L. Marty, Nano-aptasensing in mycotoxin analysis: Recent updates and progress, *Toxins*, 2017, **9**(11), 349.

18. J. Das and H. N. Mishra, Recent advances in sensors for detecting food pathogens, contaminants, and toxins: a review, *Eur. Food Res. Technol.*, 2022, 1–24.

19. A. Alshannaq and J.-H. Yu, Occurrence, toxicity, and analysis of major mycotoxins in food, *Int. J. Environ. Res. Public Health*, 2017, **14**(6), 632.

20. P. Visciano, M. Schirone, M. Berti, A. Milandri, R. Tofalo and G. Suzzi, Marine biotoxins: occurrence, toxicity, regulatory limits and reference methods, *Front. Microbiol.*, 2016, **7**, 1051.

21. W. Ye, T. Liu, W. Zhang, M. Zhu, Z. Liu, Y. Kong and S. Liu, Marine toxins detection by biosensors based on aptamers, *Toxins*, 2020, 12(1), 1.

22. Y. Zhao, L. Li, X. Yan, L. Wang, R. Ma, X. Qi, S. Wang and X. Mao, Emerging roles of the aptasensors as superior bioaffinity sensors for monitoring shellfish toxins in marine food chain, *Journal of Hazardous Materials*, 2022, **421**, 126690.

23. H. Daguer, R. B. Hoff, L. Molognoni, C. R. Kleemann and L. V. Felizardo, Outbreaks, toxicology, and analytical methods of marine toxins in seafood, *Curr. Opin. Food Sci.*, 2018, **24**, 43–55.

24. M. E. E. Alahi and S. C. Mukhopadhyay, Detection methodologies for pathogen and toxins: a review, *Sensors*, 2017, **17**(8), 1885.

25. D. Pimbley and P. Patel, A review of analytical methods for the detection of bacterial toxins, *J. Appl. Microbiol.*, 1998, **84**, 98S.

26. E. Yamasaki, R. Sakamoto, T. Matsumoto, F. Morimatsu, T. Kurazono, T. Hiroi, G. B. Nair and H. Kurazono, Development of an immunochromatographic test strip for detection of cholera toxin, *Biomed Res. Int.*, 2013, **2013**, 679038.

27. J. W.-F. Law, N.-S. Ab Mutalib, K.-G. Chan and L.-H. Lee, Rapid methods for the detection of foodborne bacterial pathogens: principles, applications, advantages and limitations, *Front. Microbiol.*, 2015, **5**, 770.
28. J. Brunt, M. D. Webb and M. W. Peck, Rapid affinity immunochromatography column-based tests for sensitive detection of Clostridium botulinum neurotoxins and Escherichia coli O157, *Appl. Environ. Microbiol.*, 2010, **76**(13), 4143–4150.
29. X. Wang, R. Niessner, D. Tang and D. Knopp, Nanoparticle-based immunosensors and immunoassays for aflatoxins, *Anal Chim Acta*, 2016, **912**, 10–23.
30. Z. Samra, L. Madar-Shapiro, M. Aziz and J. Bishara, Evaluation of a new immunochromatography test for rapid and simultaneous detection of Clostridium difficile antigen and toxins, *Isr. Med. Assoc. J.*, 2013, **15**(7), 373–376.
31. I. Bazin, S. A. Tria, A. Hayat and J.-L. Marty, New biorecognition molecules in biosensors for the detection of toxins, *Biosens. Bioelectron.*, 2017, **87**, 285–298.
32. I. Bobrinetskiy, M. Radovic, F. Rizzotto, P. Vizzini, S. Jaric, Z. Pavlovic, V. Radonic, M. V. Nikolic and J. Vidic, Advances in nanomaterials-based electrochemical biosensors for foodborne pathogen detection, *Nanomaterials*, 2021, **11**(10), 2700.
33. R. Sharma, K. V. Ragavan, M. S. Thakur and K. S. M. S. Raghavarao, Recent advances in nanoparticle based aptasensors for food contaminants, *Biosens. Bioelectron.*, 2015, **74**, 612–627.
34. H. Shan, X. Li, L. Liu, D. Song and Z. Wang, Recent advances in nanocomposite-based electrochemical aptasensors for the detection of toxins, *J. Mater. Chem. B*, 2020, **8**(27), 5808–5825.
35. M. Li, T. Chen, J. J. Gooding and J. Liu, Review of carbon and graphene quantum dots for sensing, *ACS Sens.*, 2019, **4**(7), 1732–1748.
36. X. Pei, B. Zhang, J. Tang, B. Liu, W. Lai and D. Tang, Sandwich-type immunosensors and immunoassays exploiting nanostructure labels: A review, *Anal. Chim. Acta*, 2013, **758**, 1–18.
37. R. Gao, Z. Zhong, X. Gao and L. Jia, Graphene oxide quantum dots assisted construction of fluorescent aptasensor for rapid detection of Pseudomonas aeruginosa in food samples, *J. Agric. Food Chem.*, 2018, **66**(41), 10898–10905.
38. B. Jurado-Sánchez, M. Pacheco, R. Maria-Hormigos and A. Escarpa, Perspectives on Janus micromotors: Materials and applications, *Applied Materials Today*, 2017, **9**, 407–418.
39. B. Jurado-Sánchez, M. Pacheco, J. Rojo and A. Escarpa, Magnetocatalytic graphene quantum dots Janus micromotors for bacterial endotoxin detection, *Angew. Chem. Int. Ed.*, 2017, **56**(24), 6957–6961.

40. M. Pacheco, B. Jurado-Sánchez and A. Escarpa, Sensitive monitoring of enterobacterial contamination of food using self-propelled Janus microsensors, *Anal. Chem.*, 2018, **90**(4), 2912–2917.
41. M. Veissi, S. Maktabi, Z. Ramezani and M. Khosravi, Highly Sensitive Fluorescence Assay of Enterotoxin A in Milk Using Carbon Quantum Dots as a Fluorophore, *Food Anal. Methods*, 2021, **14**(9), 1815–1825.
42. D. Liu, W. Li, C. Zhu, Y. Li, X. Shen, L. Li, X. Yan and T. You, Recent progress on electrochemical biosensing of aflatoxins: A review, *TrAC, Trends Anal. Chem.*, 2020, 115966.
43. K.-Y. Xing, S. Shan, D.-F. Liu and W.-H. Lai, Recent advances of lateral flow immunoassay for mycotoxins detection, *TrAC, Trends Anal. Chem.*, 2020, 116087.
44. D. Bueno, R. Muñoz and J. L. Marty, Fluorescence analyzer based on smartphone camera and wireless for detection of Ochratoxin A, *Sens. Actuators, B*, 2016, **232**, 462–468.
45. R. Sivakumar and N. Y. Lee, Recent progress in smartphone-based techniques for food safety and the detection of heavy metal ions in environmental water, *Chemosphere*, 2021, 130096.
46. L. Anfossi, C. Baggiani, C. Giovannoli, G. D'Arco and G. Giraudi, Lateral-flow immunoassays for mycotoxins and phycotoxins: a review, *Anal. Bioanal. Chem.*, 2013, **405**(2), 467–480.
47. X. Yan-Jun, Y. Ying, K. Wei-Jun, Y. Shi-Hai and Y. Mei-Hua, Application of nanoparticle probe-based lateral flow immunochromatographic assay in mycotoxins detection, *Chinese Journal of Analytical Chemistry*, 2015, **43**(4), 618–628.
48. P. Li, Z. Zhang, Q. Zhang, N. Zhang, W. Zhang, X. Ding and R. Li, Current development of microfluidic immunosensing approaches for mycotoxin detection via capillary electromigration and lateral flow technology, *Electrophoresis*, 2012, **33**(15), 2253–2265.
49. M. Yang, Y. Zhang, M. Cui, Y. Tian, S. Zhang, K. Peng, H. Xu, Z. Liao, H. Wang and J. Chang, A smartphone-based quantitative detection platform of mycotoxins based on multiple-color upconversion nanoparticles, *Nanoscale*, 2018, **10**(33), 15865–15874.
50. M. Rai, P. S. Jogee and A. P. Ingle, Emerging nanotechnology for detection of mycotoxins in food and feed, *Int. J. Food Sci. Nutr.*, 2015, **66**(4), 363–370.
51. I. S. Oliveira, A. G. da Silva Junior, C. A. S. de Andrade and M. D. L. Oliveira, Biosensors for early detection of fungi spoilage and toxigenic and mycotoxins in food, *Curr. Opin. Food Sci.*, 2019, **29**, 64–79.
52. Y. Alhamoud, D. Yang, S. S. F. Kenston, G. Liu, L. Liu, H. Zhou, F. Ahmed and J. Zhao, Advances in biosensors for the detection of ochratoxin A: Bio-receptors, nanomaterials, and their applications, *Biosens. Bioelectron.*, 2019, **141**, 111418.

53. N. M. Danesh, H. B. Bostan, K. Abnous, M. Ramezani, K. Youssefi, S. M. Taghdisi and G. Karimi, Ultrasensitive detection of aflatoxin B1 and its major metabolite aflatoxin M1 using aptasensors: A review, *TrAC, Trends Anal. Chem.*, 2018, **99**, 117–128.
54. R. Eivazzadeh-Keihan, P. Pashazadeh, M. Hejazi, M. de la Guardia and A. Mokhtarzadeh, Recent advances in nanomaterial-mediated bio and immune sensors for detection of aflatoxin in food products, *TrAC, Trends Anal. Chem.*, 2017, **87**, 112–128.
55. X. Lin and X. Guo, Advances in biosensors, chemosensors and assays for the determination of fusarium mycotoxins, *Toxins*, 2016, **8**(6), 161.
56. C. Wang, J. Qian, K. Wang, M. Hua, Q. Liu, N. Hao, T. You and X. Huang, Nitrogen-doped graphene quantum dots@SiO_2 nanoparticles as electrochemiluminescence and fluorescence signal indicators for magnetically controlled aptasensor with dual detection channels , *ACS Appl. Mater. Interfaces*, 2015, **7**(48), 26865–26873.
57. J. Tian, W. Wei, J. Wang, S. Ji, G. Chen and J. Lu, Fluorescence resonance energy transfer aptasensor between nanoceria and graphene quantum dots for the determination of ochratoxin A, *Anal. Chim. Acta*, 2018, **1000**, 265–272.
58. H. Bhardwaj, C. Singh, R. K. Kotnala and G. Sumana, Graphene quantum dots-based nano-biointerface platform for food toxin detection, *Anal. Bioanal. Chem.*, 2018, **410**(28), 7313–7323.
59. H. Bhardwaj, M. K. Pandey and G. Sumana, Electrochemical Aflatoxin B1 immunosensor based on the use of graphene quantum dots and gold nanoparticles, *Microchim. Acta*, 2019, **186**(8), 1–12.
60. H. Bhardwaj, C. A. Marquette, P. Dutta and G. Sumana, Integrated graphene quantum dot decorated functionalized nanosheet biosensor for mycotoxin detection, *Anal. Bioanal. Chem.*, 2020, **412**(25), 7029–7041.
61. Q. Wang, F. Zhao, Q. Yang and W. Wu, Graphene oxide quantum dots based nanotree illuminates AFB1: Dual signal amplified aptasensor detection AFB1, *Sens. Actuators, B*, 2021, **345**, 130387.
62. R. Shadjou, M. Hasanzadeh, M. Heidar-poor and N. Shadjou, Electrochemical monitoring of aflatoxin M1 in milk samples using silver nanoparticles dispersed on α-cyclodextrin-GQDs nanocomposite, *J. Mol. Recognit.*, 2018, **31**(6), e2699.
63. L. Luo, S. Ma, L. Li, X. Liu, J. Zhang, X. Li, D. Liu and T. You, Monitoring zearalenone in corn flour utilizing novel self-enhanced electrochemiluminescence aptasensor based on NGQDs-NH_2-Ru@ SiO_2 luminophore, *Food Chem.*, 2019, **292**, 98–105.
64. M. Shao, M. Yao, S. D. Saeger, L. Yan and S. Song, Carbon quantum dots encapsulated molecularly imprinted fluorescence quenching

particles for sensitive detection of zearalenone in corn sample, *Toxins*, 2018, **10**(11), 438.

65. S. Li, J. Wang, W. Sheng, W. Wen, Y. Gu and S. Wang, Fluorometric lateral flow immunochromatographic zearalenone assay by exploiting a quencher system composed of carbon dots and silver nanoparticles, *Microchim. Acta*, 2018, **185**(8), 1–9.
66. S. Mandal and P. Das, Ultrasensitive visual detection of mycotoxin citrinin with yellow-light emitting carbon dot and Congo red, *Food Chem.*, 2020, **312**, 126076.
67. F. Rahimi, H. Roshanfekr and H. Peyman, Ultra-sensitive electrochemical aptasensor for label-free detection of Aflatoxin B1 in wheat flour sample using factorial design experiments, *Food Chem.*, 2021, **343**, 128436.
68. H. Singh, S. Singh, S. K. Bhardwaj, G. Kaur, M. Khatri, A. Deep and N. Bhardwaj, Development of carbon quantum dot-based lateral flow immunoassay for sensitive detection of aflatoxin M1 in milk, *Food Chem.*, 2022, 133374.
69. L. Zhao, Y. Huang, Y. Dong, X. Han, S. Wang and X. Liang, Aptamers and aptasensors for highly specific recognition and sensitive detection of marine biotoxins: recent advances and perspectives, *Toxins*, 2018, **10**(11), 427.
70. J. Peng, Z. Zhao, M. Zheng, B. Su, X. Chen and X. Chen, Electrochemical synthesis of phosphorus and sulfur co-doped graphene quantum dots as efficient electrochemiluminescent immunomarkers for monitoring okadaic acid, *Sens. Actuators, B*, 2020, **304**, 127383.
71. H. Gu, L. Hao, H. Ye, P. Ma and Z. Wang, Nuclease-assisted target recycling signal amplification strategy for graphene quantum dot-based fluorescent detection of marine biotoxins, *Microchim. Acta*, 2021, **188**(4), 1–9.
72. M. Gholami, M. A. Salmasi, E. Sohouli, B. Torabi, M. R. Sohrabi and M. Rahimi-Nasrabadi, A new nano biosensor for maitotoxin with high sensitivity and selectivity based fluorescence resonance energy transfer between carbon quantum dots and gold nanoparticles, *J. Photochem. Photobiol., A*, 2020, **398**, 112523.
73. C. Wang, R. Xiao, S. Wang, X. Yang, Z. Bai, X. Li, Z. Rong, B. Shen and S. Wang, Magnetic quantum dot based lateral flow assay biosensor for multiplex and sensitive detection of protein toxins in food samples, *Biosens. Bioelectron.*, 2019, **146**, 111754.
74. M. G. Warner, J. W. Grate, A. Tyler, R. M. Ozanich, K. D. Miller, J. Lou, J. D. Marks and C. J. Bruckner-Lea, Quantum dot immunoassays in renewable surface column and 96-well plate formats for the fluorescence detection of botulinum neurotoxin using high-affinity antibodies, *Biosens. Bioelectron.*, 2009, **25**(1), 179–184.

75. M. Gharaat, R. H. Sajedi, M. Shanehsaz, N. Jalilian, M. Mirshahi and M. Gholamzad, A dextran mediated multicolor immunochromatographic rapid test strip for visual and instrumental simultaneous detection of Vibrio cholera O1 (Ogawa) and Clostridium botulinum toxin A, *Microchim. Acta*, 2017, **184**(12), 4817–4825.
76. A. Sharma, V. K. Rao, D. V. Kamboj, R. Gaur, S. Upadhyay and M. Shaik, Relative efficiency of zinc sulfide (ZnS) quantum dots (QDs) based electrochemical and fluorescence immunoassay for the detection of Staphylococcal enterotoxin B (SEB), *Biotechnol. Rep.*, 2015, **6**, 129–136.
77. R. Li, Y. Wen, L. Yang, A. Liu, F. Wang and P. He, Dual quantum dot nanobeads-based fluorescence-linked immunosorbent assay for simultaneous detection of aflatoxin B1 and zearalenone in feedstuffs, *Food Chem.*, 2022, **366**, 130527.
78. B. Jia, X. Liao, C. Sun, L. Fang, L. Zhou and W. Kong, Development of a quantum dot nanobead-based fluorescent strip immunosensor for on-site detection of aflatoxin B_1 in lotus seeds, *Food Chem.*, 2021, **356**, 129614.
79. T. B. Nguyen, T. B. Vu, H. M. Pham, C. S. Tran, H. H. Le Thi and N. T. Vo Thi, Detection of aflatoxins B1 in maize grains using fluorescence resonance energy transfer, *Applied Sciences*, 2020, **10**(5), 1578.
80. C. Wang, J. Qian, K. An, X. Lu and X. Huang, A semiconductor quantum dot-based ratiometric electrochemical aptasensor for the selective and reliable determination of aflatoxin B1, *Analyst*, 2019, **144**(16), 4772–4780.
81. X. Lu, C. Wang, J. Qian, C. Ren, K. An and K. Wang, Target-driven switch-on fluorescence aptasensor for trace aflatoxin B1 determination based on highly fluorescent ternary CdZnTe quantum dots, *Anal. Chim. Acta*, 2019, **1047**, 163–171.
82. R. Li, C. Meng, Y. Wen, W. Fu and P. He, Fluorometric lateral flow immunoassay for simultaneous determination of three mycotoxins (aflatoxin B1, zearalenone and deoxynivalenol) using quantum dot microbeads, *Microchim. Acta*, 2019, **186**(12), 1–9.
83. Y. Shao, H. Duan, L. Guo, Y. Leng, W. Lai and Y. Xiong, Quantum dot nanobead-based multiplexed immunochromatographic assay for simultaneous detection of aflatoxin B1 and zearalenone, *Anal. Chim. Acta*, 2018, **1025**, 163–171.
84. R. Renuka, J. Achuth, V. Mudili and S. Poda, Development of a FRET-based fluorescence aptasensor for the detection of aflatoxin B1 in contaminated food grain samples, *RSC Adv.*, 2018, **8**(19), 10465–10473.
85. N. V. Beloglazova, A. Foubert, A. Gordienko, M. D. Tessier, T. Aubert, E. Drijvers, I. Goryacheva, Z. Hens and S. De Saeger, Sensitive QD@SiO2-based immunoassay for triplex determination of cereal-borne mycotoxins, *Talanta*, 2016, **160**, 66–71.

86. W. Jiang, N. V. Beloglazova, P. Luo, P. Guo, G. Lin and X. Wang, A dual-color quantum dots encoded frit-based immunoassay for visual detection of aflatoxin M1 and pirlimycin residues in milk, *J. Agric. Food Chem.*, 2017, **65**(8), 1822–1828.
87. N. Gan, J. Zhou, P. Xiong, F. Hu, Y. Cao, T. Li and Q. Jiang, An ultrasensitive electrochemiluminescent immunoassay for Aflatoxin M1 in milk, based on extraction by magnetic graphene and detection by antibody-labeled CdTe quantumn dots-carbon nanotubes nanocomposite, *Toxins*, 2013, **5**(5), 865–883.
88. O. A. Goryacheva, N. V. Beloglazova, I. Y. Goryacheva and S. De Saeger, Homogenous FRET-based fluorescent immunoassay for deoxynivalenol detection by controlling the distance of donor-acceptor couple, *Talanta*, 2021, **225**, 121973.
89. O. A. Goryacheva, C. Guhrenz, K. Schneider, N. V. Beloglazova, I. Y. Goryacheva, S. De Saeger and N. Gaponik, Silanized Luminescent Quantum Dots for the Simultaneous Multicolor Lateral Flow Immunoassay of Two Mycotoxins, *ACS Appl Mater Interfaces*, 2020, **12**(22), 24575–24584.
90. A. Foubert, N. V. Beloglazova and S. De Saeger, Comparative study of colloidal gold and quantum dots as labels for multiplex screening tests for multi-mycotoxin detection, *Anal. Chim. Acta*, 2017, **955**, 48–57.
91. N. V. Beloglazova, A. M. Sobolev, M. D. Tessier, Z. Hens, I. Y. Goryacheva and S. De Saeger, Fluorescently labelled multiplex lateral flow immunoassay based on cadmium-free quantum dots, *Methods*, 2017, **116**, 141–148.
92. M. Jia, B. Jia, X. Liao, L. Shi, Z. Zhang, M. Liu, L. Zhou, D. Li and W. Kong, A CdSe@ CdS quantum dots based electrochemiluminescence aptasensor for sensitive detection of ochratoxin A, *Chemosphere*, 2022, **287**, 131994.
93. Z. Tang, X. Liu, B. Su, Q. Chen, H. Cao, Y. Yun, Y. Xu and B. D. Hammock, Ultrasensitive and rapid detection of ochratoxin A in agro-products by A nanobody-mediated FRET-based immunosensor, *J. Hazard. Mater.*, 2020, **387**, 121678.
94. L. Liu, Z. I. Tanveer, K. Jiang, Q. Huang, J. Zhang, Y. Wu and Z. Han, Label-Free Fluorescent Aptasensor for Ochratoxin—A Detection Based on CdTe Quantum Dots and (N-Methyl-4-pyridyl) Porphyrin, *Toxins*, 2019, **11**(8), 447.
95. J. Gao, Z. Chen, L. Mao, W. Zhang, W. Wen, X. Zhang and S. Wang, Electrochemiluminescent aptasensor based on resonance energy transfer system between CdTe quantum dots and cyanine dyes for the sensitive detection of Ochratoxin A, *Talanta*, 2019, **199**, 178–183.
96. H. Duan, Y. Li, Y. Shao, X. Huang and Y. Xiong, Multicolor quantum dot nanobeads for simultaneous multiplex immunochromatographic

detection of mycotoxins in maize, *Sens. Actuators, B*, 2019, **291**, 411–417.

97. C. Wang, J. Qian, K. An, X. Huang, L. Zhao, Q. Liu, N. Hao and K. Wang, Magneto-controlled aptasensor for simultaneous electrochemical detection of dual mycotoxins in maize using metal sulfide quantum dots coated silica as labels, *Biosens. Bioelectron.*, 2017, **89**, 802–809.
98. Y. Liang, X. Huang, R. Yu, Y. Zhou and Y. Xiong, Fluorescence ELISA for sensitive detection of ochratoxin A based on glucose oxidase-mediated fluorescence quenching of CdTe QDs, *Anal. Chim. Acta*, 2016, **936**, 195–201.
99. Q. Li, Y. Kang, S. Yin, Y. Qian, Y. Cai and Z. Yang, Graphene Oxide Synergy with the Conjugation of DNA and Quantum Dots for the Sensitive Detection of Ochratoxin A, *Food Anal. Methods*, 2021, 1–8.
100. Z. Lu, X. Chen and W. Hu, A fluorescence aptasensor based on semiconductor quantum dots and MoS2 nanosheets for ochratoxin A detection, *Sens. Actuators, B*, 2017, **246**, 61–67.
101. L. Wang, W. Chen, W. Ma, L. Liu, W. Ma, Y. Zhao, Y. Zhu, L. Xu, H. Kuang and C. Xu, Fluorescent strip sensor for rapid determination of toxins, *Chem. Commun.*, 2011, **47**(5), 1574–1576.
102. T. Lu, S. Zhan, Y. Zhou, X. Chen, X. Huang, Y. Leng, Y. Xiong and Y. Xu, Fluorescence elisa based on cat-regulated fluorescence quenching of cdte qds for sensitive detection of fb 1, *Anal. Methods*, 2018, **10**(48), 5797–5802.
103. S.-J. Li, W. Sheng, W. Wen, Y. Gu, J.-P. Wang and S. Wang, Three kinds of lateral flow immunochromatographic assays based on the use of nanoparticle labels for fluorometric determination of zearalenone, *Microchim. Acta*, 2018, **185**(4), 1–8.
104. S. Zhan, X. Huang, R. Chen, J. Li and Y. Xiong, Novel fluorescent ELISA for the sensitive detection of zearalenone based on H2O2-sensitive quantum dots for signal transduction, *Talanta*, 2016, **158**, 51–56.
105. H. Duan, X. Chen, W. Xu, J. Fu, Y. Xiong and A. Wang, Quantum-dot submicrobead-based immunochromatographic assay for quantitative and sensitive detection of zearalenone, *Talanta*, 2015, **132**, 126–131.
106. H.-W. Yu, J. Lee, S. Kim, G. H. Nguyen and I. S. Kim, Electrochemical immunoassay using quantum dot/antibody probe for identification of cyanobacterial hepatotoxin microcystin-LR, *Anal. Bioanal. Chem.*, 2009, **394**(8), 2173–2181.
107. Z. Zhang, X. Yu, J. Zhao, X. Shi, A. Sun, H. Jiao, T. Xiao, D. Li and J. Chen, A fluorescence microplate assay based on molecularly imprinted silica coated quantum dot optosensing materials for the separation and detection of okadaic acid in shellfish, *Chemosphere*, 2020, **246**, 125622.

108. X. Weng and S. Neethirajan, Paper-based microfluidic aptasensor for food safety, *J. Food Saf.*, 2018, **38**(1), e12412.
109. N. Xu, Y. Liu, Y. Li, B. Tang, X. Liang, Y. Yang, M. Liu, X. Liu and Y. Zhou, Rapid Quantum Dot Nanobead-mAb Probe-Based Immunochromatographic Assay for Antibody Monitoring of *Trichinella spiralis* Infection, *Int. J. Nanomedicine*, 2021, **16**, 2477.

CHAPTER 9

Carbon-based Quantum Dots from Food Waste: Synthesis to Application in Food Safety

S. SADEGH HASSANI*[a], M. DARAEE[b], S. H. NAVAZANI[c] AND F. RAHIMI[c]

[a]Catalysis Research Division, Research Institute of Petroleum Industry (RIPI), Tehran, Iran; [b]School of chemical, Gas and Petroleum Engineering, Semnan University, Semnan, Iran; [c]Department of Energy Systems, Faculty of Mechanics, K. N. Toosi University of Technology, Tehran, Iran
*E-mail: nsadeghhassani@yahoo.co.uk, sadeghs@ripi.ir

9.1 Introduction

Food waste is an important issue on a global scale: the Food and Agriculture Organization (FAO) estimates that billions of tons of this waste are generated annually, with serious ecological, economic, and social impacts.[1,2] Food waste can, however, be a valuable resource as it is a potential component of valuable chemicals and fuels. One such chemical produced from food waste is a new type of carbon-based material known as carbon-based quantum dots (CQDs). CQDs were discovered by Xu *et al.* during the synthesis and purification of single-walled carbon nanotubes. They are semiconductor NPs that exhibit quantum mechanical behavior due to their small size (below 10 nm). CQDs have superior stability owing to their sp^2 covalent carbon core structure. They have different

Detection Science Series No. 22
Quantum Dots in Bioanalytical Chemistry and Medicine
Edited by Michael Thompson and Zahra Ramezani

Published by the Royal Society of Chemistry, www.rsc.org

types of oxygen-containing functional groups, such as carboxyl, hydroxyl, and aldehyde.[3–7] Recently, CQDs have attracted a great deal of attention due to their features such as good biocompatibility, chemical resistance, water solubility, easy functional modification, low toxicity, high chemical inertness, optical properties, and visible fluorescence.[3,4]

Various natural biological materials (including food waste) can be used as carbon sources for preparing biomass CQDs, including vegetable- and fruit-derived biomass and animal resources. Potential sources include carrot juice, mushrooms, sweet potatoes, pomegranates, bananas, apples, grass, bamboo leaves, black tea, soy milk, carbohydrates, and turtle shells.[8]

In the last few decades, CQDs' synthesis routes, modification methods, and applications have expanded. They can be synthesized through "bottom-up" and "top-down" techniques. In the top-down approach, immense molecules such as carbon nanotubes and graphite are broken down or cut into very small molecules (including CQDs) *via* physical and/or chemical techniques. In the bottom-up approach, in contrast, CQDs are formed through chemical reactions from various molecules. Bottom-up synthesis is more time consuming than the top-down approach and, therefore, requires optimization; however, it is more environmentally friendly, simple, and cost-effective.

CQDs' sizes are affected by the source materials, preparation method, reaction time, and temperature.[5,9] Their properties vary with size, which allows them to be used in a variety of contexts (*e.g.*, they can be used in optical or humidity sensors, photocatalysis, biomedicine, bio/cell imaging, medical diagnosis, energy conversion/storage, solar cells, drug delivery systems, and LED devices).[4,5,10–13] Their unique fluorescence properties also make them very effective in the field of food safety analysis, where they can be used to detect additives and food product contamination.[14]

Characterization techniques are important to determine the structural and photophysical properties of synthesized CQDs. Transmission electron microscopy (TEM), infrared (IR), X-ray diffraction (XRD), scanning electron microscopy (SEM), UV-Visible spectroscopy (UV-VIS), and photoluminescence (PL) are some of the most frequently used techniques in the characterization of CQDs.

This chapter provides an overview of food waste resources that can be used to synthesize CQDs. We also review synthesis methods, properties of synthesized CQDs, and characterization techniques. Finally, we discuss CQDs' applications with a focus on their potential uses in food safety contexts.

9.2 Food Waste for Preparation of Carbon-based Quantum Dots

Food waste is a significant issue worldwide, but its effects can be mitigated if uses can be found for it. Recently, researchers have explored the possibility of using food waste as a carbon source for the preparation of CQDs. Plants, animal food, and food processing by-products are three important categories of food waste that have been used as carbon sources for CQDs. CQDs are environmentally friendly fluorescent nanomaterials with potential applications in biomedical research and other nanotechnology purposes.[5,15]

9.2.1 Preparation of CQDs from Fruit and Vegetable Waste

Beetroot (*Beta vulgaris*) waste has been used to synthesize CQDs through two procedures: a one-pot hydrothermal method and an acidic treatment method using ortho-phosphoric acid (Figure 9.1). In both methods, the product was purified *via* repeated centrifugation at 1500, 2500, and 4000 rpm for 10 min in order to remove unreacted natural plant ingredients, heavy organic residues, and inorganic residues.[16] CQDs prepared using these methods are spherical in shape with average particle sizes of 4–8 nm. They are photo-stable and biocompatible in aqueous media. These CQDs were first used in *C. elegans* soil nematodes, where they showed stable and consistent fluorescence in the worms' bodies without any side effects. Knowing this, the as-prepared CQDs with proper biocompatibility could potentially be used for diagnostic and therapeutic purposes. In another study, fluorescent CQDs were successfully synthesized from red beetroot using a simple hydrothermal method. The resulting CQDs ranged between 5 and 7 nm in size. They were used as selective probes for the detection of Pd^{2+} in an aqueous solution.[17]

Lemon juice and lemon peel waste have also been used as carbon sources for CQDs using a heat decomposition process. In one study, lemon waste was obtained from a retail shop and decomposed using a heating plate at 175 °C. The non-polar residues were extracted using chloroform in a separating funnel.[18] In another study, water-soluble fluorescent CQDs were synthesized from lemon peel waste and functionalized with TiO_2 for use as a photocatalyst. The TiO_2 introduced emissive energy traps to promote the recombination of electron–carbon pairs, which in turn resulted in a further increase in quantum yield (QY). In this work, lemon peel waste was used as a carbon source *via* hydrothermal reaction to produce water-soluble fluorescent CQDs (Figure 9.1). The initial materials were transferred to an autoclave and heated for 12 h at 200 °C. The resulting dark brown CQD-containing product was washed using dichloromethane. Centrifugation at 10 000 rpm for 30 min separated the CQDs from larger particles. TEM imaging of CQDs confirmed that they were spherical in shape with uniform dispersion and an average size of 3 nm.[19]

Purple perilla is another vegetable whose waste can be used to prepare CQDs. This plant can be found worldwide but is especially prevalent in some parts of Asia. In a typical preparation method, CQDs are synthesized *via* a one-step hydrothermal treatment. The purple perilla is smashed and added to distilled water, then heated for 5 h at 260 °C in an autoclave to produce CQDs. Next, the light brown suspension is filtered through a 0.22

μm micro-pore film filter to remove the unreacted raw materials. Finally, CQDs are obtained when the product is dialyzed against ultrapure water using a dialysis membrane.[20–22]

Miscanthus, a perennial grass, is another resource that can be used to prepare CQDs and doped CQDs using an economically feasible and green one-step heating process (Figure 9.1). In one study, *Miscanthus* fibers were placed into deionized water. The mixture was heated to 80 °C on a hot plate and stirred at 500 rpm. After vacuum filtration, the recovered filtrate was used as a precursor to produce CQDs using conventional heating methods. The aqueous extract of *Miscanthus* grass contained cellulose, lignin, ash, pentosans, and extractives. These materials undergo an intermolecular dehydration process, leading to the formation of polymer NPs. Further intramolecular dehydration caused aromatic clusters to form, producing CQDs after a nuclear burst. The pH of *Miscanthus* extract was adjusted to 11 using KOH followed by a hydrothermal process at 180 °C for 4 h. Finally, it was cooled to room temperature and CQDs were obtained by centrifuging the samples at high speed to remove the insoluble and large particles. The average size of the CQDs was between 4 and 8 nm, with QYs of 4.71 and 11.65 for CQDs and nitrogen-doped CQDs (N-CQDs), respectively.[10]

Fruit shells are another potential CQD source. Expired passion fruit shells have been used successfully for this purpose with the help of a hydrothermal method with two major steps. First, expired passion fruit shells were heated at 100 °C for 24 h and then ground into a powder and transferred to an autoclave to be heated for 3 h at 180 °C. The final product was then obtained *via* 2000 Dalton dialysis. TEM results showed sphere-shaped CQDs that were less than 5 nm in size. The CQDs had a weak graphite phase structure and produced a QY of 1.8%.[21]

9.2.2 Preparation of CQDs from Microorganisms

Cyanobacteria is a microorganismused for synthesizing photoluminescent CQDs with a QY of 5.30%. Transforming cyanobacteria into CQDs is an interesting way to remedy cyanobacteria pollution, and their fluorescence has many potential applications; for example, it is useful in biological imaging. As-prepared CQDs were brown and bright blue under visible and UV light, respectively. These materials were used as ideal nanofillers to reinforce composites in combination with polyvinyl alcohol and cellulose nanofibers to produce fluorescent films.[22]

9.2.3 Preparation of CQDs from Seeds

Seeds are among the most important carbon sources for CQDs from a waste reduction standpoint: in many industries, seeds end up as unused by-products. Cotton linter is a type of short fiber that is removed from seeds during the processing of cotton and is used in certain applications including paper making. It has also been used to produce CQDs with an average size of 10.14 nm in acidic media using a closed microwave system with a preparation time of 5 min.[23]

Wheat, rice, *Pennisetum glaucum*, and sorghum, though not waste material themselves, are other seeds that have been used for CQD synthesis.[24] As shown in Figure 9.1, synthesis was performed by thermally calcinating seeds at 400 °C for 2 h under a nitrogen atmosphere. The resulting supernatant was filtered using a filter membrane and centrifuged at 17 000 rpm to produce a solid CQD extract. A lyophilization step was used to remove all moisture from the samples, thus producing pure CQDs. The as-prepared CQDs were used in a chemo-sensing study to identify Cr^{3+} ions in aqueous media – a simple, rapid, cost-effective, and harmless means of identifying metal ion impurities with high selectivity and sensitivity.[24]

In another study, N and S co-doped CQDs were prepared from palm powder in a green and sustainable process. Palm powder was treated with triflic acid and then heated in an oven at 150 °C to obtain a char. The char was subsequently washed with water and a 2% solution of $NaHCO_3$ and dried. The triflic acid-activated palm shell char was ultrasonicated in 50 mL conductivity water and filtered through Whatman filter paper to obtain the CQDs. Finally, the filtrate was centrifuged at 10,000 rpm to eliminate large particles, and the CQD solution was purified using a dialysis membrane.[6]

In yet another study, *Phoenix dactylifera* (commonly known as date palm fruit) was used to produce CQDs *via* microwave-assisted pyrolysis. In this synthesis procedure, whole date palm fruit was crushed and dissolved in autoclaved distilled water and filtered using a muslin cloth to obtain an aqueous extract that was pyrolyzed by heating in a microwave. The pure CQDs were obtained by dialysis and the prepared solution was stored at 4 °C in a refrigerator.[25]

9.2.4 Preparation of CQDs from Animal Products

Animal products are another green CQD precursor that is connected to economical green-chemistry purposes. In one study, cow milk was used to synthesize water-soluble CQDs for simultaneous imaging and drug delivery in cancer therapy. To produce these CQDs, commercially pasteurized cow milk was heated in a domestic microwave, and the resulting residue was added to a solution of NaOH and ethanol. The obtained solution showed UV illumination at an excitation wavelength of 365 nm and this illuminated fraction contained graphene-based quantum dots (GQDs).

The solution was subjected to sucrose density gradient centrifugation producing layers that were examined for UV illumination. The UV active layer was collected carefully by puncturing the centrifugation tube with a syringe and centrifuging it at 16 770 rpm followed by a slow evaporation step in a vacuum oven for 4 h to obtain a black-colored GQD powder. The obtained GQDs were used as a carrier platform to deliver the anti-cancer drug berberine hydrochloride (BHC); Cysteamine hydrochloride was used to attach the drug to the nanocarriers, forming a GQDs@Cys-BHC complex. The as-prepared GQDs were spherical in shape with a lateral size of 5 nm and multifluorescent properties. The GQDs@Cys-BHC complex had a high drug-loading capacity (88%) and a potent cytotoxic effect on various cancerous cell lines. As a result, GQDs have been demonstrated to be a suitable drug delivery system as well as nanoprobes for dual-fluorescent detection of cancerous cells at low concentration.[26]

Animal shells are another green CQD precursor. Turtle shells, for instance, have been successfully used to fabricate highly fluorescent NPs using a simple pyrolysis method. The main component of turtle shells is collagen, which contains functional groups such as amidogen, carboxyl, and oxhydryl. These groups have been demonstrated to aid in the formation of fluorescent CQDs, which can be used in multi-signal anti-counterfeiting, coding, and optoelectronic devices. In one study, researchers removed impurities from turtle shells, then dried and milled them to form a powder that was transferred to a tube furnace and heated for 5 h at 500 °C under a nitrogen atmosphere. Centrifugation of the supernatant at 8000 rpm produced a brownish-colored CQD powder.[27]

9.2.5 Preparation of CQDs from Tea and Coffee

Household waste, such as used tea leaves and peanut shells, is a low-cost material to produce CQDs with a simple and less-toxic synthesis method (Figure 9.1). Each year, large amounts of discarded tea leaves are produced worldwide; they can be used to synthesize water-soluble CQDs with high fluorescent properties. In one study, CQDs obtained from tea leaves using this method exhibited strong blue fluorescence under UV light with a wavelength of 365 nm. The many hydrophilic groups on the surfaces of the as-prepared CQDs led to interesting fluorescence, showing a red shift as the excitation light intensity was increased. These CQDs were spherical in shape, approximately 7–9 nm in size, and homogenous in water.[28] As in most research, the amount of carbon source used in the reaction, in this case, used tea leaves, is an essential factor in CQDs production.

In another study, manganese-doped carbon-based quantum dots (Mn-CQDs) with blue emissions were prepared *via* a one-pot hydrothermal method using green tea waste for use as magneto-fluorescent photodynamic therapy (PDT) agents.[29] In a third study, green tea leaf-derived CQDs with stability to the buffer's pH and ionic strength and ultraviolet radiation fluorescence. The CQDs were used to sense gefitinib.[30] As these CQDs were sensitive to environmental temperatures, they could be quenched by gefitinib *via* the inner-filter effect and successfully applied as a sensor for gefitinib detection in urine samples.

Coffee grounds are another type of food waste that can be used as a carbon source for CQDs. The use of coffee grounds not only reduces waste but also allows researchers to leverage the structural and luminescent properties of CQDs obtained from this specific source, including enhanced band-edgePL with single exponential decay. Coffee ground-derived CQDs have been used for bioimaging and as sensing probes to detect Fe^{3+} and other heavy metals in water.[31,32]

9.2.6 Preparation of CQDs from Animal Food (Meat, Egg, Seafood)

Egg yolk oil (EYO) is a traditional medicine in China. It is derived from the cooked yolks of *Gallus gallus domesticus* Brisson eggs and is used to treat burn wounds and acute and chronic eczema. EYO was used to synthesize CQDs with an almost spherical shape and an average size of less than 10 nm. The resulting EYO-CQDs were studied for their potential hemostatic effects. The effect of these materials depended on the dosage; the researchers claimed that the EYO-CQDs stimulated the intrinsic blood coagulation system and activated the fibrinogen system.[33]

9.2.7 Preparation of CQDs from Industrial Waste and By-products

Finding uses for food waste can both decrease environmental pollution and create economic benefits. As food waste is rich in carbon, it is an inexpensive and abundant material for CQDs. Plant by-products, animal food by-products, and food processing by-products have several potential food waste-based carbon sources.[5]

Hydrothermal synthesis is a green and sustainable approach for CQD preparation. However, one of the major issues is the low yield of CQDs because the traditional hydrothermal treatment produces a lot of hydrochar as a by-product. This hydrochar and carbonized biomass can be used for large-scale synthesis of CQDs in mild oxidation conditions using sodium hydroxide and hydrogen peroxide. One advantage of this method is that it can be used to prepare CQDs with a uniform size (approximately 2.4 nm). CQDs obtained using this method have a good quantum yield of between 4% and 23% depending on reaction conditions, which can be increased by further surface treatment. The abundant conjugated double bond, carbonyl group, and quinoid groups in the hydrochar are oxidized during the treatment, which produces CQDs with abundant carboxyl and hydroxyl groups on the surfaces. These groups allow as-prepared CQDs to be used as sensors for Pb^{2+} detection.[34]

N-doped CQDs derived from food waste have also been used for Fe^{3+} sensing and cell imaging purposes. Moreover, the reaction feedstocks contain crude protein, crude fat, crude fiber, and crude ash. CQDs can be prepared by grounding the feedstocks to fine powder and drying followed

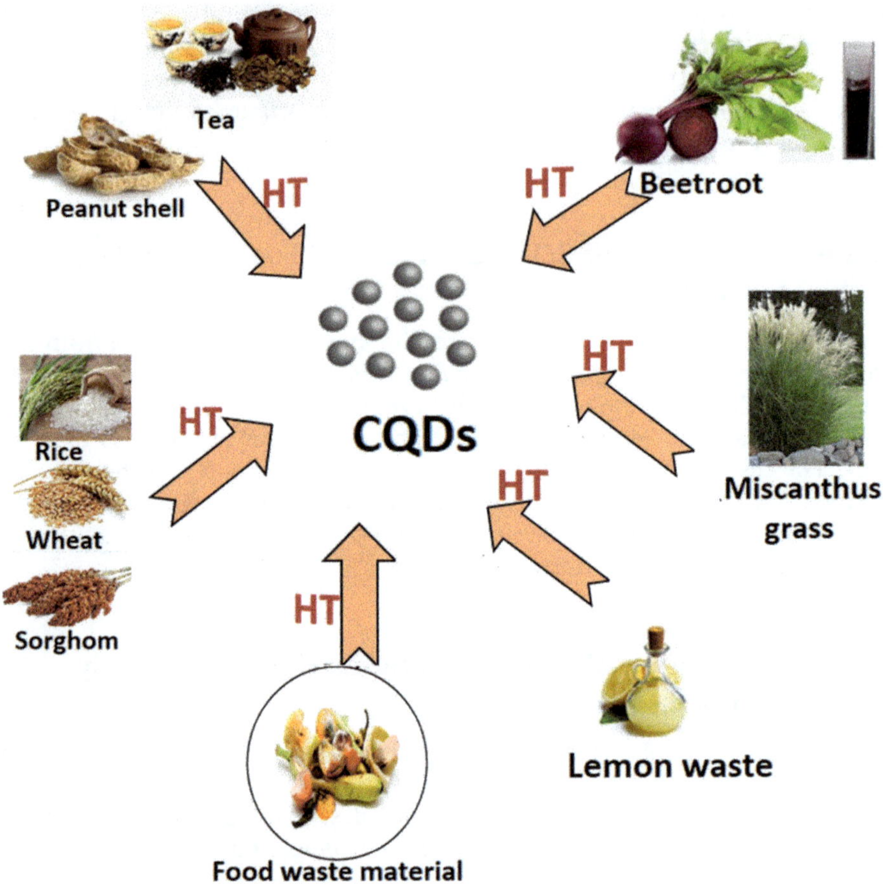

Figure 9.1 Preparation of CQDs from various waste materials.

by hydrothermal carbonization and filtration using a dialysis bag as shown in Figure 9.1. Extraordinarily fluorescent CQDs with a high QY of 28% were prepared in this study. As the PL of the food waste-derived CQDs are quenched by Fe^{3+} even in the presence of other metal ions, it can be applied to fluorescence detection for Fe^{3+} with high selectivity.[35]

9.3 Methods of CQD Synthesis

CQDs are fabricated using a variety of approaches. Methods of synthesis include pyrolysis, ultrasonic synthesis, microwave irradiation, hydrothermal/solvothermal methods, chemical oxidation, electrochemical techniques, and laser ablation.[36] Various precursors can be used to fabricate CQDs, including thiourea, ethylenediaminetetraacetic acid (EDTA), phytic acid, phenylenediamine, benzene, ethylene glycol, ammonium citrate, ascorbic acid, and citric acid.[5,37–39]

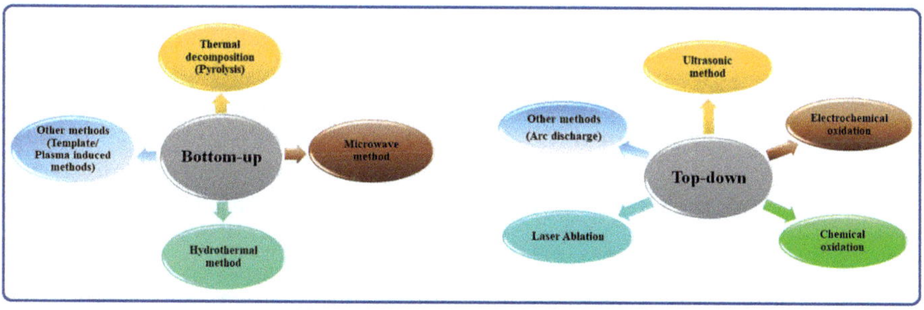

Figure 9.2 Classification of CQDs preparation methods.

The subject of green chemistry is a significant one in current approaches to CQD synthesis. By choosing a green approach to synthesis, researchers aim to prevent or reduce the use of toxic substances and produce minimal or no waste.[40] Green synthesis reduces the environmental impact of fabrication and may have additional benefits (*e.g.*, green methods may be more efficient and produce higher-quality results).[41]

In the last ten years, CQDs have attracted widespread attention among researchers due to their impressive resistance to photobleaching, ease of production and operation, biocompatibility, and excitation wavelength-dependent and size-commensurate luminescence. QDs are prepared using chemical, physical, or electrochemical methods that can be classified as "bottom-up" and "top-down" (Figure 9.2). In top-down methods, large carbon structures are broken down into smaller ones using techniques such as electrochemistry, laser ablation, and arc discharge. In bottom-up methods, CQDs are produced from molecular precursors using techniques such as reverse micelle, pyrolytic methods, oxidation, thermal methods, hydrothermal and aqueous processes, and microwave-assisted synthesis.[42]

The advantages of bottom-up techniques are a better adherence to green chemistry and admirable sustainability, which some of the relevant methods for mass production of CQDs include thermal/chemical oxidation of molecular precursors are in this category. Of all the bottom-up techniques mentioned above, most scientific attention is focused on the hydrothermal method owing to its simplicity, repeatability, and low cost, especially for the majority of carbon sources. It is also possible to increase the PL efficiency of CQDs during production using this method. The natural resources used to fabricate CQDs lead to high PL efficiency and adjustable emissions because of doping heteroatoms (such as P, N, and S) and self-passivation. Moreover, it is inexpensive and easy to add other elements using natural resources. The QYs of various precursors provide a broad range of PL values (between 2% and 70%), depending on the process used for synthesis.[43] In the next section, we will review several commonly used methods of CQD production.

9.3.1 Microwave Synthesis

One of the fastest and least expensive ways to produce CQDs is through microwave irradiation of organic compounds (Figure 9.3a). It takes 1 min to produce green luminescent CQDs from diethylene glycol (DEG, the reaction medium) and sucrose (the carbon source) under microwave irradiation. Proper water dispersion and a clear appearance were achieved for DEG-stabilized CQDs (DEGCQDs) using this method.[44] Microwaves, as electromagnetic waves, have a large range of wavelength between 1 mm and 1 m, thereby presenting intensive energy for cleaving chemical bonds in the structure of a substance. Microwave-based approaches may, therefore, reduce the duration of reaction times and provide concurrent and homogenous heating, which can result in QDs with an even size distribution.[45] Zhu *et al.* developed the first facile and affordable microwave pyrolysis method for preparing stable and brightly luminescent CQDs with proper water dispersion in just a few minutes.[46,47]

Microwave-based methods are commonly used for green CQD fabrication due to the ease and speed of reactions with repetitive temperature settings. Various carbon sources can be used to produce green CQDs with a PL efficiency equal to or greater than 15%.[48] Green CQDs have been fabricated using microwave-based methods from diverse other precursor materials, including rose petals, eutrophic algae flowers, chitin-rich crab shells containing Eu^{3+}, Mn^{2+}, and Gd^{3+}, winged animal feathers containing N (16.3%), O (33.3%), and S (1.90%) by mass, silkworm chrysalis containing 5.72% N by mass (46% QY), garlic (5% QY), flour (11% QY), onion peel, quince fruit (8.6% QY), and glucose and cotton linter waste with 5% s and

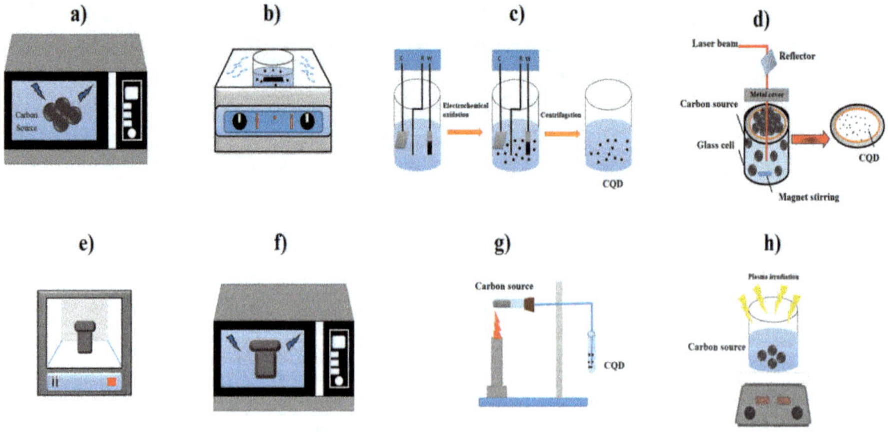

Figure 9.3 Schematic of (a) microwave synthesis method, (b) ultrasonic supported synthesis, (c) electrochemical oxidation method, (d) laser ablation method, (e) hydro/solvo thermal method, (f) microwave hydrothermal method, (g) pyrolysis method, and (h) plasma-induced method.

N by mass. CQDs fabricated from such precursors have greater QY values compared to those synthesized using conventional methods with additives of F and N (QY = 1.2%).[36,49–51]

9.3.2 Ultrasonic-supported Synthesis

Ultrasound can be used to produce alternating high- and low-pressure waves in aqueous solutions, which leads to the formation and collapse of tiny vacuum bubbles (Figure 9.3b). The results of such cavitation are deagglomeration, robust hydrodynamic shear forces, and high-speed impinging liquid jets. In fact, the energy emitted by ultrasonic waves is capable of cleaving materials containing macroscopic carbon into CQDs at the nanoscale. In a study by Kang *et al.*, a single-step ultrasonic treatment mediated by hydrogen peroxide (H_2O_2) was employed to fabricate water-soluble CQDs from activated carbon.[52] Their prepared CQD solution was stable for six months without precipitation of NPs.[47]

9.3.3 Extraction Method

Physical extraction is a new approach to CQD fabrication. It was first introduced in a study by Sk *et al.*, in which amorphous carbon NPs were extracted from food caramels. Subsequently, multiple attempts were made to extract natural CQDs from natural ingredients or foods such as beer, coffee, humic substances, and commercial beverages.[53–57] Also, the extracts of tropical fruits may be associated with difficulty in commercialization, due to failure to export or consume the residue. Thus, the fruit leftovers can be used as additional utilization to produce cost-effective CQDs.[57] It should be noted that there are several disadvantages to this approach: samples are difficult to prepare, separation and purification are time consuming, and the QY is low.[58]

9.3.4 Calcination Approach

Calcination is a two-step method where first, the precursor materials are calcinated in mesoporous silicon templates or spheres, and second, etching is used to remove the templates and produce the nanoscale CQDs. Zong *et al.* introduced a method of synthesis that used mesoporous silicon spheres as rigid templates for producing CQDs. The sphere templates were saturated with a mixture of complex salts and citric acid. By calcination of mesoporous templates, the CQDs' photostability, along with its monodispersion, caused them to exhibit appreciable luminescent activities.[59] Using the calcination method, the crystalline degrees, adjustable sizes, and compositions can be achieved for CQDs, with excellent stability, up-conversion PL, and a PL efficiency of 3.3–4.7%. The soft–hard templates can be used to eliminate the difficulties caused by aggregate formation, and it is possible to make CQDs with a limited size distribution because of the confinement of size.[42]

9.3.5 Electrochemical Oxidation

Electrochemical exposure is a top-down approach for CQD preparation from bulk carbon materials (Figure 9.3c). However, only a limited number of reports are available on the electrochemical carbonization of tiny molecules to CQDs.

In this method, large carbon-based materials such as carbon fiber, multi-walled carbon nanotubes (MWCNTs), graphene, and graphite may be used.[60] Its advantages include simplicity, availability of raw materials, cost-effectiveness, feasibility of mass production, and non-toxicity. Electrochemically fabricated CQDs exhibit uniform particle size, with the ability to cooperate with other routine electrocatalysts in one-pot production using green chemistry.[61] One of this technique's shortcomings is the long purification process required for as-fabricated particles.[48]

Generally, in a two-electrode system, graphite rods act as both cathodes and anodes, while combined organic and inorganic solvents act as electrolytes. The intensity of the current used for electrolysis is in the range of 10–200 mA cm^{-2}. In a three-electrode system, CQDs are fabricated through the electrochemical carbonization of low-molecular-weight alcohols. In one study, researchers used two Pt sheets (as working and counter electrodes) and a calomel electrode (as the reference electrode) installed on a freely tunable Luggin capillary. In a basic environment, the alcohols were transformed into CQDs after electrochemical carbonization.[62]

According to some experiments in the presence of acids as electrolytes (such as H_2SO_4/EtOH), an alkaline environment can be a pivotal parameter for the effective fabrication of CQDs during such electrochemical oxidation. Moreover, the pH values and applied potentials have a significant effect on the production of CQDs.[47]

9.3.6 Laser Ablation

Laser ablation is a simple and green method that can be used to fabricate carbon-based materials such as CQDs with adjustable surfaces of particles. In this method, a high-energy laser pulse irradiates the target surface, creating high pressure and raising the surface's temperature (Figure 9.3d). The atoms of the target surface heat up rapidly and evaporate into plasma; subsequently, the vapor crystallizes to create NPs. Laser ablation can be used to produce fluorescent CQDs with acceptable water solubility and limited size distribution.

In a study by Li *et al.*, a simple protocol was introduced for the preparation of CQDs through the laser irradiation of a carbon precursor, resulting in CQD dispersion in various routine organic solvents. The as-fabricated CQDs exhibited visible and adjustable PL. In a work published by Hu and colleagues, CQDs' surfaces were modified using specific organic solvents while performing laser irradiation; these modifications enabled the adjustment of as-prepared CQDs' PL features. Accordingly, a routine solvent acts as a liquid medium (*i.e.*, water, ethanol, or acetone) and carbon-containing material as a carbon precursor (organic/inorganic).[44,48,62] This method's use is limited due to its cost and difficulty of use.[61]

9.3.7 Hydro/Solvo-thermal Method

This is one of the most common CQD fabrication methods. It is easy to perform and produces particles with relatively uniform sizes and a high QY. Typically, water or an organic solvent is used to dissolve tiny organic molecules and/or polymers to obtain the precursor, which is transferred to a Teflon-lined stainless-steel autoclave (Figure 9.3e). Next, organic molecules and/or polymers at fairly high temperature are integrated as carbon seeding cores, followed by growing into CQDs with a mean particle size of <10 nm.[61]

Hydrothermal/solvothermal synthetic techniques are among the best candidates for green CQD fabrication, especially due to product uniformity and appropriate monitoring of conditions, allowing for higher yields and less waste. Some of the natural precursors, such as fruit peels, juices, and plant leaves, are found as optimal substances for the hydrothermal synthesis of green CQDs. Studies have reported the production of high-QY CQDs using various plant leaves, including *Tamarindus indica* (47%), *Pinellia ternata* (21%), black pepper (44%), turmeric (38%), red chili (27%), cinnamon (36%), *Azadirachta indica* (27%), and *Aegle marmelos* (22%). In addition, low PL efficiency has been obtained for the synthesized CQDs applying the plant leaves including purple perilla (9%), waste green tea (12%), Maojian green tea (13%), and Chinese medicine (12%). Other examples of natural precursors used in various studies include willow leaves, scindapsus, scallion, *Prosopis juliflora*, pipe tobacco, magnolia, paulownia, ginkgo biloba, orange peel, *Ocimum tenuiflorum*, jujube, grass, and crown daisy. Other natural precursors used specifically for hydrothermal CQD fabrication include sweet potato, willow bark, soy milk, *Enteromorpha prolifera*, cornstalk, corncob, cherry tomatoes, broccoli, beetroot extract, and potato.[19,34,63,64]

It should be noted that high PL efficiency has been observed in CQDs prepared hydrothermally from fruit juices. Examples of fruit juices with relative PL efficiency include grape juice (14%), bitter orange juice (20%), orange juice (26%), banana juice (32%), and lemon juice (79%). Hydrothermally produced CQDs with a QY value of 11% have been obtained using precursors such as sugar cane juice, strawberry juice, watermelon juice, *Carica papaya* juice, *Eleocharis dulcis* juice, and apple juice.[19,65,66]

Food waste is another precursor that can be used for green hydrothermal CQD synthesis. Relatively high PL efficiency has been reported for CQDs fabricated hydrothermally from precursors such as glycine and rice residue (24%), oil palm waste (25%), onion waste (28%), and orange peel waste (36%). Natural CQD precursors with relatively low PL efficiency include lemon peel, grapefruit peel, grape peel, orange waste peel, orange pericarp, and pomelo peels.

CQDs have also been fabricated from insects, animals, and animal products; common and abundant precursors with high PL efficiency include milk (12%), bee pollen (13%), *Bombyx mori* silk (14%), spider silk (18%), honey (20%), pigskin (24%), hair (25%), cocoon silk (38%), and dried shrimp (54%). In addition, chicken blood, eggshell membrane, milk, cyanobacteria biomass, and pig liver have been used as natural precursors for CQDs prepared through hydrothermal methods; these precursors yielded CQDs with relatively low QY efficiency.[5,34,67]

The QY of hydrothermally generated CQDs is quite low compared to that of CQDs fabricated with conventional precursors (*e.g.*, citric acid, glucose, glucosamine). For example, citric acid has been used as a carbon source to produce 2.8 nm CQDs with a QY of 52% and a maximum emission of 410 nm. QY values for CQDs produced using various precursors with acceptable PL values were 50 and 29% for citric acid alone, 36% for urea precursor and citric acid (N-CQDs), 35% for ammonia and citric acid (N-CQDs), and 31% for boric acid and citric acid (B-CQDs). Stable CQDs with >10% QY have also been prepared using common precursors such as tetrahydrofuran, polycyclic aromatic hydrocarbon, BBr_3 and hydroquinone, ammonia, and graphene oxide.[36,44,62]

9.3.8 Microwave Hydrothermal Technique

The microwave-supported hydrothermal method is a significant bottom-up CQD production method. Its advantages include energy- and cost-efficiency and a lower reaction time relative to other methods (Figure 9.3f). Microwave-assisted hydrothermal synthesis can be used to produce CQDs from various chemicals (*e.g.*, graphite, citric acid, glucose) or natural substances (*e.g.*, bergamot, garlic, bamboo leaves).[68]

A routine hydrothermal CQD synthesis procedure takes about 24 h to perform; in contrast, it takes about 1–10 min to perform microwave-based hydrothermal CQD synthesis.[23] This highly effective and ultra-fast method can produce CQDs from a mixture of carbon sources and solvents inside an autoclave using a closed microwave system within about 1–10 min.[69] There are numerous reports on the fabrication of carbon dots from various organic waste products, including wine lees, waste paper, cotton linter, onion, lemon peel, rice husk, food leftovers, palm shell powder, whey, wheat straw, kitchen waste, and orange waste.[62]

9.3.9 Chemical Oxidation

Fluorescent particles with ultrahigh PL efficiency can be produced easily and properly using chemical oxidation.[70] Green CQDs have been synthesized using this approach from numerous precursors, including *Trapa bispinosa* peel, plant soot, and waste frying oil. QY values for CQDs produced using this method have QY values of 1–4% and a particle size of 1–7 nm. Other relevant examples of CQDs produced *via* chemical oxidation include cow manure (65% QY) and hair fibers (11% QY and λ_{em} = 383 nm).[36]

In this method, small organic molecules are carbonized by strong oxidizing acids to form carbonaceous substances. These substances subsequently cleave to small sheets under controlled oxidation. This method generally involves drastic processes and harsh conditions due to the strong oxidizing acids used. In a study by Peng and Travas-Sejdic, an easy protocol was developed for the fabrication of luminescent CQDs in an aqueous solution through carbohydrate dehydration *via* concentrated H_2SO_4. The resulting carbonaceous substances were subsequently cut into separate CQDs *via* HNO_3, and eventually inactivated *via* 4,7,10-Trioxa-1,13-tridecanediamine (amine-terminated compounds).[71] In order to produce PL CQDs using this method, surface passivation is required. A branched cationic polyelectrolyte such as polyethylenimine (PEI) is used concurrently as a passivating agent and carbon source under the oxidation of HNO_3 to fabricate PL CQDs. It is possible to adjust these CQDs' emission wavelength by altering the initial carbon source and the nitric acid treatment duration. CQDs produced using this method are especially useful in life science, as researchers in this field can take advantage of their nontoxicity and multicolor emission potencies. The PL of these CQDs is very sensitive to pH, unlike the usual non-pH sensitive CQDs, such that increasing the pH from 2–12 reduced the intensity of PL. Therefore, in these CQDs, PL has an inverse relationship with pH.[44]

9.3.10 Pyrolysis (Thermal Decomposition)

Pyrolysis (thermal decomposition) is a method that can be used to produce various magnetic and semiconductor nanostructures (Figure 9.3g). Recently, external heat was suggested for the dehydration and carbonization of organic materials to obtain the CQDs. This method has several advantages: it is simple and solvent-free, can be used with a broad variety of precursors, the reaction time is short, and production is scalable and cost-effective.[47]

Xu *et al.* were the first group to use combustion/thermal oxidation to prepare CQDs; since then, many other researchers have followed their example. Li *et al.*, for instance, fabricated fluorescent GQDs *via* citric acid combustion followed by functionalization using carboxyl groups *via* the conjugation of acetic acid moieties under exposure to high temperatures.

The resulting GQDs exhibited a uniform particle size (8.5 nm) and abundant surface carboxyl groups. These oxygen-bearing moieties provide water adsorption properties, a useful trait for the electrocatalytic process in aqueous media.[61]

Waste biomass has been used for many years to produce CQDs through pyrolysis and carbonization pyrolysis methods.[72] These methods have garnered much attention because of their low cost and provision of large quantities of carbon; they are also easy to perform, and, unlike several other methods, do not require post-synthesis processes or expensive devices. It is possible to simply produce heteroatom-doped CQDs from various precursors *via* carbonization. Several green-CQDs obtained by carbonization exhibited almost identical size (approximately 2 nm), QY values (approximately 10%), and the highest emission wavelength of approximately 450 nm.[73] When used with appropriate precursors, pyrolysis can produce highly luminescent green CQDs. It can also produce low-QY CQDs.[36]

9.3.11 Microwave-assisted Pyrolysis

Bottom-up microwave-assisted pyrolysis is a method with fast production and the potential for commercialization. In a study by Zhu *et al.*, a simple microwave-assisted pyrolysis method was used to synthesize CQDs using a combination of saccharide (*i.e.*, fructose and glucose) and poly(ethylene glycol) (PEG200) in water. The resulting transparent solution was subsequently heated using a microwave oven to obtain CQDs.[46] The as-fabricated CQDs exhibited excitation-based PL features. This is an easy, rapid, and green method for synthesizing CQDs with numerous oxygen-based functional groups.[61]

9.3.12 Plasma-mediated Approaches

Thermal activation can be used to accelerate chemical reactions, but it consumes a great deal of energy and results in unnecessary waste and carbon dioxide emissions.[74,75] Injection of radicals and electrons is considered an alternative approach that can be used with low-temperature non-equilibrium plasmas (Figure 9.3h). Non-thermal controls can also be achieved using plasma-based methods. Surface bombardment by plasma may direct numerous reactions in (bio)chemical media (such as the generation of extensive reactive oxygen/nitrogen species). The combination of micro-plasma with chemical methods in the fabrication of CQDs presents extra control of surface functionalization and defects, the same as other plasmas, providing further options to tailor C-dot features and photoemission.[76]

The direct advantage of micro-plasma incorporation is to inject energetic electrons with ≥1 eV energy from the plasma and plasma-driven radicals into the volume of the reaction, thereby accelerating reaction processes and forming smaller blue luminescent C-dots. Fourier-transform infrared

(FTIR) spectroscopy also confirmed that CQD samples produced using this plasma-mediated method contained various functional groups not present in dots produced using chemical methods.[75]

A fast and easy plasma-mediated method was used to produce CQDs using chicken egg yolk as the carbon source. The egg yolk was irradiated by intense plasma beams for 3 min to produce black products containing CQDs. Then, extraction is applied to separate the CQDs in the presence of water and some other polar solvents. CQDs produced using this method had a uniform distribution with a mean particle size of 2.15 nm. The CQD solution can be considered a fluorescence carbon ink.[62]

9.3.13 Templating Method

Limited pyrolysis of organic precursors using templates can produce CQDs with uniformly distributed and adjustable sizes (Figure 9.4). This approach consists of three stages: (a) impregnation of silica spheres as templates in the presence of carbon precursor under capillary force, (b) limited pyrolysis of the carbon precursor, and (c) removal of the templates to produce template-free CQDs.[61] In this technique, the diameters of the templates' pores determine the size and size distribution of the CQDs. Porous silica is routinely used to make templates because of its adjustability, thermal stability, texture accessibility, and easy elimination. The best alternative template candidates are polymeric core–shell NPs, as they have thermally removable shells and thermally cross-linkable cores.[48]

9.3.14 Arc Discharge Method

The first recorded use of arc discharge to prepare CQDs was in 2004. In a study by Xu *et al.* on the production of Single-Wall Carbon Nanotubes (SWNTs), this method yielded three types of carbon NPs with diverse relative molecular mass and fluorescence features.[7] In a study by Arora and Sharma, arc discharge was used to reorganize carbon atoms decomposed from bulk carbon precursors in an anodic electrode by gas plasma in a sealed reactor.[76] The temperature of the reactor can be high (4000K)

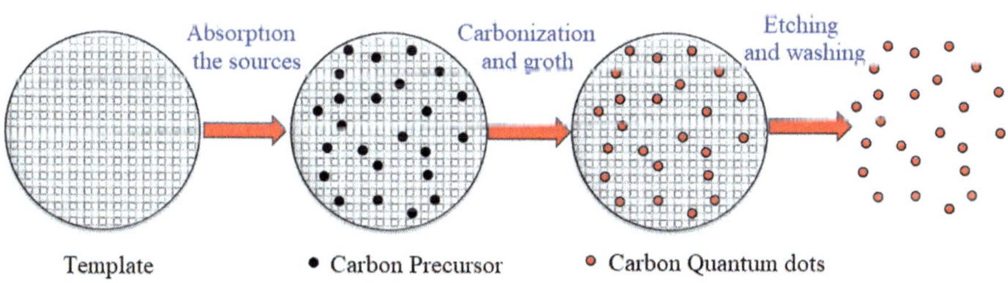

Figure 9.4 Templating method.

with electric current for the creation of high-energy plasma. The resulting carbon vapors aggregate in the gas phase and drift toward the cathode for the formation of CQDs. Yellow, blue-green, and orange fluorescence can be emitted from the as-fabricated CQDs at an excitation wavelength of 365 nm. Based on the evidence, the CQD surface is linked with the hydrophilic carboxyl group. The resultant CQDs have acceptable water solubility and a distribution of large particle sizes in terms of various carbon particle sizes formed in the discharge process. The surface area of CQDs is reduced due to the large particle size, probably limiting the active reaction sites for any electrocatalytic reaction.[61]

9.4 CQD Characterization Techniques

CQDs' excellent and stable performance is attributed to their small size. However, the exact methods used to prepare them can critically affect their nano- or microstructures and their physical and chemical properties; these differences can, in turn, affect their performance. The ability to measure and characterize CQDs according to size and other specific properties is, therefore, crucial in many disciplines such as food sciences. Various methods have been used to study CQDs' structural and photophysical properties at various resolutions, including microscopy, spectroscopy, and diffraction techniques. Examples of common methods include FTIR, nuclear magnetic resonance (NMR), TEM, SEM, X-ray photoelectron spectroscopy (XPS), and XRD.

CQDs' surface morphology, nanostructure, and particle size can be studied using TEM and SEM. Using XPS allows researchers to investigate CQDs' elemental composition, chemical state, the density of their electronic states, and their electronic structure. XRD can be used to determine the crystalline structure and phase purity of synthesized CQDs.

9.4.1 Spectroscopic Evaluations

CQDs' chemical bonds and functional groups can be evaluated using FTIR. The FTIR spectrum is usually used to determine the existence of specific surface functionalities on CQDs.[11] This method is mainly used to identify organic compounds; carbonyl, amine, amide, and hydroxyl groups; and atoms doped on CQDs. NMR provides detailed information about the physical and chemical properties of atoms or molecules, the electronic structure of molecules, functional groups on the surface of a sample, and the reaction state and chemical environment of molecules. can be used to study radio frequency radiation absorption by the nucleus of a CQD in the range of 4–1000 MHz. It can also be used to identify synthesized CQDs' structure and functional groups; this use is especially common in the analysis of food and drug samples. In one study, CQD NPs were produced from D-lactose and then grafted with polyethylene glycol (PEG).

NMR results indicated that the hydroxyl groups in the ethylene glycol molecules were attached to the CQDs' surfaces through covalent bonds between PEG and the CQDs.[77]

The characteristics of CQDs' surfaces affect how they interact with the external environment and with other materials, so it is important that they are accurately identified. One key tool for the surface analysis of CQD is XPS. It can be used to identify CQDs' elemental composition (and the relative percentage of elements), flag contaminants on a sample's surface, and investigate the functional groups on their surfaces. In one study, the XPS spectra of a CQD sample were investigated; the study found that the sample contained mainly C (typical binding energy: *ls* = 284.8 eV), N (typical binding energy: *ls* = 400 eV), and O (typical binding energy: *ls* = 532 eV). S and P were also present; the atomic ratio in the sample was C1s:O1s:N1s:S2p:P2P = 61.0:29.6:8.5:0.5:0.4. The high-resolution C 1s spectra fitted by a binding energy C *ls* revealed three chemical bonds C–C/C=C at 284.7 eV, C–O at 286.08 eV, and C=O at 287.86 eV present in the CQD.[3]

XRD is a technique that can be used to determine the crystallographic structure of materials. In this technique, the sample is irradiated with incident X-rays, and the intensities and scattered angles of the X-rays that leave the material are measured. When the X-rays collide sharply with atoms or molecules, the rays may be scattered under certain conditions, which can give information about the geometric properties of these structures. One study showed that synthesized CQDs with amorphous structures exhibited a large diffraction peak at $2\theta = 23.8°$. This peak can be related to the (002) diffraction pattern of graphite in the CQDs.[78] Paul *et al.* studied the XRD pattern of nitrogen-doped CQDs (N-CQDs) obtained from jackfruit peel and tamarind peel as carbon sources. Both CQD types showed a single-wide peak at about $2\theta = 20°$. The peaks of these amorphous structures could be assigned to the (002) plane of graphitic carbon.[11] According to Bragg's equation, the interlayer spacing of the CQDs is 0.403 nm and 0.492 nm for JFP- and TP-based particles, respectively, which is larger than that of graphite (0.34 nm). The increase in interlayer distance in comparison to graphite may be attributed to the presence of functional groups and N-doping in the lattices of the CQDs.[79] Moreover, Pal *et al.* improved the crystalline properties of CQDs by incorporating phosphorus inside it, as revealed by XRD.[80]

UV-Vis, fluorescence spectroscopy, and PL can be used to determine CQDs' optical properties. UV-Vis absorption spectra and PL, specifically, can survey the optical behavior of CQDs in the range of 200–800 nm and 200–1000 nm, respectively. The surface modification affects CQDs' optical and absorption properties. Usually, in these materials, peaks of optical absorption in the UV-Vis region are considered as π–π^* transition of sp^2-conjugated carbon and n–π^* transition of hybridization with some heteroatom such as nitrogen and sulfur.

In one paper, the absorption spectra of CQD solutions with similar features in different solvents were investigated. The absorption spectra contained a shoulder at 350 nm and an absorption tail at about 450 nm. The shoulder could be attributed to n–π^* electronic transitions of conjugated C=N or C=O bonds and a weak absorption shoulder at about 450 nm is reasonably assigned to the $n \rightarrow \pi^*$ transition of the conjugated C=N bonds. The PL of CQDs changed significantly in the blue region (from 400 nm to 450 nm) with the changing of solvent; CQD aggregation led to enhanced red emissions and reduced blue emissions. According to the Förster resonance energy transfer (FRET) mechanism, when CQDs absorb light significantly in the blue region, the aggregation of CQDs causes enhanced energy transfer among the CQDs and reabsorption of their blue light emissions. However, longer-wavelength light was less affected. In these CQDs, blue emissions came mainly from C sp^2 domains and the green emission (525 nm) originated from surface C = N states. The aggregation of CQDs due to π–π attraction among C sp^2 domains caused the blue emissions to be quenched due to the delocalization of excited carriers in C sp^2, while excited carries at surface states were stabilized *via* photorecombinations.[81]

9.4.2 Electronic Microscopy Evaluations

Transmission electron microscopy is a special tool for nano-structural studies; it can be used to determine the structure and morphology of CQDs with high resolution and very high magnification. The resolution of TEM images is thousand times greater than that of images produced by an optical microscope. TEM can also be used to study CQDs' symmetry, orientation, and crystal defects. Recently, high-resolution transmission electron microscope (HRTEM) has been used as an important tool in many research of different fields, especially food science.

Basoglu *et al.* characterized the surface morphology and size of as-synthesized CQDs using TEM and showed that the agglomerated synthesized CQDs had a size distribution of 5–100 nm.[11] In another study, as-prepared CQDs were investigated using TEM, which showed that they were well-dispersed, quasi-spherical structures without any apparent aggregation. From the statistical results, the particle size and their average are estimated between 3 and 9 nm and 6.4 and 5.3 nm, for two types of starting materials: jackfruit peel and tamarind peel.[82]

SEM is a powerful and efficient microscopy tshape, approximatelyechnique for CQD analysis. In this method, an electron beam interacts with the material and emitted secondary electrons from the sample surfaces are recorded. The high magnifications possibly allow CQD morphology to be obtained using SEM.

9.4.3 Application of CQDs in Food Safety

Due to such qualities, as readily available from abundant carbon sources, low toxicity, environmental friendliness, great biocompatibility, and other unique properties, CQDs have many applications in fields related to food safety, including food component identification and contaminant detection. They can also be used as fluorescent probes for food safety detection.[5,83,84]

Figure 9.5 shows various food safety applications for CQDs. CQDs can be used, for example, to detect pathogenic bacteria, antibiotics, anions and cations, and additives and colorants. They can also be used to disinfect wastewater and can act as bio- and nano-sensors.

9.4.3.1 Chemical Detection Applications

9.4.3.1.1 Metal Ions and Molecules. Toxic, non-biodegradable metal ions, such as copper, chrome, mercury, aluminum, silver, lead, and iron, may enter the food chain, where they aggregate in animals and plants and harm human health.[85,86] Due to their optical and luminescent properties, CQDs can be used as fluorescent probes to detect several metal ions.

The existence of a number of cations such as Fe^{3+} ions in the biological process must be controlled because this ion has an important role in the chelation of various components such as proteins and is considered as an

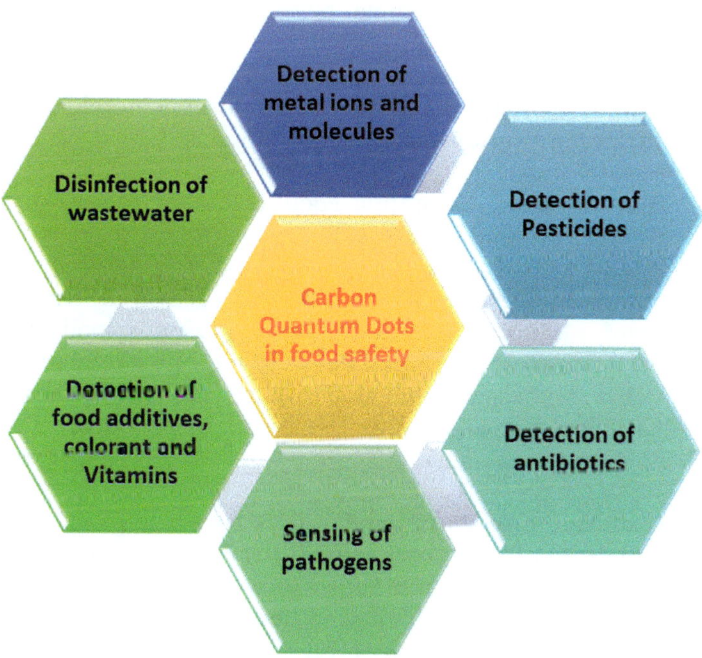

Figure 9.5 Applications of CQDs in food safety.

important element in biological systems. Fluorescent CQDs offer a simple, environmentally friendly, non-time-consuming, and inexpensive approach to cation determination: their functional groups (such as hydroxyls and carbonyls) enable them to interact with metals and especially with cations. CQDs can also detect anions such as ClO^-, $PO4^{3-}$, I^-, and F^-. Much research has been conducted on treated CQDs (N-CQD or S-CQDs) as probes for the detection of various ions.[11,85,86]

9.4.3.1.2 Pesticides. Much research has been conducted on the use of CQD-based sensors for the selective detection of small inert molecules, such as those found in pesticides used in the food industry. Generally, the surface state of CQDs plays an important role in sensing these molecules, as there is no effective energy transfer between inert molecules and the sensor.[87]

In one study, CQDs and N-CQDs were synthesized from *Moringa oleifera* roots and successfully used to detect sulcotrione pesticide in real samples (detection limit = 2 μg mL^{-1}). The results of this study also showed that the N-CQDs performed well at inhibiting the pathogenicity of *Corynespora cassiicola* and *Phytophtora nicotianae* with 82.8% and 75.3%, respectively.[88]

Another study explored the use of quantum dots to improve the performance of optical biosensors for quick and accurate pesticide detection. Naturally, fluorescent and non-toxic CQDs were conjugated with acetylcholinesterase to produce a fluorescent biosensor, which was used to detect pure chlorpyrifos and profenofos. This is one of the few works that have reported the detection of commercial pesticide formulations with promising results.[89]

Fluorescent CQDs with silver nanoparticles (AgNPs) have also been used to identify pesticides such as propanyl, parathion, dimethoate, chlorpyrifos, and pyrimicarb in rice, carrot, orange, and pepper samples.[90]

9.4.3.1.3 Harmful Food Colorants. Food colorants can be harmful to human health if consumed in excess. For example, tartrazine is a common colorant in beverages, dairy products, and baked goods,[91] but research has shown that its overconsumption can lead to harmful health effects.[92] Controlling and monitoring the amount of tartrazine in food samples is a serious issue; fortunately, CQDs can detect it and other colorants easily. In one study, CQDs synthesized from aloe without chemical surface modification were used for the sensitive and selective detection of tartrazine in foods collected from local markets.[93]

9.4.3.1.4 Disinfection of Wastewater. CQDs have a large specific surface area (as high as 1600 m^2 g^{-1}) and a small particle size (less than 10 nm); these features make them suitable for use as adsorbents for the removal of organic and inorganic pollutants or heavy metals from wastewater.[94] Research has shown that they have a high capacity to adsorb heavy metal ions such as Cd^{2+} and Pb^{2+} from water.[95] The polarity and surface area of adsorbent materials play an important role in the adsorption process. The adsorption of metal ions can be controlled by electrostatic adhesion and chemical chelation as well as by physisorption between metal ions and the adsorbent material.[94]

In one study, a CQDs/TiO_2 composite was used as a highly efficient photocatalyst for the degradation of various water-soluble dyes, such as methylene blue (a common and highly toxic wastewater pollutant). The CQDs/TiO_2 composite, which was derived from waste rice noodles, had high photocatalytic performance: its degradation rates were 81.84% within 10 min and 99.87% within 80 min. Moreover, this study found that the degradation rate of the CQDs/TiO_2 composite was affected by the CQDs' loading dosage, with higher carbon content resulting in higher degradation rates. This composite can be recycled and reused without any significant decrease in the degradation rate, which maintained more than 96.5% effectiveness after seven photocatalytic cycles.[96]

9.4.3.1.5 Detection of Food Additives. There are some limitations in using additives in the world, so the use of different natural and synthetic additives in foods must be monitored according to a country's law.[5,83] Natural additives in comparison to synthetic additives are largely used in the food industry because they have good stability and excellent water solubility and are inexpensive. However, they show concerns about human health. CQDs have been proposed as a solution for detecting harmful additives in food products because CQDs have fewer limitations compared to conventional methods such as UV-Vis, chromatography, high-performance liquid chromatography (HPLC), and fluorescence spectrometry, including small size and low cost.[5,83]

Carmine is a colorant additive that can enhance genetic damage in cells and can cause tumors. Su *et al.* successfully produced CQDs from dried lemon peels and used them to detect carmine in beverages.[97] Thiourea is an additive that is used in fertilizers and fungicides. It can be used to ripen some fruits, but it is toxic to humans. In one study, highly fluorescent N-CQDs with excellent chemiluminescence (CL) were used to measure the amount of thiourea in orange juice. The study produced satisfactory results even in the juice's acidic environment.[98]

9.4.3.1.6 Detection of Food Toxins. Toxic chemicals can be absorbed by plants, which may be used in food products and subsequently consumed by humans; their detection is, therefore, an important concern. In one study, CQDs with optical properties were synthesized from spoiled milk and used as fluorescent probes to detect food toxins such as heavy metals and pathogens.[99]

9.4.3.1.7 Detection of Vitamins. Traditional methods for detecting vitamins in food samples are expensive and time consuming. CQDs modified with chitosan can be used as an electrochemical bio-sensing platform to detect vitamin D2.[100] CQDs have also been used as probes or sensors to detect amino acids, proteins, vitamins, sugar, and glucose – additives that can cause severe harm to human health if under- or over-consumed.[100]

In one study, CQDs dually doped with phosphorus and nitrogen (PN-CQDs) were used to determine vitamin B12 in drug samples and human cells. The response to vitamin B12 was linear in the range of 2.0–31 μM with a detection limit of 3.0 nM.[101]

9.4.4 Biological Detection Applications

A wide variety of CQD-based nano/biosensors have been used to detect targets in the food industry, including pathogenic microorganisms, antibiotics, and drugs.

9.4.4.1 Sensing of Pathogens

Bacterial pathogens can easily live under specific conditions and rapidly reproduce and increase their numbers. Consequently, the presence of these bacterial pathogens in food can provide a serious danger to human health. Food poisoning and food pathogen identification are known as a significant issue in the world and in many of countries, resulting in their serious regulation.[83] Therefore, rapid and sensitive identification of these pathogens is essential in the prevention of the threat such as *Escherichia coli* (*E. coli*), *Staphylococcus aureus* (*S. aureus*), *Desulfovibrio desulfuricans*, *Staphylococcus sciuri*. CQDs can be used to detect, monitor, and even capture pathogenic bacteria in foods.[102]

CQDs can be used as fluorescent probes for bacteria and fungus cell imaging; owing to their non-toxicity and biocompatibility, they can be used both *in vitro* and *in vivo*. These probes detect bacteria based on fluorescence changes. The detection mechanism is a result of interactions with bacterial cell walls, enzyme reactions for intracellular compounds, or peptidoglycan synthesis reactions.[103]

CQDs can also function as excellent antibacterial agents due to their controlled size, good dispersal in water, and good antimicrobial properties.[84] One study explored the antimicrobial effects of N-CQDs on pathogens such as *Staphylococcus*, *S. aureus*, *S. epidermidis*, and Methicillin-resistant *Staphylococcus aureus* (MRSA).[104–106] In another study, highly fluorescent CQDs were prepared from sugarcane bagasse pulp using a hydrothermal technique. The CQDs' antibacterial effects on aquatic Gram-positive (*S. aureus* and *Benthesicymus cereus*) and Gram-negative (*Vibrio cholerae*, *E. coli*, and *Pseudomonas aeruginosa*) microbial organisms were studied. The results indicated that sugarcane waste CQDs are promising materials for the functioning of NLO devices, bioimaging, and pharmaceutical applications.[107]

9.4.4.2 Detection of Antibiotics

Antibiotics are commonly used in both medical and agricultural contexts. They can affect human and environmental health, which makes their detection especially important. CQDs' applications and limitations in the detection of antibiotics and other medical drugs have been widely studied. CQDs with their promising optical properties can be used to track antibiotics. They make excellent biosensors due to their high sensitivity and specificity to antibiotics[108]; they have been used to investigate antibiotics such as tetracyclines,[109] macrolide,[110] aminoglycosides,[111] and quinolones.[112]

Oxytetracycline (OTC) is a tetracycline that is used to treat infections caused by a variety of Gram-positive and Gram-negative microorganisms, including *Mycoplasma pneumoniae* and *Pasteurella pestisand*. In one investigation, N-CQDs with strong fluorescence were prepared from natural biomass waste (including orange and watermelon peels) using a hydrothermal method. Efficient fluorescent probes based on the fluorescence-quenching responses of N-CQDs were constructed for the sensitive and selective detection of OTC.[113]

9.5 Future Challenges and Perspectives

As the world's population increases, so too does the need for improvements in the fields of food quality and food safety. The future applications of CQDs are broad due to the expanded range of applications for this group of carbon-based materials, especially in food safety and industry, in addition to the green methods of producing them from waste materials.

Research has shown that CQDs excel at detecting a variety of pathogenic bacteria, antibiotics, ions, molecules, additives, and colorants. They can also be used to remove heavy metals and other substances from wastewater, and can be incorporated into nano/bio sensors. Also, other fields of science have benefited from these progresses; for instance bio-imaging, drug delivery, catalysis, optronics, and rocket fuels as discussed in other chapters.

The application of CQDs in viral and bacterial investigation, especially in different foods, is important as it can be useful to understand how they act in various environments. CQDs could help researchers to manage viruses and bacteria for human protection, treatment of disease and providing proper food safety for a better future.

In this regard, improving the performance of nano/bio sensors based on CQDs, enhancing the selectivity and sensitivity of the sensing materials, and facilitating the fabrication methods and design should be targeted in the future investigations.

The use of CQDs in sensors does, however, pose some challenges. First, sensor disposal can affect human health. Second, sensors must be made more user-friendly – for example, by developing better read-out systems that are applicable in the real-world applications. Soon, CQDs and modified sensors will be able to enhance food safety in a green way, helping protect the environment and earth.

9.6 Conclusion

CQDs are inexpensive, abundant, environmentally friendly, biocompatible materials that can be produced using easy methods of synthesis. They have extraordinary optical properties, including excellent and adjustable PL, which give them unique advantages over conventional semiconductor materials for food safety and sensing applications. Various waste materials and by-products (including vegetable, animal, and industrial waste) have been used as carbon sources for CQD synthesis. Among the wide variety of available synthesis methods, the low-cost hydrothermal method has received the most attention. Despite their low cost and environmental friendliness, these waste materials' carbon sources have a low yield.

CQDs' QY can be tuned by changing the composition of the raw materials used as precursors. Therefore, by changing the synthetic parameters with the same starting material and the same synthesis method, distinct effects on CQD's QY can be observed. Moreover, heteroatom addition and surface passivation can enhance CQDs' PL efficiency.

CQDs obtained from food waste using green synthesis methods demonstrate excellent sensitivity and selectivity toward metal ions and anions, and can be used in real-life detection and elimination of these compounds. It should be noted that conventional (*i.e.*, non-green) methods of CQD preparation are still mostly preferred; however, recent literature has shown that CQDs obtained through green synthesis routes can perform acceptably in various sensing and bioimaging applications. It is, therefore, expected that future research will focus on greener CQD production methods and developing ideas for large-scale production of this interesting group of materials.

References

1. C. Martin-Rios, C. Demen-Meier, S. Gössling and C. Cornuz, *Waste Manag.*, 2018, **79**, 196.
2. H. Kumar, K. Bhardwaj, R. Sharma, E. Nepovimova, K. Kuča, D. S. Dhanjal, R. Verma, P. Bhardwaj, S. Sharma and D. Kumar, *Molecules*, 2020, **25**(12), 2812.
3. J. Zhang, H. Wang, Y. Xiao, J. Tang, C. Liang, F. Li, H. Dong and W. Xu, *Nanoscale Res. Lett.*, 2017, **12**, 611.
4. B. P. de Oliveira and F. O. M. da Silva Abreu, *Mater. Lett.*, 2021, **282**, 128764.

5. H. Fan, M. Zhang, B. Bhandari and C. Yang, *Trends Food Sci. Technol.*, 2020, **95**, 86.
6. H. Soni and P. Sudhakar, *Mater. Res. Bull.*, 2018, **108**, 250.
7. X. Xu, R. Ray, Y. Gu, H. J. Ploehn, L. Gearheart, K. Raker and W. A. Scrivens, *J. Am. Chem. Soc.*, 2004, **126**, 12736.
8. T. Wang, H. Luo, X. Jing, J. Yang, M. Huo and Y. Wang, *Molecules*, 2021, **26**, 1246.
9. S. Y. Park, H. U. Lee, E. S. Park, S. C. Lee, J. W. Lee, S. W. Jeong, C. H. Kim, Y. C. Lee, Y. Suk Huh and J. Lee, *ACS Appl. Mater. Interfaces*, 2014, **6**, 3365.
10. M. Picard, S. Thakur, M. Misra and A. K. Mohanty, *RSC Adv.*, 2019, **9**, 8628.
11. A. Başoğlu, Ü. Ocak and A. Gümrükçüoğlu, *J. Fluoresc.*, 2020, **30**(3), 515.
12. X. Y. Du, C. F. Wang, S. Wu and G. Chen, *Angew. Chem. Int. Ed.*, 2021, **60**(16), 8585.
13. L. Jiang, H. Ding, M. Xu, X. Hu, S. Li, M. Zhang, Q. Zhang, Q. Wang, S. Lu, Y. Tian and H. Bi, *Small*, 2020, **16**(9), 2000680.
14. X. Zhu, W. Jiang, Y. Zhao, H. Liu and B. Sun, Single, dual and multi-emission carbon dots based optosensing for food safety, *Trends Food Sci. Technol.*, 2021, **111**, 388.
15. C. Kang, Y. Huang, H. Yang, X. F. Yan and Z. P. Chen, *Nanomaterials*, 2020, **10**(11), 2316.
16. V. Singh, K. S. Rawat, S. Mishra, T. Baghel, S. Fatima, A. A. John, N. Kalleti, D. Singh, A. Nazir, S. Rath and A. Goel, *J. Mater. Chem. B*, 2018, **6**(20), 3366.
17. N. Hashemi and M. H. Mousazadeh, *J. Photochem. Photobiol., A*, 2021, **421**, 113534.
18. F. Calì, V. Cantaro, V. Fichera, L. Ruffino, R. Trusso Sfrazzetto, G. Li-Destri and G. N. Tuccitto, *Chemosensors*, 2021, **9**(8), 202.
19. A. Tyagi, K. M. Tripathi, N. Singh, S. Choudhary and R. K. Gupta, *RSC Adv.*, 2016, **6**, 72423.
20. X. Zhao, S. Liao, L. Wang, Q. Liu and X. Chen, *Talanta*, 2019, 201, 1.
21. H. Yang, B. Zhou, Y. Zhang, H. Liu, Y. Liu, Y. He and S. Xia, *Waste Biomass Valorization*, 2020, **12**(4), 2109.
22. L. Xu, Y. Li, S. Gao, Y. Niu, H. Liu, C. Mei, J. Cai and C. Xu, *Polymers*, 2020, 12 (5), 1143.
23. H. Eskalen, S. Uruş, S. Cömertpay, A. H. Kurt and Ş. Özgan, *Ind. Crops Prod.*, 2020, **147**, 112209.
24. S. Chaudhary, S. Kumar, B. Kaur and S. K. Mehta, *RSC Adv.*, 2016, **6**(93), 90526.

25. K. Tungare, M. Bhori, K. S. Racherla and S. Sawant, *3 Biotech.*, 2020, 10(12), 540.
26. M. Thakur, A. Mewada, S. Pandey, M. Bhori, K. Singh, M. Sharon and M. Sharon, *Mater. Sci. Eng. C Mater. Biol. Appl.*, 2016, **67**, 468.
27. J. Guo, H. Li, L. Ling, G. Li, R. Cheng, X. Lu, A.-Q. Xie, Q. Li, C.-F. Wang and S. Chen, *ACS Sustainable Chem. Eng.*, 2019, **8**(3), 1566.
28. J. Zhu, F. Zhu, X. Yue, P. Chen, Y. Sun, L. Zhang, D. Mu and F. Ke, *J. Nanomater.*, 2019, **2019**, 1.
29. N. Irmania, K. Dehvari, G. Gedda, P. J. Tseng and J. Y. Chang, *J. Biomed. Mater. Res. B Appl. Biomater.*, 2020, **108**(4), 1616.
30. Z. Hu, X.-Y. Jiao and L. Xu, *Microchem. J.*, 2020, **154**, 104588.
31. D. M. A. Crista, A. El Mragui, M. Algarra, J. C. G. Esteves da Silva, R. Luque and L. Pinto da Silva, Turning Spent Coffee Grounds into Sustainable Precursors for the Fabrication of Carbon Dots, *Nanomaterials*, 2020, **10**(6), 1209.
32. L. Wang, W. Li, B. Wu, Z. Li, S. Wang, Y. Liu, D. Pan and M. Wu, *Chem. Eng. J.*, 2016, **300**, 75.
33. Y. Zhao, Y. Zhang, X. Liu, H. Kong, Y. Wang, G. Qin, P. Cao, X. Song, X. Yan and Q. Wang, *Sci. Rep.*, 2017, **7**(1), 4452.
34. S. Jing, Y. Zhao, R. C. Sun, L. Zhong and X. Peng, *ACS Sustainable Chem. Eng.*, 2019, **7**(8), 7833.
35. J. Ahn, Y. Song, J. E. Kwon, S. H. Lee, K. S. Park, S. Kim, J. Woo and H. Kim, *Mater. Sci. Eng. C Mater. Biol. Appl.*, 2019, **102**, 106.
36. M. O. Caglayan, F. Mindivan and S. Şahin, *Crit. Rev. Anal. Chem.*, 2020, 1.
37. X. Li, H. Wang, Y. Shimizu, A. Pyatenko, K. Kawaguchi and N. Koshizaki, *Chem. Commun.*, 2010, **47**, 932.
38. K. J. Mintz, Y. Zhou and R. M. Leblanc, Recent development of carbon-quantum dots regarding their optical properties, photoluminescence mechanism, and core structure, *Nanoscale*, 2019, **11**, 4634–4652.
39. T. Yuan, T. Meng, P. He, Y. Shi, Y. Li, X. Li, L. Fan and S. Yang, *J. Mater. Chem. C*, 2019, **7**, 6820.
40. H. Duan, D. Wang and Y. Li, Green chemistry for nanoparticle synthesis, *Chem. Soc. Rev.*, 2015, **44**, 5778.
41. P. Gupta and A. Mahajan, *RSC Adv.*, 2015, **5**, 26686.
42. I. Singh, R. Arora, H. Dhiman and R. Pahwa, *Turk. J. Pharm. Sci.*, 2018, **15**, 219.
43. P.-C. Hsu, Z.-Y. Shih, C.-H. Lee and H.-T. Chang, *Green Chem.*, 2012, **14**, 917.
44. Y. Wang and A. Hu, *J. Mater. Chem. C*, 2014, **2**, 6921.
45. Q. Wang, X. Liu, L. Zhang and Y. Lv, *Analyst*, 2012, **137**, 5392.

46. H. Zhu, X. Wang, Y. Li, Z. Wang, F. Yang and X. Yang, *Chem. Commun.*, 2009, 5118.
47. R. Wang, K.-Q. Lu, Z.-R. Tang and Y.-J. Xu, *J. Mater. Chem. A*, 2017, **5**, 3717.
48. M. Farshbaf, S. Davaran, F. Rahimi, N. Annabi, R. Salehi and A. Akbarzadeh, *Artif. Cells, Nanomed. Biotechnol.*, 2018, **46**, 1331.
49. Y. Hu and Z. Gao, *J. Hazard. Mater.*, 2020, **382**, 121048.
50. R. K. Singh, R. Kumar, D. P. Singh, R. Savu and S. A. Moshkalev, *Mater. Today Chem.*, 2019, **12**, 282.
51. W. L. Ang, C. A. L. B. Mee, N. S. Sambudi, A. W. Mohammad, C. P. Leo, E. Mahmoudi, M. Ba-Abbad and A. Benamor, *Sci. Rep.*, 2020, **10**, 1.
52. H. Li, X. He, Y. Liu, H. Yu, Z. Kang and S.-T. Lee, *Mater. Res. Bull.*, 2011, **46**, 147.
53. M. P. Sk, A. Jaiswal, A. Paul, S. S. Ghosh and A. Chattopadhyay, *Sci. Rep.*, 2012, **2**, 1.
54. C. Jiang, H. Wu, X. Song, X. Ma, J. Wang and M. Tan, *Talanta*, 2014, 127, 68.
55. Y. Dong, L. Wan, J. Cai, Q. Fang, Y. Chi and G. Chen, *Sci. Rep.*, 2015, **5**, 1.
56. S. Mandani, D. Dey, B. Sharma and T. K. Sarma, *Carbon*, 2017, 119, 569.
57. C. A. Franco, C. H. Candela, J. Gallego, J. Marin, L. E. Patiño, N. Ospina, E. Patiño, M. Molano, F. Villamil and K. M. Bernal, *Ind. Eng. Chem. Res.*, 2020, **59**, 11359.
58. B. Han, M. Yu, T. Pen, Y. Li, X. Hu, R. Xiang, X. Hou and G. He, *New J. Chem.*, 2017, **41**, 5267.
59. J. Zong, Y. Zhu, X. Yang, J. Shen and C. Li, *Chem. Commun.*, 2011, **47**, 764.
60. Y.-P. Sun, B. Zhou, Y. Lin, W. Wang, K. A. S. Fernando, P. Pathak, M. J. Meziani, B. A. Harruff, X. Wang and H. Wang, *J. Am. Chem. Soc.*, 2006, **128**, 7756.
61. X. Wang, Y. Feng, P. Dong and J. Huang, *Front. Chem.*, 2019, **7**, 671.
62. R. Das, R. Bandyopadhyay and P. Pramanik, Carbon quantum dots from natural resource: A review, *Mater. Today Chem.*, 2018, **8**, 96–109.
63. M. Sabet and K. Mahdavi, *Appl. Surf. Sci.*, 2019, **463**, 283.
64. T. Shen, Q. Wang, Z. Guo, J. Kuang and W. Cao, *Ceram. Int.*, 2018, **44**, 11828.
65. D. Kumar, K. Singh, V. Verma and H. S. Bhatti, *J. Bionanoscience*, 2014, **8**, 274.
66. M. R. Hasan, N. Saha, T. Quaid and M. T. Reza, *Energies*, 2021, **14**, 986.
67. C. Wang, H. Shi, M. Yang, Y. Yan, E. Liu, Z. Ji and J. Fan, *Mater. Res. Bull.*, 2020, **124**, 110730.

68. Q. Liang, Y. Wang, F. Lin, M. Jiang, P. Li and B. Huang, *Anal. Methods*, 2017, **9**, 3675.
69. N. R. Pires, C. M. W. Santos, R. R. Sousa, R. Paula, P. L. R. Cunha and J. Feitosa, *J. Braz. Chem. Soc.*, 2015, **26**, 1274.
70. S. Hu, R. Tian, L. Wu, Q. Zhao, J. Yang, J. Liu and S. Cao, *Chem. Asian J.*, 2013, **8**, 1035.
71. H. Peng and J. Travas-Sejdic, *Chem. Mater.*, 2009, **21**, 5563.
72. B. Thangaraj, P. R. Solomon and S. Ranganathan, *Curr. Pharm. Des.*, 2019, **25**, 1455.
73. B. Thangaraj, P. R. Solomon, S. Chuangchote, N. Wongyao and W. Surareungchai, *ChemBioEng. Rev.*, 2021, **8**, 265.
74. D. Vollath, Plasma synthesis of nanopowders, *J. Nanopart. Res.*, 2008, **10**, 39.
75. X. Huang, Y. Li, X. Zhong, A. E. Rider and K. Ostrikov, *Plasma Process. Polym.*, 2015, **12**, 59.
76. N. Arora and N. N. Sharma, *Diam. Relat. Mater.*, 2014, **50**, 135.
77. B. B. Campos, D. Mutavdžić, M. Stanković, K. Radotić, J. M. Lázaro-Martínez, J. C. G. Esteves da Silva, R. Contreras-Cáceres, M. S. Pino-González, E. Rodriguez-Castellón and M. Algarra, *New J. Chem.*, 2017, **41**, 4835.
78. J. B. Essner, C. H. Laber, S. Ravula, L. Polo-Parada and G. A. Baker, *Green Chem.*, 2016, **18**(1), 243.
79. A. Mewada, S. Pandey, S. Shinde, N. Mishra, G. Oza, M. Thakur, M. Sharon and M. Sharon, *Mater. Sci. Eng., C*, 2013, **33**, 2914.
80. A. Pal, G. Natu, K. Ahmad and A. Chattopadhyay, *J. Mater. Chem. A*, 2018, **6**, 4111.
81. X.-D. Mai, Y. T. H. Phan and V.-Q. Nguyen, *Adv. Mater. Sci. Eng.*, 2020, 1.
82. A. Paul and M. Kurian, *Cleaner Engineering and Technology*, 2021, 100103.
83. W. Zhang, H. Zhong, P. Zhao, A. Shen, H. Li and X. Liu, *Food Control*, 2022, 133, 108591.
84. J. H. Qu, Q. Wei and D. W. Sun, *Crit. Rev. Food Sci. Nutr.*, 2018, **58**(14), 2466.
85. H. Abdelsalam, N. H. Teleb, I. S. Yahia, H. Y. Zahran, H. Elhaes and M. A. Ibrahim, *J. Phys. Chem. Solids*, 2019, **130**, 32.
86. J. Dou, D. Gan, Q. Huang, M. Liu, J. Chen, F. Deng, X. Zhu, Y. Wen, X. Zhang and Y. Wei, *Int. J. Biol. Macromol.*, 2019, **136**, 476.
87. E. C. Hett, R. E. Kyne, A. Gopalsamy, M. A. Tones, H. Xu, G. L. Thio, E. Nolan and L. H. Jones, *ACS Comb. Sci.*, 2016, **18**(10), 611.
88. Z. Wang, Q. Liu, J. Leng, H. Liu, Y. Zhang, C. Wang, W. An, C. Bao and H. Lei, *J. Saudi Chem. Soc.*, 2021, **25**, 12.

89. M. I. Gaviria, K. Barrientos, J. P. Arango, J. B. Cano and G. A. Peñuela, *Front. Environ. Sci.*, 2022, **10**, 825112.
90. S. V. Carneiro, V. H. R. de Queiroz, A. A. C. Cruz, L. M. U. D. Fechine, J. C. Denardin, R. M. Freire, R. F. do Nascimento and P. B. A. Fechine, *Sens. Actuators, B*, 2019, **301**, 127149.
91. X. L. Ye, Y. L. Du, D. B. Lu and C. M. Wang, *Anal. Chim. Acta*, 2013, **779**, 22.
92. K. A. Amin, H. A. Hameid and A. H. Abd Elsttar, *Food Chem. Toxicol.*, 2010, **48**, 2994.
93. T. Chatzimitakos, A. Kasouni, L. Sygellou, A. Avgeropoulos, A. Troganis and C. Stalikas, *Talanta*, 2017, **175**, 305.
94. U. A. Rani, L. Y. Ng, C. Y. Ng and E. Mahmoudi, *Adv. Colloid Interface Sci.*, 2020, **278**, 102124.
95. A. Azimi, A. Azari, M. Rezakazemi and M. Ansarpour, *ChemBioEng. Rev.*, 2017, **4**, 37.
96. X. Jin, R. Che, J. Yang, Y. Liu, X. Chen, Y. Jiang, J. Liang, S. Chen and H. Su, *Nanomaterials*, 2022, **12**, 472.
97. A. Su, D. Wang, X. Shu, Q. Zhong, Y. Chen, J. Liu and Y. Wang, *Chem. Res. Chin. Univ.*, 2018, **34**, 164.
98. Y. Liu and S. Han, *Food Anal. Methods*, 2017, **10**, 3398.
99. A. Nayab, A. M. Najabat and T. J. Khan, *Front. Mater.*, 2021, 8.
100. T. Sarkar, H. B. Bohidar and P. R. Solanki, *Int. J. Biol. Macromol.*, 2017, **109**, 687.
101. Z. Li, W. Huiping, H. Qin and D. Chuan, *Microchim. Acta*, 2019, **186**(506), 1.
102. C. Zhao, X. Wang, L. Wu, W. Wu, Y. Zheng, L. Lin, S. Weng and X. Lin, *Colloids Surf. B Biointerfaces*, 2019, **179**, 17.
103. O. Moradlou, Z. Rabiei and N. Delavari, *J. Photochem. Photobiol., A*, 2019, **379**, 144.
104. E. Mechcatie and K. Rosenberg, *Am. J. Nurs.*, 2018, **118**(11), 56.
105. H. Du, X. Wang, Q. Yang and W. Wu, *Trends Food Sci. Technol.*, 2021, **110**, 1.
106. Q. Xu, T. Kuang, Y. Liu, L. Cai, X. Peng, T. S. Sreeprasad, P. Zhao, Z. Yu and N. Li, *J. Mater. Chem. B*, 2016, **4**, 7204.
107. S. Pandiyan, L. Arumugam, S. P. Srirengan and G. Vinitha, *ACS Omega*, 2020, **5**(47), 30363.
108. F. Sun, J. Zhang, Q. Yang and W. Wu, *Food Qual. Saf.*, 2021, 5, fyab019.
109. R. Daghrir and P. Drogui, *Environ. Chem. Lett.*, 2013, **11**, 209.
110. G. V. Sanchez, D. J. Shapiro and A. L. Hersh, *Open Forum. Infect. Dis.*, 2017, 4(4), ofx220.
111. L. Jiafeng, X. Fu and Z. Chang, *Sci. Rep.*, 2015, **5**, 14247.

112. W. Simonson, *Geriatr. Nurs.*, 2017, **38**, 152.

113. R. Gao, Z. Wu, L. Wang, J. Liu, Y. Deng, Z. Xiao, J. Fang and Y. Liang, *Nanomaterials*, 2020, **10**(8), 1561.

Subject Index